VOLUME ONE HUNDRED TWENTY NINE

ADVANCES IN
APPLIED MICROBIOLOGY

Fungal Stress Mechanisms and
Responses

VOLUME ONE HUNDRED TWENTY NINE

Advances in
APPLIED MICROBIOLOGY
Fungal Stress Mechanisms and Responses

Edited by

DRAUZIO EDUARDO NARETTO RANGEL

Inbioter—Institute of Biotechnology Rangel,
Itatiba, SP, Brazil

ACADEMIC PRESS

An imprint of Elsevier

Academic Press is an imprint of Elsevier
125 London Wall, London, EC2Y 5AS, United Kingdom
50 Hampshire Street, 5th Floor, Cambridge, MA 02139, United States
525 B Street, Suite 1650, San Diego, CA 92101, United States

First edition 2024

Notices
Knowledge and best practice in this field are constantly changing. As new research and experience broaden our understanding, changes in research methods, professional practices, or medical treatment may become necessary.

Practitioners and researchers must always rely on their own experience and knowledge in evaluating and using any information, methods, compounds, or experiments described herein. In using such information or methods they should be mindful of their own safety and the safety of others, including parties for whom they have a professional responsibility.

To the fullest extent of the law, neither the Publisher nor the authors, contributors, or editors, assume any liability for any injury and/or damage to persons or property as a matter of products liability, negligence or otherwise, or from any use or operation of any methods, products, instructions, or ideas contained in the material herein.

ISBN: 978-0-443-21660-2
ISSN: 0065-2164

For information on all Academic Press publications
visit our website at https://www.elsevier.com/books-and-journals

Publisher: Zoe Kruze
Acquisitions Editor: Leticia M. Lima
Editorial Project Manager: Devwart Chauhan
Production Project Manager: James Selvam
Cover Designer: Gopalakrishnan Venkatraman

Typeset by MPS Limited, India

Working together
to grow libraries in
developing countries

www.elsevier.com • www.bookaid.org

Contents

Contributors

Alexandre Melo Bailão
Universidade Federal de Goiás (UFG), Goiânia, GO, Brazil

Xianxian Cheng
MOE Key Laboratory of Biosystems Homeostasis & Protection, Institute of Microbiology, College of Life Science, Zhejiang University, Hangzhou, P.R. China

Weiguo Fang
MOE Key Laboratory of Biosystems Homeostasis & Protection, Institute of Microbiology, College of Life Science, Zhejiang University, Hangzhou, P.R. China

Ming-Guang Feng
Institute of Microbiology, College of Life Sciences, Zhejiang University, Hangzhou, P.R. China

Marina Fomina
Zabolotny Institute of Microbiology and Virology, National Academy of Sciences of Ukraine, Kyiv, Ukraine

Geoffrey Michael Gadd
Geomicrobiology Group, School of Life Sciences, University of Dundee, Dundee, Scotland, United Kingdom; State Key Laboratory of Heavy Oil Processing, Beijing Key Laboratory of Oil and Gas Pollution Control, College of Chemical Engineering and Environment, China University of Petroleum, Beijing, P.R. China

Olena Gromozova
Zabolotny Institute of Microbiology and Virology, National Academy of Sciences of Ukraine, Kyiv, Ukraine

Erika Kothe
Friedrich Schiller University Jena, Institute of Microbiology, Jena, Germany

Katrin Krause
Friedrich Schiller University Jena, Institute of Microbiology, Jena, Germany

Dayane Moraes
Universidade Federal de Goiás (UFG), Goiânia, GO, Brazil

Drauzio Eduardo Naretto Rangel
Inbioter—Institute of Biotechnology Rangel, Caixa Postal 5, Itatiba, SP, Brazil

Mirelle Garcia Silva-Bailão
Universidade Federal de Goiás (UFG), Goiânia, GO, Brazil

Lea Traxler
Friedrich Schiller University Jena, Institute of Microbiology, Jena, Germany

Xinru Wang
MOE Key Laboratory of Biosystems Homeostasis & Protection, Institute of Microbiology, College of Life Science, Zhejiang University, Hangzhou, P.R. China

Sheng-Hua Ying
Institute of Microbiology, College of Life Sciences, Zhejiang University, Hangzhou, P.R. China

Qing Zhen
MOE Key Laboratory of Biosystems Homeostasis & Protection, Institute of Microbiology, College of Life Science, Zhejiang University, Hangzhou, P.R. China

Preface

Volume 129 of *Advances in Applied Microbiology*, entitled *Fungal Stress Mechanisms and Responses*, explores the adaptive strategies and biotechnological applications of fungi under stress conditions. The various reviews by experts explore different aspects of fungal stress responses. The first three chapters focus on stress in insect-pathogenic fungi, while Chapters 4–6 address the impact of stressful environmental conditions on fungi used for bioremediation. The last chapter investigates the molecular aspects of copper homeostasis in human fungal pathogens.

In Chapter 1, Drauzio E. N. Rangel from Inbioter – Institute of Biotechnology, Brazil, discusses how higher stress tolerance can be induced in entomopathogenic fungi, which is crucial for their application for biological control of insects in agriculture. Stress tolerance and cross-protection are reviewed with emphasis on the responses of *Metarhizium robertsii* to nutritive stress, osmotic stress, heat-shock stress, oxidative stress, and chemical stress. Conditions, including visible light, magnetic, and electric fields, are discussed. Furthermore, the chapter relates how a productive scientific career, including instigating and organizing the International Symposium on Fungal Stress – ISFUS, can be achieved with much enthusiasm, hard work, and a great mentor.

In Chapter 2, Sheng-Hua Ying from Zhejiang University, China, provides an in-depth exploration of the lifecycle of filamentous entomopathogen fungi to identify opportunities for increasing their virulence and, thus, their efficacy as biocontrol agents in modern agriculture. The biogenesis and functions of organelles and other subcellular structures are reviewed, focusing on recent advances. A mechanistic overview is provided about the life cycle of these eukaryotes from a cell biology perspective. This chapter also discusses the current state of knowledge about the biological roles and regulatory mechanisms of organelles and subcellular structures in the physiology of entomopathogenic fungi, as well as suggestions for further investigation.

In Chapter 3, Ming-Guang Feng, also from Zhejiang University, China, investigates how fungal UV tolerance depends on mechanisms such as nucleotide excision repair or photorepair of UV-induced DNA lesions to recover UV-impaired cells in the darkness or the light. The genetic/molecular basis for photorepair-dependent photoreactivation, which serves as a primary anti-UV mechanism in insect-pathogenic fungi, is reviewed, focusing on the methodology established to quantify fungal responses to solar UV radiation. Emphasis is placed on their anti-UV mechanisms compared to those documented in the model yeast system and principles for properly timing the application of a fungal pesticide to improve pest control during the summer months. An overview about the progress of anti-UV roles and mechanisms in fungal entomopathogens elucidated in the past two decades is provided.

In Chapter 4, Lea Traxler from Friedrich Schiller University Jena in Germany, together with Katrin Krause and Erika Kothe, present innovative approaches to the problem of reusing large areas often co-contaminated with both metals and organic pollutants and provide technical solutions, including mycoremediation. Fungi cope particularly well with heterogeneous conditions due to their adaptability and large hyphal network. This chapter summarizes the advantages of basidiomycetes, focusing on the ability of wood rotting fungi to tolerate metals, radionuclides, and organic contaminants such as polycyclic aromatic hydrocarbons. It also explains how these fungi can reduce the toxicity of contaminants to other organisms, including plants, to help restore land use. The processes based on diverse molecular mechanisms are introduced, and their use for mycoremediation is discussed.

In Chapter 5, Marina Fomina from the National Academy of Sciences of Ukraine, along with Olena Gromozova and Geoffrey M. Gadd, discuss the morphological adaptations of filamentous fungi when exposed to environmental stress. Growth conditions elicit diverse morphological responses in filamentous fungi, that are coupled with fungal biogeochemical activity, and can ameliorate hostile conditions. However, our understanding of the diverse range of morphological responses exhibited by filamentous fungi under different stressors is still limited. The diverse morphological responses to environmental stressors manifested by filamentous fungi in surface and submerged mycelial growth conditions are summarized. Emphasis is placed on the morphological flexibility of filamentous fungi in response to metal- and mineral-rich conditions that ensure successful adaptation to stressful environments. Such knowledge will contribute to both fundamental mycology and applied microbiology, enabling better control of fungal-based biotechnological processes.

In Chapter 6, Weiguo Fang from Zhejiang University, China, explores the role of plant symbiotic fungi in mitigating pollution from toxic metals and metalloids. Anthropogenic activities have dramatically accelerated the release of toxic metals and metalloids into soil and water, which can subsequently accumulate in plants and animals, threatening biodiversity, human health, and food security. Mycorrhizal and endophytic fungi represent an excellent strategy for remediation compared to physical and chemical treatments. The current knowledge of how plant symbiotic fungi remediate metal(loid)-polluted soil, and the underlying cellular, molecular, biochemical, and evolutionary mechanisms are reviewed. Emphasis is placed on the removal of methylmercury pollution by endophytic species of *Metarhizium*, which could be acquired from bacteria via horizontal gene transfer. These fungi and others, including *Trichoderma* sp., can be cost-effectively produced at large scale, and seem to be well-placed for field trials for remediation of metal(loid)-polluted soil and water on a large scale.

In Chapter 7, Alexandre Melo Bailão and his group from the Universidade Federal de Goiás, Brazil, provide a detailed analysis of copper homeostasis in fungi. Copper homeostasis in fungi is a tightly regulated process crucial for cellular functions, including respiration and antioxidant defense. Excessive copper can be toxic, promoting cell damage mainly due to oxidative stress and metal displacements. This chapter summarizes the mechanisms involved in copper homeostasis in representative fungi, including *Saccharomyces cerevisiae*, *Schizosaccharomyces pombe*, *Aspergillus fumigatus*, *Cryptococcus neoformans*, and *Candida albicans*. The biological function and characteristics of proteins, such as transporters, ferrireductases, metallothioneins, antioxidant metallochaperones, and transcription factors involved in the transport, storage, and detoxification of copper, are analyzed. These mechanisms enable fungi to balance copper levels, ensuring proper cellular function while preventing toxicity. Understanding copper homeostasis in fungi is not only essential for fungal biology but also critical for various applications, including biotechnology and antifungal drug development.

This volume offers a comprehensive collection of findings that help our understanding of fungal stress responses and their applications in agriculture, medicine, and the environment.

ALENE ALDER-RANGEL

Alder's English Services,
São José dos Campos, SP, Brazil

HUMBERTO R. MEDINA

Tecnológico Nacional de México,
Celaya, Gto., Mexico

DRAUZIO E. N. RANGEL

Inbioter – Institute of Biotechnology Rangel,
Itatiba, SP, Brazil

How *Metarhizium robertsii's* mycelial consciousness gets its conidia Zen-ready for stress

Drauzio Eduardo Naretto Rangel*,1
Inbioter—Institute of Biotechnology Rangel, Caixa Postal 5, Itatiba, SP, Brazil
*Corresponding author. e-mail address: drauzio@live.com

Contents

Abstract

This memoir takes a whimsical ride through my professional adventures, spotlighting my fungal stress research on the insect-pathogenic fungus *Metarhizium robertsii*, which transformed many of my wildest dreams into reality. Imagine the magic of fungi meeting science and me, a happy researcher, arriving at Utah State University ready to dive deep into studies with the legendary insect pathologist, my advisor Donald W. Roberts, and my co-advisor Anne J. Anderson. From my very first "Aha!" moment in the lab, I plunged into a vortex of discovery, turning out research like a mycelium on a mission. Who knew 18 h/day,

1 https://drauzioeduardo.wordpress.com

Advances in Applied Microbiology, Volume 129
ISSN 0065-2164, https://doi.org/10.1016/bs.aambs.2024.07.002

seven days a week, could be so exhilarating? I was fueled by an insatiable curiosity, boundless creativity, and a perhaps slightly alarming level of motivation. Years later, I managed to bring my grandest vision to life: the International Symposium on Fungal Stress—ISFUS. This groundbreaking event has attracted 162 esteemed speakers from 29 countries to Brazil, proving that fungi can be both fun and globally fascinating. ISFUS is celebrating its fifth edition in 2024, a decade after its 2014 debut.

1. Introduction

Everything started soon after I arrived at Utah State University on October 5, 2000, just two months after defending my master's degree in agricultural microbiology at the Universidade Estadual Paulista (UNESP), advised by Antônia do Carmo Barcelos Correia, at the age 42. My professor Donald W. Roberts (Alder-Rangel, 2021) brought several boxes of house crickets (*Acheta domesticus*, Orthoptera: Gryllidae), waxworms (*Galleria mellonella*, Lepidoptera: Pyralidae), and giant mealworms (*Zophobas morio*, Coleoptera: Tenebrionidae) to the laboratory. He handed them to me and gave me instructions. At that time, I could read well in English, but my speaking and listening skills were worse than poor. Therefore, I did not understand anything he said. The fact, that Dr. Roberts had suffered from tongue cancer, could make his speech unclear even to native speakers (Alder-Rangel, 2021). Therefore, I just smiled from ear to ear and nodded my head. Every day when he arrived in the laboratory, Dr. Roberts would ask me how the experiments with the crickets, waxworms, and mealworms were (Fig. 1), and again, I smiled and nodded my head (Fig. 2). Although I do not know if Dr. Roberts understood what my smiling responses really meant, his wife Mae Roberts (Fig. 3) figured out very quickly that I really did not understand when I responded that way.

This situation continued for two or three weeks until I was very ashamed, and I asked a colleague in the laboratory to ask Dr. Roberts what I needed to do. Therefore, my colleague went to his office and came back to tell me that I needed to infect the insects *G. mellonella* and *Z. morio* with *Metarhizium robertsii* isolates ARSEF 23, ARSEF 2575, and *A. domesticus* with *Metarhizium acridum* ARSEF 324. Then after the death of the insects, I should pick up conidia from the insect cadavers and evaluate the conidial tolerances of the conidia produced on insect cadavers with the tolerance of the conidia produced on potato dextrose agar medium supplemented with yeast extract (PDAY). Once I understood the task, I was very grateful to Dr. Roberts and excited to do my own first project in the laboratory. By that time, all the insects that Dr. Roberts had purchased were dying of old

Fig. 1 Donald W. Roberts arriving at the laboratory at Utah State University asking me how the experiments with insects are progressing. Unpublished.

Fig. 2 Drauzio E. N. Rangel, Donald W. Roberts, and Mae Roberts in front of their house's porch Summer 2002. Unpublished.

Fig. 3 I was already ashamed that I did not understand Dr. Roberts's English. Unpublished.

Fig. 4 Larvae of *Zophobas morio* (Coleoptera: Tenebrionidae) mummified with *Metarhizium robertsii* (ARSEF 2575). Unpublished.

age, so I ashamedly asked him to purchase new insects. The next day, the insects arrived, and I started inoculating dry conidia of *M. robertsii* (ARSEF 23 and ARSEF 2575) on their backs with a cotton swab. I transferred the waxworm and giant mealworm larvae to high relative humidity chambers and waited for their demise. Then, I transferred the dead insects to Petri dishes with humid filter paper to allow conidial production on the insect cadavers (Fig. 4). Subsequently, I made a suspension of the conidia from insect cadavers and another conidial suspension from PDAY medium from both isolates and evaluated their UV-B radiation tolerance for 1, 2, 3, and 4 h. To my surprise, the conidia of the isolates ARSEF 23 and ARSEF 2575 collected from insects were less tolerant than the conidia produced on the PDAY medium. Furthermore, ARSEF 2575 conidia from *G. mellonella* were more tolerant than conidia from *Z. morio* (Rangel, Braga, Flint, Anderson, & Roberts, 2004).

After these first experiments were done, I started working with house crickets (*Acheta domesticus* Orthoptera: Gryllidae), infecting them with *Metarhizium acridum* (ARSEF 324) due to its host specificity is limited to Orthoptera (Rangel et al., 2022); however, these insects died and started smelling a foul odor from bacteria. Then Dr. Roberts asked Dr. Stefan T. Jaronski from the United States Department of Agriculture, Sidney, MT, USA, to send some grasshoppers (*Austracris guttulosa*, Orthoptera: Acrididae) from his rearing, and I started my second project infecting these insects with *M. acridum* (ARSEF 324). Although I did not observe any difference in tolerance to UV-B radiation when conidia were produced on grasshopper or PDAY medium (Rangel, Braga, Anderson, & Roberts, 2005a), the conidia produced on grasshoppers germinated slower than conidia produced on PDAY (Rangel et al., 2005a). In addition, conidia produced on grasshoppers were much smaller than conidia produced on PDAY (Rangel et al., 2005a).

These results broadened my horizons and really got my juices flowing. Studying the fascinating phenomenon of stress tolerance and phenotypic plasticity, I found that when microorganisms grow under harsh stress conditions, they may become more tolerant to other stresses. For example, bacterial spores collected from the Sonora Desert are two-fold more tolerant than the same bacterial spores produced on laboratory medium (Nicholson & Law, 1999). Furthermore, I read articles from Naresh Magan's group, which explain that tolerance of *Metarhizium* to osmotic stress may be improved by physiological manipulation of the endogenous trehalose and mannitol in conidia (Hallsworth & Magan, 1994a, 1994b, 1994c; Magan, 2001). Then I focused on articles published by Anita Panek's and Johan Thevelein's groups, which explain that endogenous yeast trehalose protects against heat, ethanol, oxidative stress, frost, and dehydration (Coutinho, Bernardes, Felix, & Panek, 1988; Panek, 1962, 1963; Thevelein & Hohmann, 1995; Thevelein, 1984; van Dijck, Colavizza, Smet, & Thevelein, 1995) as does the fungus *Aspergillus nidulans* (Fillinger et al., 2001). Then my eyes lit up at the possibility of applying this information to improve tolerance to UV-B radiation and heat in *Metarhizium*.

After Dr. Roberts suggested this first research project, he let me free in the lab to start exploring my own ideas and gave me wings to fly. The project funding he had was on UV-B radiation tolerance of *M. anisopliae* now *M. robertsii* (Project: Improving tolerance to UV-B radiation of an insect-biocontrol fungus, *Metarhizium anisopliae*, by Donald W. Roberts and Anne J. Anderson from the United States Department of Agriculture). However, he did not give me orders to work on exactly what he wanted done. I was free in the laboratory, mostly by myself, for several years working on my own projects. This is how Dr. Roberts preferred to work. For his biography, he explained that he feels that researchers are more productive if they have buy-in. "You just have to give them enough rope, where they can finally get themselves tangled. Then they decide that they better come in at night and watch that experiment or whatever has to get done… It is a lot more fun to work on your stuff than it is on his stuff" (Alder-Rangel, 2021).

Dr. Roberts was not very happy with me when he visited my laboratory in Brazil in 2010, in which I told the students what to do and closely supervised their experiments. I really enjoy the laboratory, and I love to see the results; thus, I love to work together with my students. Yes, I agree with him that encouraging the students to be independent would be ideal; however, every person is different. All of Dr. Roberts's postdocs and students also recalled that he gave them the freedom to explore their own projects (Alder-Rangel, 2021).

Dr. Roberts gave me more than just the freedom to explore my own research but is the mentor who really gave me the opportunity to become a scientist. Although I had always dreamed of being a researcher, after finishing my undergraduate in biology, I was not given the opportunity to continue my studies. I returned to university at 40 for my master's degree. Then Dr. Roberts brought me to the USA to intern in his lab and eventually advised me through my PhD.

He was a leading insect pathologist internationally known for his work on insect-fungal pathogens and biological control. Although Dr. Roberts worked on many different insect pathogens, his favorite was *Metarhizium*. Thus, when *Metarhizium anisopliae* was divided into several new species in 2008, one was named *Metarhizium robertsii* after him. This is also the fungus that I have used in most of my studies.

During my time in Dr. Roberts' lab, I came to understand that the ability to induce higher stress tolerance of entomopathogenic fungi is vital for their application for biological control of insects in agriculture (Rangel et al., 2015c; Rangel, Finlay, Hallsworth, Dadachova, & Gadd, 2018).

Consequently, if the insect-pathogenic fungus *M. robertsii* is more tolerant to stress, it is apt to infect insects in the field. Could conidial tolerance to UV-B radiation and other stress conditions be improved by cross-protection? In the next sections, I discuss the stress conditions that induced cross-protection and those that did not.

Biological control of insects using insect-pathogenic fungi is crucial in Brazil, given its vast agricultural landscape. These fungi, which naturally infect and kill insects, offer an environmentally friendly alternative to chemical pesticides. Employing insect pathogens helps sustain agricultural productivity, maintain ecological balance, and promote sustainable farming practices (Li et al., 2010). The development of fungal insect-pathogenic isolates displaying higher stress tolerance is important to improve the use of these fungi to control insects. Before use in the field, these fungi must endure a short shelf life (Faria, Hotchkiss, & Wraight, 2012; Krell, Jakobs-Schoenwandt, Persicke, & Patel, 2018) and be transported sometimes at high temperatures for delivery (Faria et al., 2012). When applied in agricultural fields, fungal conidia are subjected to solar UV and heat radiation (Dias et al., 2018; Feng, 2024; Santos, Dias, Ferreira, Pasin, & Rangel, 2011; Souza, Azevedo, Lobo, & Rangel, 2014). In the soil, these fungi are under biotic stress caused by necrotrophic mycoparasites, including *Trichoderma* and *Clonostachys* species (Costa, Rangel, Morandi, & Bettiol, 2013; Karlsson, Atanasova, Jensen Dan, & Zeilinger, 2017; Medina, Oliveira, Medina, & Rangel, 2020). Chemical fungicides also greatly influence the survival of insect pathogens in the field (Rangel, Dettenmaier, Fernandes, & Roberts, 2010). After infecting the host, fungal insect pathogens encounter oxidative and osmotic stresses in the host hemocoel (Bergin, Reeves, Renwick, Wientjes, & Kavanagh, 2005; Zhang et al., 2009). Therefore, insect-pathogenic fungi must be able to endure various types of stress, and a fungal screening of isolates to a variety of stress conditions is vital to improve the biological control of insects (Araújo et al., 2018, 2020; Azevedo, Souza, Braga, & Rangel, 2014; Dias et al., 2018; Licona-Juárez et al., 2023a; Lima et al., 2021; Rangel, Braga, Anderson, & Roberts, 2005b; Souza et al., 2014).

2. Cross-protection

Soon after I discovered the importance of trehalose and mannitol on fungal stress tolerance, I started working on a review about bacterial stress tolerance and cross-protection. The regulation of sigma S in the general resistance conferred in *Escherichia coli* has been well studied; thus, I read articles related to this bacterium (Jenkins, Schultz, & Matin, 1988; Lynch & Matin, 2005; Lynch, Brodie, & Matin, 2004; Matin & Lynch, 2005; McCann, Fraley, & Matin, 1993). However, after a 14-year break from the biological field, working in my family's printing industry, my knowledge was very poor. I humbly asked my co-advisor, Anne J. Anderson, to read my literature review. She said that the mechanisms that fungi and bacteria use for cross-protection are very different; nevertheless, she liked my review. Although I was frustrated because I first thought that bacteria and fungi have the same stress response mechanisms, her information was very important for me to improve my understanding of fungal stress responses. A few years later when Dr. Anderson was in Brazil in 2010 for the First Brazilian Symposium about the Effects of Increased UV Radiation on Agriculture (UVRAG) that I organized (Rangel & Alder-Rangel, 2020), I gave her my manuscript entitled "Stress induced cross-protection against environmental challenges on prokaryotic and eukaryotic microbes" and I saw tears coming out her eyes (which make me very emotive as well). She said that she always asked her master's and PhD students to publish their literature reviews, but I was the only student who did this without being asked. She was able to read the manuscript during the night, and the next day she said that she really enjoyed reading the review, which was published in 2011 (Rangel, 2011).

Fungi have Core Environmental Response genes that are commonly induced in response to diverse types of environmental stresses, which also induce a second set of genes that are commonly repressed in response to these conditions (Brown, Larcombe, & Pradhan, 2020). My objective in this review is to analyze the stress responses in *Metarhizium robertsii* that I have personally studied, for further information about fungal priming to stress and its mechanisms, I strongly suggest reading the following articles (Brown, Barton, Pan, Buck, & Wigneshweraraj, 2014; Brown et al., 2020; Deveau et al., 2007; Emri, Forgacs, & Pocsi, 2022; Gasch & Werner-Washburne, 2002; Gasch et al., 2000; Harish & Osherov, 2022; Hilker et al., 2016; Mitchell & Pilpel, 2011; Mitchell et al., 2009; Morano, Grant, & Moye-Rowley, 2012; Rangel, 2011; Sessa, Pedrini, Altier, & Abreo, 2022).

Cross-protection, also known as priming, adaptive response, acclimation, or acquired stress resistance, occurs when an organism is exposed to mild stress, resulting in a more robust and protective response to future stress (Harish & Osherov, 2022). For example, a mild heat stress of 35 °C induced priming to 40 °C in eight of 19 soil filamentous fungi tested (Andrade-Linares, Veresoglou, & Rillig, 2016). However, one stress can induce priming to several other stress conditions, in *Saccharomyces cerevisiae* a mild heat shock confers protection against subsequent oxidative, osmotic, or freeze-thaw stress (Herdeiro, Pereira, Panek, & Eleutherio, 2006; Lewis, Learmonth, & Watson, 1995; Park, Grant, Attfield, & Dawes, 1997).

In *M. robertsii,* many types of stresses induce cross-protection, including nutritive stress, osmotic stress, heat-shock stress, oxidative stress, and many chemicals. Furthermore, some conditions, such as visible light, magnetic, and electric fields, which do not necessarily induce stress on the fungus, can still lead to fungus that produces conidia that are more tolerant to stress.

In the following sections, I discuss each type of stress condition during growth that induces higher conidial tolerance.

3. Nutritive stress as a potent inducer to cross-protection

An oligotrophic fungus is defined as a fungus that can grow at low levels of essential nutrients or even in their apparent absence (Wainwright, 1993). Oligotrophy seems to be the rule among common soil fungi (Wainwright, 1993), including *Metarhizium* species (Bidochka, Kasperski, & Wild, 1998; Bidochka, Kamp, Lavender, Dekoning, & De Croos, 2001). Fungal growth under nutritive stress has been more intensively studied because this stress causes greater cross-protection to a variety of stress conditions, such as oxidative stress, heat, osmotic stress, freeze-thaw tolerance, gamma radiation, and UV radiation (Hohmann & Mager, 2003; Rangel, 2011).

Growth under nutritive stress (Czapek medium without sucrose) is known to induce the greater stress tolerance in *Beauveria bassiana, Cordyceps fumosorosea, Metarhizium anisopliae, M. brunneum, M. robertsii, T. cylindrosporum,* and *T. inflatum* (Dias, Souza, Pupin, & Rangel, 2021).

In January 2003 during my first year as a PhD student, I started working with nutritive stress by growing *M. robertsii* (ARSEF 2575) on minimum medium (MM), which is Czapek medium (Czapek, 1901) without sucrose, and comparing tolerance to UV-B radiation and heat of conidia produced on PDAY medium. Conidia produced on MM were twofold more tolerant to UV-B radiation and heat than conidia produced on PDAY medium (Rangel, Anderson, & Roberts, 2006, 2008b). In addition, conidia produced on MM accumulated twice as much trehalose and mannitol as conidia produced on MM supplemented with glucose (Rangel et al., 2006). Later, my group also discovered that genes encoding heat-shock proteins (*hsp30* and *hsp101*) and the gene encoding superoxide dismutase (*sod2*) of conidia produced on MM were upregulated compared to conidia produced on (PDA) potato dextrose agar medium (Silva, Pedrini, Pupin, Roberts, & Rangel, 2023).

How did conidia produced on MM accumulate more trehalose and mannitol without any carbon source to grow than conidia produced on MM supplemented with 3% glucose? My hypothesis is that newer hyphae dissolve older mycelia by autophagy, which I have observed through a microscope and is discussed in this book by Ying (2024). Autophagic events are associated with the whole lifecycle of insect-pathogenic fungi, including pathogenic and saprotrophic growth. Autophagy plays an important role in various processes of the infection cycle of insect-pathogenic fungi, including nutritional starvation/shift, development, stress response, and virulence (Lin, Wang, Feng, & Ying, 2019; Ying & Feng, 2019). Although I could not prove that autophagy was the reason *M. robertsii* was able to accumulate more trehalose and mannitol, other authors have confirmed that autophagy is important for stress response (Lin et al., 2019; Ying & Feng, 2019). Nevertheless, fungi can grow oligotrophically without apparent involvement of the lysis or utilization of preformed hyphae (Jennings, 1993).

On the other hand, growth on MM produces 90 million fewer conidia than growth on PDAY medium (Rangel et al., 2006), which makes it completely impractical for mass production. Because MM produces conidia with the highest tolerances of all other stresses, I have used this condition as a positive control for all other priming experiments.

4. Carbohydrates that induce cross-protection

In addition to nutritive stress caused by MM, I also found that growth on MM supplemented with some carbon sources induced high conidial tolerance to UV-B radiation. This was true for MM supplemented with 3% arabinose, fructose, galactose, inositol, lactose, and mannitol. In addition, conidia produced on MM supplemented with lactose accumulated three times more trehalose and mannitol than conidia produced on MM plus 3% glucose (Rangel et al., 2006).

5. Heat-shock stress

Yeast cells subjected to heat shock demonstrated increased tolerance to heat, freezing, and salt stress (Lewis et al., 1995). Based on this report,

I started experiments with heat-shock in May 2003 to test weather heat shock confers protection against a subsequent UV-B and heat stress. I exposed mycelium of *M. robertsii* ARSEF 2575 on days two, three, or four of growth to two heat-shock conditions, heated by convection in an incubator at 50 °C (40 min) or heated by visible light and infrared radiation under an Oriel solar simulator (25 min). In both heat treatments, the final temperature was approximately 45 °C. One set of cultures was also exposed three times (days 2, 3, and 4) to high temperatures. Conidia produced on mycelia heat-shocked by both methods on the third day of growth were more tolerant to UV-B radiation and heat, and they accumulated two times more trehalose and mannitol than conidia produced in the other treatments (Rangel et al., 2008b). In conclusion, both heat-shock methods produced similar results, but the time window in which mycelia were shocked determined the level of this tolerance.

6. Osmotic stress

Salt stress induces cross-protection of *S. cerevisiae* to heat, salt, and freezing (Lewis et al., 1995). Accordingly, in May 2004, I started experiments to test osmotic stress for cross-protection against UV-B radiation and heat. *M. robertsii* ARSEF 2575 was grown on PDAY (control) or PDAY supplemented with potassium chloride or sodium chloride to 0.4, 0.6, 0.8, or 1.0 M in the media. The tolerance to UV-B increased in direct proportion to the salt concentration, and the concentration of 1.0 M of both salts induced the highest UV-B tolerance. In contrast, heat tolerance significantly increased at lower salt concentrations but decreased as the salt concentration rose (Rangel et al., 2008b). Conidia produced on sodium chloride-supplemented PDAY exhibited higher UV-B tolerance compared to those produced on potassium chloride-augmented PDAY. However, thermotolerance was higher for conidia produced on PDAY with potassium chloride than with sodium chloride (Rangel et al., 2008b).

7. Oxidative stress

Mild oxidative stress induces cross-protection to higher oxidative stress, heat, and UV radiation (Jamieson, 1992; Kapoor & Sveenivasan, 1988; Verma & Singh, 2001). In May 2004, I also started experiments to test oxidative stress for cross-protection against UV–B radiation and heat. This study employed menadione (0.1 mM), hydrogen peroxide, or UV–A irradiation to induce oxidative stress. Varying hydrogen peroxide concentrations (0, 5, 10, and 20 mM) and different exposure methods (supplemented in medium or flooding the mycelium with H_2O_2) were used to evaluate the response of conidia to UV–B or heat treatments. Additionally, conidia from mycelia treated with UV–A (for 0, 1, 2, 4, and 6 h on the third day of mycelial growth) or produced on PDAY medium supplemented with menadione were assessed for their relative germination following UV–B or heat exposure.

M. *robertsii* conidia generated under oxidative stress conditions (mycelia exposed to UV–A irradiation or hydrogen peroxide) did not exhibit increased tolerance to UV–B radiation or heat compared to conidia produced under normal conditions (PDAY medium) (Rangel et al., 2008b). However, conidia produced on PDAY supplemented with menadione showed UV–B tolerance similar to the tolerance of conidia produced on PDAY, but their thermotolerance increased to levels comparable to conidia produced on MM (Rangel et al., 2008b). Therefore, similar to yeast (Herdeiro et al., 2006), not all oxidative stress agents confer cross-protection against other stress conditions in M. *robertsii*.

8. Toxic stress

During my Plant Pathology class with Professor Anne J. Anderson in September 2002, I learned that salicylic acid (SA) released by plant cells is associated with systemic acquired resistance (SAR) in plants, which can induce defenses of plants against pathogens (Agrios, 2005). After the class, I asked Dr. Anderson if she had a little bit of SA to test if M. *robertsii* conidia produced on medium supplemented with SA induces higher tolerance to UV–B radiation and heat. She was very excited about my curiosity and gave me three tubes with different salicylic acids. However, I started these experiments only in May 2006 just before my PhD defense.

Conidia were produced on PDAY medium with and without supplements of 1, 2, 4, or 8 mM SA (2-hydroxybenzoic acid sodium salt). The medium pH was adjusted to 6.9 to avoid an acid response. On the one hand, producing M. *robertsii* conidia on medium supplemented with SA did not increase their UV–B radiation tolerance, which was similar to conidia produced on PDAY (control). On the other hand, growth on PDAY with SA (at concentrations of 1, 2, and 4 mM) increased conidial thermotolerance to a level comparable to that of conidia produced on MM. However, the highest concentration of SA (8 mM) did not induce higher tolerance to heat. Growth of *M. robertsii* on PDAY + SA reduced conidial yields, but not nearly as much as MM (Rangel, Fernandes, Anderson, & Roberts, 2012).

Congo red is another toxic stress that may also induce cross-protection. It induces cell wall stress (Csillag, Emri, Rangel, & Pócsi, 2023; Lima et al., 2021). In this study, my group evaluated the capacity of Congo red to induce priming against heat, oxidative, osmotic, and UV radiation in *M. robertsii*. Conidia were produced on three different media: PDA (control), PDA supplemented with 200 mg/mL of Congo red, or MM. Congo red did not induce priming against heat, osmotic, or oxidative stress in *M. robertsii*. For these three stress conditions, conidial germination was similar when conidia were produced on either PDA medium or PDA supplemented with Congo red. Congo red only induced higher conidial tolerance against UV-B radiation in *M. robertsii* at 2 h of exposure, but not at 3 h of exposure (Licona-Juárez et al., 2023b).

9. Alkaline and acid stress

Alkaline medium (NaOH 1 mM) induces priming to UV-C radiation in *S. cerevisiae* (Verma & Singh, 2001), and mildly acidic conditions induce priming to heat, salt, and crystal violet in *Salmonella typhimurium*; furthermore, acid adaptation increases cell surface hydrophobicity (Leyer & Johnson, 1993). To discover if acid or alkaline medium can induce priming in *M. robertsii*, in May 2006, I tested the tolerance of conidia produced either on PDAY at pH 6.90 or on aerial conidia from surface cultures grown on potato dextrose broth supplemented with 0.1% yeast extract (PDBY), with the pH adjusted to 4.59, 5.85, 6.60, 6.90, 8.04, or 9.45. Additionally, conidia were produced on PDAY supplemented with 10 mM of the organic acids: sorbic, maleic, propionic, and taurocholic. Conidia produced on PDBY adjusted to pH 8.04 and 9.45 exhibited higher UV-B tolerance compared to those produced on PDAY medium (control), while their heat tolerance remained similar to the control. Conidia produced on low-pH media showed UV-B tolerance comparable to those produced at neutral pH (Rangel et al., 2015c).

Additionally, weak acids such as sorbic acid and propionic acid slightly improved heat tolerance, but the UV-B radiation tolerance was similar to the control at neutral pH (Rangel et al., 2015c).

10. White, blue, green, and red light

An amazing dream six months before my dissertation defense was inspirational. In this dream, I cultured *M. robertsii* under the light and in the dark. Upon waking up, I rushed to the laboratory and prepared two Petri dishes, growing one of them in an incubator under fluorescent light and the other in the same incubator but in the dark. Fourteen days later, I tested the conidia produced on PDA medium under the light and in the dark, as well as conidia produced on MM in the dark to UV-B radiation and heat. To my surprise, conidia produced under the light had similar tolerance to UV-B radiation and heat as conidia produced on MM, which were more tolerant to UV-B radiation and heat than conidia produced on PDAY medium in the dark. Moreover, conidial production in the dark and under the light were similar (Rangel, Fernandes, Braga, & Roberts, 2011). I did not believe the results, so I ran the test four times with four different cultures at different times; all of them produced the same results. Interestingly, all my previous studies growing the fungus under stress were, fortunately, performed in the dark! Many papers had been published before this study on photo-reactivation after UV radiation (Berrocal-Tito, Rosales-Saavedra, Herrera-Estrella, & Horwitz, 2000), but in the case of photo-reactivation, white light is shined on the conidia after the UV radiation, for the enzyme photolyase to repair the damage in the conidial DNA. Conversely, in my study (Rangel et al., 2011), the conidia were produced under fluorescent light and then the harvested conidia were exposed to UV-B irradiation or heat.

I was very interested in continuing this study after returning to Brazil. Using the light conditions in my laboratory incubator, I obtained similar results: conidia produced under light were more tolerant to UV-B radiation and heat than conidia produced in the dark (Rangel et al., 2015c). I was relieved by this result because I was still skeptical of the first results published in 2011 that light during mycelial growth induces higher conidial tolerance to UV-B radiation and heat. Moreover, my delight in these results made me interested enough to study the effect of light on the conidial tolerance of other insect and plant pathogens. We initially worked with ten fungal insect pathogens (i.e., *Aschersonia aleyrodis*, *B. bassiana*, *C. fumosorosea*, *Lecanicillium aphanocladii*, *M. anisopliae*, *M. brunneum*, *M. robertsii*, *Simplicillium lanosoniveum*, *T. cylindrosporum*, and *T. inflatum*) grown under the light and in the dark. Then, we tested the conidia produced under light and dark to five stress conditions: heat stress, oxidative stress (menadione), osmotic stress (KCl), UV radiation, and genotoxic stress (4-NQO). Several fungal species demonstrated greater stress tolerance when conidia were produced under white light than in the dark. For instance, white light induced higher tolerance of *A. aleyrodis* to KCl and 4-NQO; *B. bassiana* to KCl and 4-NQO; *C. fumosorosea* to UV radiation; *M. anisopliae* to heat and menadione; *M. brunneum* to menadione, KCl, UV radiation, and 4-NQO; *M. robertsii* to heat, menadione, KCl, and UV radiation; and *T. cylindrosporum* to menadione and KCl. However, conidia of *L. aphanocladii*, *S. lanosoniveum*, and *T. inflatum* produced under white light exhibited similar tolerance as conidia produced in the dark (Dias et al., 2021).

In addition, we tested during mycelial growth different white light irradiances (2.2, 5.9, 9.6, 15.5, and 27.6 W m^{-2}) produced by Panasonic Plant Growth Chamber MLR-352H-PA, which can be adjusted to five different white light irradiances. Then we evaluated the tolerance of conidia produced under these light irradiances to osmotic and oxidative stress. For both stress conditions, growth under any of the tested irradiances did not differ in stress tolerance, thus conidia produced in any light intensity were more tolerant than conidia produced in the dark (Dias, Pupin, Roberts, & Rangel, 2022).

Blue, green, and red light during mycelial growth influenced differently conidial stress tolerance. Conidia produced under white, blue, and green lights were more tolerant to osmotic stress than conidia produced in the dark. However, conidia produced under red light were the least tolerant to osmotic stress. Furthermore, conidia produced under white and blue light were more tolerant to UV radiation than conidia produced in the dark. Conidia produced under green and red light were less tolerant to UV radiation than conidia produced in the dark (Dias et al., 2020).

The plant pathogens *Colletotrichum acutatum* and *Fusarium fujikuroi* were also grown on PDA in the dark (control) or on PDA under continuous white, blue, green, or red light. *C. acutatum* were more tolerant to UV radiation when conidia were produced under white light than under blue or green light and more tolerant than conidia produced in the dark. Conidia of *C. acutatum* produced under red light were the least tolerant.

F. fujikuroi conidia produced under white or blue light were more UV tolerant than conidia produced in the dark or under green or red light (Costa et al., 2021). Interestingly mycelial growth under red light produced conidia of *M. robertsii*, *C. acutatum*, and *F. fujikuroi* less tolerant than the conidia produced in the dark (Costa et al., 2021; Dias et al., 2020).

The results from the studies Dias et al. (2020) and Costa et al. (2021) with blue, green, and red light made me recall my childhood. When I was eight years old and showed interest in science, my parents, Maria Hilda Naretto Rangel and Drauzio Taveiros Rangel, gave me a small microscope, which is still my office. Later, when I was 10 years old, my parents purchased some science kits at the newspaper stand called "The Scientists", which were sold inside small styrofoam boxes. The collection of kits produced in the late 1960s and early 1970s was published by the defunct FUNBEC (Brazilian Foundation for the Development of Science Teaching) in partnership with the publisher Editora Abril. These kits contained simple experiments inspired by the works of historically renowned scientists. There were 50 kits, one released every two weeks. Each kit contained a scientist's biography, one experimental manual, and the material to conduct related experiments. Although my parents were not wealthy and could not buy all these boxes, I clearly remember receiving "Einstein—The Photoelectric Effect" (Machado, 2009), which contained three plastic filters with the colors blue, green, and red. I conducted many experiments in physics (for which the kit was prepared) and biology using these filters. In one experiment, I grew seeds of *Ceiba speciosa* (Malvales: Malvaceae), a common tree in Brazil, under white, blue, green, or red light, and I found that seeds growing under blue light grew quickly and produced more mucilage during germination. These experiments were so vivid that these results stuck in my mind forever; consequently, the experiments published with light fondly revived the memories of my childhood.

11. Hypoxia and anoxia

Oxygen limitation restricts glucose metabolism, leading to reduced glucose transport and subsequent nutrient limitation of cellular metabolism (Griffin, 1996), also causing nutritive stress to the fungus. During my last year at Utah State University, I also discovered that conidia produced in Petri dishes sealed with three layers of Parafilm, which causes hypoxic stress, produced conidia more tolerant to heat but not to UV-B radiation (Rangel et al., 2015c). My group continued this study later, finding that growth under continuous hypoxia or under transient anoxia induced higher conidial stress tolerance of *M. robertsii* than conidia produced in normoxia (in a normal atmosphere). Conidia produced under hypoxia were more tolerant to four stress conditions—heat, oxidative stress, osmotic stress, and UV radiation (Silva et al., 2023). For transient anoxia, 24-h-old-mycelia of *M. robertsii* exposed to transient anoxia for five days grew as soon as it was returned to optimal atmospheric conditions. Presumably, it became dormant and then resumed active growth once the anoxia condition was eliminated. Transient anoxia, on the other hand, induced higher conidial stress tolerance to only two stress conditions—osmotic stress and UV radiation (Silva et al., 2023).

Conidia produced under hypoxia exhibited higher UV tolerance (Silva et al., 2023) than conidia produced under white light (Dias et al., 2021). This is particularly interesting because light and oxygen are sensed by the same LOV (light, oxygen, or voltage) domain, which was named by Winslow Briggs based on sequence similarity among proteins sensing these three environmental factors (Huala et al., 1997). PAS (Per-Arnt-Sim) domains are important signaling modules that monitor changes in light, redox potential, oxygen, and overall energy level of a cell (Taylor & Zhulin, 1999). Hypoxia and anoxia are environmental factors that are sensed by PAS.

12. Biotic stress

The most renowned fungal antagonism-inducing bacterial lysis was noticed by Alexander Fleming in 1928 (Shama, 2021), which led to his discovery of penicillin (Fleming, 1946). Chemical warfare between fungal communities occurs as a constitutive defense through the modification of the territory occupied by an individual and the deposition of antimicrobial compounds within (Hiscox & Boddy, 2017). The competing mycelia undergo changes in morphology, secondary metabolite production, pigment deposition, water potential or pH reduction, reactive oxygen species accumulation, and enzyme activity alterations (Hiscox & Boddy, 2017). Fungi also emit mixtures of volatile organic compounds (VOCs) during growth, which serve as communication agents and display biological activity such as germination inhibitors (Jaddaoui, Rangel, & Bennett, 2023). All these are stressful biotic conditions created by fungi that possibly induce cross-protection to stress.

I had a dream in 2016 that the fungi *Metarhizium* and *Trichoderma* were in a boxing ring. They were boxing with red gloves, and I was the referee. After waking up, I had the inspiration of letting *Trichoderma atroviride* (IMI 206040) and *M. robertsii* (ARSEF 2575) fight it out in a Petri dish to determine the conidial stress tolerance of *M. robertsii*. *M. robertsii* was grown in dual culture with *T. atroviride* on PDA using the following treatments:

(1) *Trichoderma* inoculated at the same time with *Metarhizium* (A0); (2) *Trichoderma* inoculated two days after the inoculation of *Metarhizium* (A2); (3) *Trichoderma* inoculated four days after *Metarhizium* (A4); and (4) *Trichoderma* inoculated 6 days after *Metarhizium* (A6). Antagonism by *Trichoderma* definitely stressed *Metarhizium*. Although *Trichoderma* was not able to overrun *Metarhizium* in any dual culture (Medina et al., 2020), *Metarhizium* conidia produced at A0 and A2 dual culture had a very low viability. However, conidia of *M. robertsii* produced at A4 and A6 dual cultures were more tolerant to osmotic stress, oxidative stress, and UV-B radiation than conidia produced in the control treatment without antagonism (Medina et al., 2020). Heat stress was the only stress that the dual culture A4 and A6 did not improve tolerance, which was similar to the control (Medina et al., 2020).

13. Magnetic and electric fields

Magnetic field affects mycelial growth (Mateescu, Burunţea, & Stancu, 2011; Nagy & Fischl, 2004; Ružič, Gogala, & Jerman, 1997), germination (Albertini et al., 2003; Wittekindt, Broers, Kraepelin, & Lamprecht, 1990), conidial production (Nagy & Fischl, 2004), and virulence (Jaworska, Domanski, Tomasik, & Znoj, 2016). Some studies have also reported changes in gene expression and enzyme synthesis when fungi are cultivated under a magnetic field (Albertini et al., 2003; Manoliu et al., 2006; Potenza et al., 2012). Superoxide dismutase and catalase activities also increase in magnetized cultures of *S. cerevisiae* compared with unexposed samples, suggesting oxidative stress (Kthiri et al., 2019).

Electric field, on the other hand, induces fungal hyphae to grow in a particular direction. The direction of this galvanotropic response varies for different fungal strains and for different hyphal types (Gooday, 1995). For example, *Candida albicans* reorients its growth axis in response to an electric field (Crombie, Gow, & Gooday, 1990).

However, little is known about the phenotypic effects caused by magnetic and electric fields on fungal mycelial growth and its priming responsiveness on conidial tolerance to different stress conditions; therefore, I took on this challenge (unpublished results). In this study, conidia of the insect-pathogenic fungus *Metarhizium robertsii* were produced on (1) PDA = control, (2) under nutritional stress (MM), and on PDA medium under (3) magnetic field (MF) (Fig. 5A and C) and (4) electric field (EF)

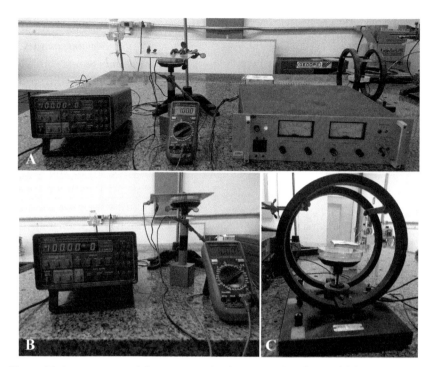

Fig. 5 (A) Equipment used for exposing the fungus to the electric field and magnetic field. (B) Parallel plate system for electric field generation. The applied electric field was 100 V/cm. By applying a potential difference (ddp) between the electrodes, it was possible to generate a uniform electric field that does not vary in time. In the experiment performed, an electric field of 100 V/cm was generated. (C) Helmholtz coil with aluminum substrate holder arranged on its central axis. The applied magnetic field was around 10 mT. For experiments with magnetism, a Helmholtz coil with 140 turns and a coil radius of 0.22 m is used, allowing the generation of magnetic fields of up to 30 mT. For both experiments, the cultures were kept on continuously for 14 days. Unpublished. *(A) For the experiments in an environment containing a uniform electric field, a set of two parallel metallic (aluminum) plates separated by a distance, d = 10 mm, was used, where one of the plates was polarized by a direct current source (Keithley brand, provided a voltage of 10–100 V) and the other was grounded. (b) To generate the magnetic field, an HP brand current source was used, which could provide values of up to 10 A.*

(Fig. 5A and B). All four treatments were incubated in the dark. The tolerances of conidia produced in these conditions were evaluated in relation to oxidative and osmotic stress, heat, and UV–B radiation. The cultures of the fungus grown on the PDA medium under magnetic and electric fields were similar to the fungus grown on the control PDA medium. *M. robertsii* conidia produced under MF and EF were more tolerant to oxidative and osmotic stress, heat, and UV–B radiation than conidia produced on PDA (control).

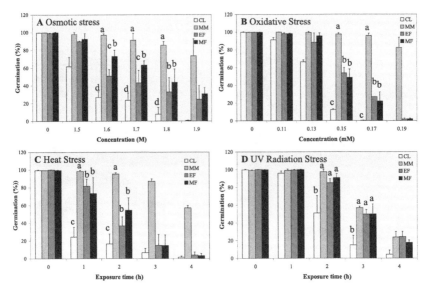

Fig. 6 Tolerance of *Metarhizium robertsii* conidia produced on (1) potato dextrose agar medium (PDA = control = CL); (2) under nutritional stress on minimal medium (MM) and on PDA medium under (3) magnetic field (MF); and (4) electric field (EF) to (A) Osmotic Stress: conidial germination on PDA medium supplemented with potassium chloride. (B) Oxidative Stress: conidial germination on PDA medium supplemented with menadione. (C) Heat Stress: germination of conidia after heat stress. (D) UV Radiation: conidial germination after exposure to UV radiation. Stress experiments were done according to Licona-Juárez et al. (2023a, 2023b). All experiments had at least three repetitions. Error bars are standard deviations from at least three independent experiments performed at different times. The effect of treatment on the percent of germination was assessed with analysis of variance of a one-way factorial. Significance levels of pair-wise mean comparisons among treatments were controlled for experiment-wise type I error using the Tukey method with overall $\alpha = 0.05$. All analyses were carried out in the free statistical program Sisvar (Ferreira, 1999, 2011). The statistical comparisons were done within each exposure time. Graph bars with the same letter are not significantly different ($P < 0.05$). Unpublished.

Both treatments—MF and EF—produced conidia with similar tolerances to all stress conditions, except for osmotic stress, where conidia from MF were more tolerant than conidia from EF (Fig. 6).

14. Some stress response mechanisms observed in *Metarhizium robertsii*

To elucidate the genetic responses of *M. robertsii* to various stresses, my group studied expression patterns using real-time PCR. For oxidative stress genes, we found that only *sod2* and *catc* genes were upregulated by nutritive stress but not by hypoxia or anoxia (Dias et al., 2020; Silva et al., 2023). On the other hand, *S. cerevisiae* exposed to anoxia experienced transient oxidative stress. The expression of *sod1* gene from yeast cells was quite variable, because first, it dropped 2–4 h after the shift from aerobic to anaerobic conditions and then increased after 5–6 h from such shift (Dirmeier et al., 2002). Studies on additional stress response genes, along with the transcription factors involved in the regulation of such genes in *M. robertsii* are needed to elucidate these complex regulatory networks.

The influence of hypoxic stress, anoxic stress, and nutritive stress on higher conidial stress tolerance is evidenced by the upregulation of heat-shock proteins 30 and 101 in these treatments (Silva et al., 2023). Anoxic stress only upregulated *hsp30* but not *hsp101* (Silva et al., 2023). In *Trichoderma reesei*, genes involved in cell protection (flavohemoglobin and heat-shock protein genes *hsp98* and *hsp30*), ergosterol synthesis (*erg1* and *erg3*), glycolysis, and the pentose phosphate pathway were upregulated during hypoxia (Bonaccorsi et al., 2006). Induction of *hsp70* under hypoxia has also been reported in *Blastocladiella emersonii* (Georg & Gomes, 2007), *C. albicans* (Setiadi, Doedt, Cottier, Noffz, & Ernst, 2006), and *Cryptococcus neoformans* (Chun, Liu, & Madhani, 2007).

Light during mycelial growth, on the other hand, upregulated *sod2*, *hsp30*, *hsp101*, and the photolyases genes *6–4phr* and *CPDphr* in *M. robertsii* (Dias et al., 2020). Light also regulated its the protease gene *pr1* and the trehalose–phosphate synthase gene *tps* (Dias et al., 2020).

15. Stress-inducing high virulence to insects

In response to stress, bacteria and fungi become more virulent to their hosts. Microgravity induces higher virulence of *Salmonlla typhimurium* in mice (Lynch & Matin, 2005). Conidia of *B. bassiana*, *M. anisopliae*, and *Paecilomyces farinosus* with elevated levels of low molecular weight polyols exhibited greater virulence against *G. mellonella* larvae than conidia that did not accumulate these compounds (Hallsworth & Magan, 1994c). Therefore, during my dissertation studies, I began studying the effects of fungal stress on insect virulence. The conidia produced on MM, MM supplemented with 3% lactose, or under osmotic stress caused by KCl or NaCl were more virulent to *Tenebrio molitor* [Coleoptera: Tenebrionidae] than conidia produced on rich medium (PDAY). In addition, conidia produced under these stress conditions germinated faster than conidia produced on

PDAY (Rangel, Alston, & Roberts, 2008a). In addition, conidia produced under white light were more virulent to *T. molitor* (Oliveira, Braga, & Rangel, 2018). Growth under white light and blue light also produced conidia that germinated faster than conidia produced in the dark (Oliveira et al., 2018). In another study, we found that growth of *M. robertsii* under transient anoxia produced conidia more virulent to *T. molitor* than conidia produced under normoxia and hypoxia (Oliveira & Rangel, 2018).

16. Fungal consciousness

One hundred and twenty two years ago, Charles Sedgwick wrote, "Consciousness is at once the oldest problem of philosophy and one of the youngest problems of science" (Minot, 1902). Indeed, it is only in recent years that we are starting to learn about the surprising world of plant consciousness. Gone are the days when plants were merely considered decorative roommates. Calvo, Gagliano, Souza, and Trewavas (2020), Segundo-Ortin and Calvo (2022), and Trewavas, Baluska, Mancuso, and Calvo (2020) have been leading the reseach about plant consciousness, and it turns out that your potted plant might be more aware of you than you are about yourself. Certainly, consciousness also proves to be a captivating and challenging subject within the realm of fungal biology. Fungal hyphae exhibit exceptional sensitivity to their environment. This responsiveness is evident at various levels, ranging from changes in hyphal structure due to alterations in exocytosis patterns to membrane excitation and wound repair mechanisms (Money, 2021, 2022). Growing hyphae expertly feel out ridges on surfaces and navigate around physical obstacles. Fungi are so sensitive that they can sense even the smallest changes in their surroundings, and they have a knack for finding openings in the leaves of their host plants. Fungal mycelia exhibit decision-making abilities and adjust their growth patterns based on interactions with other organisms (Money, 2021, 2022). Fungi have also mastered the art of communication through their chemical signals, known as volatile organic compounds (VOCs). These VOCs are the ultimate multi-taskers. They do not just gossip with neighboring fungi; they are also the enforcers of the fungal alliance. Some of these VOCs have a second job as germination inhibitors; they are the fungal approach of crowd control (Jaddaoui et al., 2023).

Recent studies have also highlighted capacity of fungi to react to stimuli, make adaptive decisions in their spatial navigation to deal with ensuing biotic and abiotic pressures (Aleklett & Boddy, 2021; Andrade-Linares et al., 2016; Fukasawa, Savoury, & Boddy, 2020), react to injury (Hernández-Oñate & Herrera-Estrella, 2015), and even exhibit indications of memory (Brown, Gow, Warris, & Brown, 2019).

Another impressive example of consciousness is found in *Phycomyces*. This fungus responds to all sorts of external stimuli. Take its phototropism, for instance—this fungus can detect light at levels as faint as 1 nW m^{-2}, basically the glow of a single star (Cerdá-Olmedo, 2001). The most mysterious sensory response of *Phycomyces* is its avoidance behavior, where the macrophores grow away from nearly any object, including threads and liquids that come close to their growing zone (Cerdá-Olmedo, 2001).

Biologists, ethologists, and geneticists are increasingly inclined to believe that mental states, awareness, consciousness—or, to generalize, sentience—are fundamental aspects of life; for example, one bacterium moving into an environment with an uncomfortably high salt content has a negative subjective state that elicits movement back toward a remembered earlier, less aversive environment (Baluška & Reber, 2019).

When I embarked on my PhD journey at Utah State University, little did I know that fungi were the silent geniuses pulling the strings! Now, I realize every experiment I did was just tapping into the mystical world of fungal consciousness. It is like I unwittingly joined the secret society of fungal whisperers!

17. The International Symposium on Fungal Stress—ISFUS

I was impressed with the world of fungal stress during his PhD days at Utah State University from 2002 to 2006, and this passion continued throughout my career. Turns out, understanding how fungi behave under pressure is essential! In medical mycology, fungal stress is about cooking up new drugs to tackle fungal diseases. In agricultural mycology, stress is like boot camp for fungi—training them to battle against insects, nematodes, and plant pathogens. Industrial mycology is the ultimate stress test, creating super fungi such as *S. cerevisiae* that can thrive under extreme osmotic, oxidative, ethanol, and heat stress.

Reading many scientific articles of many important scientists during my PhD, I kept to my subconscious an enormous desire to meet all these scientists in person. How to fulfill this aspiration came to me in a dream!

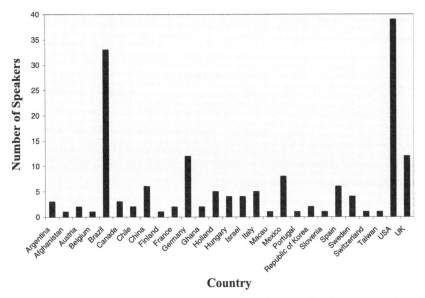

Country

Fig. 7 Country of speakers at the ISFUS 2014, ISFUS 2017, ISFUS 2019, ISFUS-IFBC 2022, and ISFUS 2024. Unpublished.

The International Symposium on Fungal Stress—ISFUS was born out of a dream, literally. On Sunday, June 16, 2013, I woke up with a eureka moment and eagerly turned to my wife, Alene Alder-Rangel, and announced, "Honey, I dreamt I organized a symposium about fungal stress!" The dream included two potential names: International Conference on Fungal Stress (ICFS) and International Fungal Stress Symposium (IFSS). However, these acronyms did not quite roll off the tongue.

Before we even got out of bed, Alene, still half-asleep, responded, "How about the International Symposium on Fungal Stress? ISFUS sounds catchy." I was, impressed by her acronym skills, but after all, she works with language editing and translating scientific articles. Then, with a sigh, Alene added, "Oh no, not another international meeting!" Remember the work of the First Brazilian Symposium about the Effects of Increased UV Radiation on Agriculture (UVRAG) which we had organized together in 2010. We both recalled the stress of organizing UVRAG but also remembered enjoying hosting the speakers and participants. Thus, ISFUS was dreamt up, acronym approved, and a new adventure in fungal stress was set in motion. From a dream and a clever wife, ISFUS came to be, proving that sometimes the best ideas are dreamt up as we sleep.

Although most of the speakers that I invited to the first ISFUS had never heard of me, they cordially accepted, and I obtained excellent funding to bring top scientists to Brazil. So far, we have organized four ISFUSs, with the fifth to occur in September 2024. The ISFUS has become well recognized in the mycological world, bringing together 162 speakers from 27 countries (Fig. 7) and cranking out five special issues. The first one made its debut in Current Genetics (Rangel et al., 2015a, 2015b), while the second, third, and fourth made their grand entrance in Fungal Biology (Alder-Rangel et al., 2018, 2020, 2023). The fifth special issue is currently in the oven, baking to perfection.

18. Conclusion

Many people have asked after my talks if all priming conditions that improved conidial stress tolerance could be applied to factory mass-production conditions. I always respond no, because most of the priming conditions lead to low conidial production. However, from a scientific standpoint, these studies offer an invaluable contribution. The extensive citation of all my publications within the field of biological control underscores their significant influence.

The trajectory of my short scientific career is an example of what can be done with very little knowledge when I started my PhD, but with a lot of hard work and the luck of having an advisor who was a great mentor, professor, father, and friend, Donald W. Roberts.

Acknowledgments

I must thank my wife, Alene Alder-Rangel, for assisting with this manuscript, helping organize ISFUS, and supporting me throughout my scientific endeavors. I would also like to extend my heartfelt gratitude to Donald Roberts and his wife, Mae Roberts, for their support and guidance throughout my PhD journey. Their dedication and encouragement have played a crucial role in my growth and deepened my understanding of mycology and insect pathology. This research was supported by grants from the National Council for Scientific and Technological Development (CNPq) of Brazil GDE 200382/02-0, PQ1D 302100/2018-0, and PQ1D 302282/2022-0.

References

Agrios, G. N. (2005). *Plant pathology* (5th ed.). San Diego: Elsevier Academic Press.

Albertini, M. C., Accorsi, A., Citterio, B., Burattini, S., Piacentini, M. P., Uguccioni, F., & Piatti, E. (2003). Morphological and biochemical modifications induced by a static magnetic field on *Fusarium culmorum. Biochimie, 85*, 963–970. https://doi.org/10.1016/j.biochi.2003.09.017.

Alder-Rangel, A. (2021). *The adventures of Donald W. Roberts*. São José dos Campos: International Insect Pathologist Inbioter.

Alder-Rangel, A., Bailão, A. M., Da Cunha, A. F., Soares, C. M. A., Wang, C., Bonatto, D., ... Rangel, D. E. N. (2018). The second International Symposium on Fungal Stress: ISFUS. *Fungal Biology, 122*, 386–399. https://doi.org/10.1016/j.funbio.2017.10.011.

Alder-Rangel, A., Bailão, A. M., Herrera-Estrella, A., Rangel, A. E. A., Gácser, A., Gasch, A. P., ... Rangel, D. E. N. (2023). The IV International Symposium on Fungal Stress and the XIII international fungal biology conference. *Fungal Biology-Uk, 127*, 1157–1179. https://doi.org/10.1016/j.funbio.2023.04.006.

Alder-Rangel, A., Idnurm, A., Brand, A. C., Brown, A. J. P., Gorbushina, A., Kelliher, C. M., ... Rangel, D. E. N. (2020). The third International Symposium on Fungal Stress—ISFUS. *Fungal Biology, 124*, 235–252. https://doi.org/10.1016/j.funbio.2020.02.007.

Aleklett, K., & Boddy, L. (2021). Fungal behaviour: A new frontier in behavioural ecology. *Trends in Ecology & Evolution, 36*, 787–796. https://doi.org/10.1016/j.tree.2021.05.006.

Andrade-Linares, D. R., Veresoglou, S. D., & Rillig, M. C. (2016). Temperature priming and memory in soil filamentous fungi. *Fungal Ecology, 21*, 10–15. https://doi.org/10.1016/j.funeco.2016.02.002.

Araújo, C. A. S., Dias, L. P., Ferreira, P. C., Mittmann, J., Pupin, B., Brancini, G. T. P., ... Rangel, D. E. N. (2018). Responses of entomopathogenic fungi to the mutagen 4-nitroquinoline 1-oxide. *Fungal Biology-UK, 122*, 621–628. https://doi.org/10.1016/j.funbio.2018.03.007.

Araújo, C. A. S., Ferreira, P. C., Pupin, B., Dias, L. P., Avalos, J., Edwards, J., ... Rangel, D. E. N. (2020). Osmotolerance as a determinant of microbial ecology: A study of phylogenetically diverse fungi. *Fungal Biology, 124*, 273–288. https://doi.org/10.1016/j.funbio.2019.09.001.

Azevedo, R. F. F., Souza, R. K. F., Braga, G. Ú. L., & Rangel, D. E. N. (2014). Responsiveness of entomopathogenic fungi to menadione-induced oxidative stress. *Fungal Biology, 118*, 990–995. https://doi.org/10.1016/j.funbio.2014.09.003.

Baluška, F., & Reber, A. (2019). Sentience and consciousness in single cells: How the first minds emerged in unicellular species. *Bioessays: News and Reviews in Molecular, Cellular and Developmental Biology, 41*, 1800229. https://doi.org/10.1002/bies.201800229.

Bergin, D., Reeves, E. P., Renwick, J., Wientjes, F. B., & Kavanagh, K. (2005). Superoxide production in *Galleria mellonella* hemocytes: Identification of proteins homologous to the NADPH oxidase complex of human neutrophils. *Infection and Immunity, 73*, 4161–4170. https://doi.org/10.1128/IAI.73.7.4161-4170.2005.

Berrocal-Tito, G. M., Rosales-Saavedra, T., Herrera-Estrella, A., & Horwitz, B. A. (2000). Characterization of blue-light and developmental regulation of the photolyase gene phr1 in *Trichoderma harzianum*. *Photochemistry and Photobiology, 71*, 662–668.

Bidochka, M. J., Kamp, A. M., Lavender, T. M., Dekoning, J., & De Croos, J. N. (2001). Habitat association in two genetic groups of the insect-pathogenic fungus *Metarhizium anisopliae*: Uncovering cryptic species? *Applied and Environmental Microbiology, 67*, 1335–1342.

Bidochka, M. J., Kasperski, J. E., & Wild, G. A. M. (1998). Occurrence of the entomopathogenic fungi *Metarhizium anisopliae* and *Beauveria bassiana* in soils from temperate and near-northern habitats. *Canadian Journal of Botany. Journal Canadien de Botanique, 76*, 1198–1204.

Bonaccorsi, E. D., Ferreira, A. J. S., Chambergo, F. S., Ramos, A. S. P., Mantovani, M. C., Farah, J. P. S., ... El-Dorry, H. (2006). Transcriptional response of the obligatory aerobe *Trichoderma reesei* to hypoxia and transient anoxia: Implications for energy production and survival in the absence of oxygen. *Biochemistry, 45*, 3912–3924. https://doi.org/10.1021/bi052045o.

Brown, A. J. P., Gow, N. A. R., Warris, A., & Brown, G. D. (2019). Memory in fungal pathogens promotes immune evasion, colonisation, and infection. *Trends in Microbiology, 27*, 219–230. https://doi.org/10.1016/j.tim.2018.11.001.

Brown, A. J. P., Larcombe, D. E., & Pradhan, A. (2020). Thoughts on the evolution of Core Environmental Responses in yeasts. *Fungal Biology, 124*, 475–481. https://doi.org/10.1016/j.funbio.2020.01.003.

Brown, D. R., Barton, G., Pan, Z., Buck, M., & Wigneshweraraj, S. (2014). Combinatorial stress responses: Direct coupling of two major stress responses in *Escherichia coli*. *Microb Cell, 1*, 315–317. https://doi.org/10.15698/mic2014.09.168.

Calvo, P., Gagliano, M., Souza, G. M., & Trewavas, A. (2020). Plants are intelligent, here's how. *Annals of Botany, 125*, 11–28. https://doi.org/10.1093/aob/mcz155.

Cerdá-Olmedo, E. (2001). Phycomyces and the biology of light and color. *FEMS Microbiology Reviews, 25*, 503–512. https://doi.org/10.1111/j.1574-6976.2001.tb00588.x.

Chun, C. D., Liu, O. W., & Madhani, H. D. (2007). A link between virulence and homeostatic responses to hypoxia during Infection by the human fungal pathogen *Cryptococcus neoformans*. *PLoS Pathogens, 3*, e22.

Costa, L. B., Rangel, D. E. N., Morandi, M. A. B., & Bettiol, W. (2013). Effects of UV-B radiation on the antagonistic ability of *Clonostachys rosea* to *Botrytis cinerea* on strawberry leaves. *Biological Control, 65*, 95–100. https://doi.org/10.1016/j.biocontrol.2012.12.007.

Costa, T. P. C., Rodrigues, E. M., Dias, L. P., Pupin, B., Ferreira, P. C., & Rangel, D. E. N. (2021). Different wavelengths of visible light influence the conidial production and tolerance to ultra-violet radiation of the plant pathogens *Colletotrichum acutatum* and *Fusarium fujikuroi*. *European Journal of Plant Pathology, 159*, 105–115. https://doi.org/10.1007/s10658-020-02146-y.

Coutinho, C., Bernardes, E., Felix, D., & Panek, A. D. (1988). Trehalose as cryoprotectant for preservation of yeast strains. *Journal of Biotechnology, 7*, 23–32. https://doi.org/10.1016/0168-1656(88)90032-6.

Crombie, T., Gow, N. A., & Gooday, G. W. (1990). Influence of applied electrical fields on yeast and hyphal growth of *Candida albicans*. *Journal of General Microbiology, 136*, 311–317. https://doi.org/10.1099/00221287-136-2-311.

Csillag, K., Emri, T., Rangel, D. E. N., & Pócsi, I. (2023). pH-dependent effect of Congo Red on the growth of *Aspergillus nidulans* and *Aspergillus niger*. *Fungal Biology-UK*. https://doi.org/10.1016/j.funbio.2022.05.006.

Czapek, F. (1901). Untersuchungen über die stickstoffgewinnung und eiweifsbildung der pflanzen" [Studies on nitrogen production and protein formation of plants]. *Beiträge zur Chemischen Physiologie und Pathologie, 1*, 538–560.

Deveau, A., Palin, B., Delaruelle, C., Peter, M., Kohler, A., & Pierrat, J. C. (2007). The mycorrhiza helper Pseudomonas fluorescens BBc6R8 has a specific priming effect on the growth, morphology and gene expression of the ectomycorrhizal fungus Laccaria bicolor S238N. *The New Phytologist, 175*. https://doi.org/10.1111/j.1469-8137.2007.02148.x.

Dias, L. P., Araújo, C. A. S., Pupin, B., Ferreira, P. C., Braga, G.Ú. L., & Rangel, D. E. N. (2018). The Xenon Test Chamber Q-SUN® for testing realistic tolerances of fungi exposed to simulated full spectrum solar radiation. *Fungal Biology, 122*, 592–601. https://doi.org/10.1016/j.funbio.2018.01.003.

Dias, L. P., Pedrini, N., Braga, G.Ú. L., Ferreira, P. C., Pupin, B., Araújo, C. A. S., ... Rangel, D. E. N. (2020). Outcome of blue, green, red, and white light on *Metarhizium robertsii* during mycelial growth on conidial stress tolerance and gene expression. *Fungal Biology, 124*, 263–272. https://doi.org/10.1016/j.funbio.2019.04.007.

Dias, L. P., Pupin, B., Roberts, D. W., & Rangel, D. E. N. (2022). Low- or high-white light irradiance induces similar conidial stress tolerance in Metarhizium robertsii. *Archives of Microbiology, 204*, 83. https://doi.org/10.1007/s00203-021-02730-8.

Dias, L. P., Souza, R. K. F., Pupin, B., & Rangel, D. E. N. (2021). Conidiation under illumination enhances conidial tolerance of insect-pathogenic fungi to environmental stresses. *Fungal Biology, 125*, 891–904. https://doi.org/10.1016/j.funbio.2021.06.003.

Dirmeier, R., O'Brien, K. M., Engle, M., Dodd, A., Spears, E., & Poyton, R. O. (2002). Exposure of yeast cells to anoxia induces transient oxidative stress: Implications for the induction of hypoxic genes. *Journal of Biological Chemistry, 277*, 34773–34784. https://doi.org/10.1074/jbc.M203902200.

Emri, T., Forgacs, K., & Pocsi, I. (2022). Biologia futura: Combinatorial stress responses in fungi. *Biologia Futura, 73*, 207–217. https://doi.org/10.1007/s42977-022-00121-8.

Faria, M., Hotchkiss, J. H., & Wraight, S. P. (2012). Application of modified atmosphere packaging (gas flushing and active packaging) for extending the shelf life of Beauveria bassiana conidia at high temperatures. *Biological Control, 61*, 78–88. https://doi.org/10.1016/j.biocontrol.2011.12.008.

Feng, M.-G. (2024). Recovery of insect-pathogenic fungi from solar UV damage: Molecular mechanisms and prospects. In D. E. N. Rangel (Ed.). *Advances in applied microbiology*. Academic Press.

Ferreira, D. F. (1999). SISVAR 4.3. Sistema de análises estatísticas, CD-ROM ed. Universidade Federal de Lavras, UFLA, Lavras, MG, Brazil.

Ferreira, D. F. (2011). Sisvar: a computer statistical analysis system. Ciência e Agrotecnologia *35*, 1039–1042.

Fillinger, S., Chaveroche, M. K., van Dijck, P., de Vries, R., Ruijter, G., Thevelein, J., & d'Enfert, C. (2001). Trehalose is required for the acquisition of tolerance to a variety of stresses in the filamentous fungus *Aspergillus nidulans*. *Microbiology-Sgm, 147*, 1851–1862.

Fleming, A. (1946). Discovery and use of penicillin. *Resenha Clinico-Cientifica, 15*, 179–186.

Fukasawa, Y., Savoury, M., & Boddy, L. (2020). Ecological memory and relocation decisions in fungal mycelial networks: Responses to quantity and location of new resources. *The ISME Journal, 14*, 380–388. https://doi.org/10.1038/s41396-019-0536-3.

Gasch, A. P., Spellman, P. T., Kao, C. M., Carmel-Harel, O., Eisen, M. B., Storz, G., ... Brown, P. O. (2000). Genomic expression programs in the response of yeast cells to environmental changes. *Molecular Biology of the Cell, 11*, 4241–4257. https://doi.org/10.1091/mbc.11.12.4241.

Gasch, A. P., & Werner-Washburne, M. (2002). The genomics of yeast responses to environmental stress and starvation. *Functional & Integrative Genomics, 2*, 181–192. https://doi.org/10.1007/s10142-002-0058-2.

Georg, R. C., & Gomes, S. L. (2007). Comparative expression analysis of members of the Hsp70 family in the chytridiomycete *Blastocladiella emersonii*. *Gene, 386*, 24–34. https://doi.org/10.1016/j.gene.2006.07.033.

Gooday, G. W. (1995). The dynamics of hyphal growth. *Mycological Research, 99*, 385–394. https://doi.org/10.1016/S0953-7562(09)80634-5.

Griffin, D. H. (1996). *Fungal physiology* (2nd ed.). New York: Wiley.

Hallsworth, J. E., & Magan, N. (1994a). Effect of carbohydrate type and concentration on polyols and trehalose in conidia of three entomopathogenic fungi. *Microbiology-Sgm, 140*, 2705–2713.

Hallsworth, J. E., & Magan, N. (1994b). Effects of KCl concentration on accumulation of acyclic sugar alcohols and trehalose in conidia of three entomopathogenic fungi. *Letters in Applied Microbiology, 18*, 8–11. 1472-765X.1994.tb00785.x.

Hallsworth, J. E., & Magan, N. (1994c). Improved biological control by changing polyols/ trehalose in conidia of entomopathogens. *Brighton Crop Protection Conference—Pests and Diseases 8D*, 1091–1096.

Harish, E., & Osherov, N. (2022). Fungal priming: Prepare or perish. *Journal of Fungi (Base), 8*. https://doi.org/10.3390/jof8050448.

Herdeiro, R. S., Pereira, M. D., Panek, A. D., & Eleutherio, E. C. A. (2006). Trehalose protects *Saccharomyces cerevisiae* from lipid peroxidation during oxidative stress. *Biochimica et Biophysica Acta (BBA)—General Subjects, 1760*, 340–346. https://doi.org/10.1016/j.bbagen.2006.01.010.

Hernández-Oñate, M. A., & Herrera-Estrella, A. (2015). Damage response involves mechanisms conserved across plants, animals and fungi. *Current Genetics, 61*, 359–372. https://doi.org/10.1007/s00294-014-0467-5.

Hilker, M., Schwachtje, J., Baier, M., Balazadeh, S., Bäurle, I., Geiselhardt, S., ... Kopka, J. (2016). Priming and memory of stress responses in organisms lacking a nervous system. *Biological Reviews, 91*, 1118–1133. https://doi.org/10.1111/brv.12215.

Hiscox, J., & Boddy, L. (2017). Armed and dangerous—Chemical warfare in wood decay communities. *Fungal Biology Reviews, 31*, 169–184. https://doi.org/10.1016/j.fbr.2017.07.001.

Hohmann, S., & Mager, W. H. (2003). *Yeast stress responses*. Berlin: Springer-Verlag.

Huala, E., Oeller, P. W., Liscum, E., Han, I. S., Larsen, E., & Briggs, W. R. (1997). Arabidopsis NPH1: A protein kinase with a putative redox-sensing domain. *Science (New York, N. Y.), 278*, 2120–2123. https://doi.org/10.1126/science.278.5346.2120.

Jaddaoui, I. E., Rangel, D. E. N., & Bennett, J. W. (2023). Fungal volatiles have physiological properties. *Fungal Biology, 127*, 1231–1240. https://doi.org/10.1016/j.funbio.2023.03.005.

Jamieson, D. J. (1992). Saccharomyces cerevisiae has distinct adaptive responses to both hydrogen peroxide and menadione. *Journal of Bacteriology, 174*, 6678–6681.

Jaworska, M., Domanski, J., Tomasik, P., & Znoj, K. (2016). Stimulation of pathogenicity and growth of entomopathogenic fungi with static magnetic field. *Journal of Plant Diseases and Protection, 123*, 295–300. https://doi.org/10.1007/s41348-016-0035-y.

Jenkins, D. E., Schultz, J. E., & Matin, A. (1988). Starvation-induced cross protection against heat or H_2O_2 challenge in *Escherichia coli*. *Journal of Bacteriology, 170*, 3910–3914. https://doi.org/10.1128/jb.170.9.3910-3914.1988.

Jennings, D. H. (1993). Stress tolerance of fungi. In P. A. Lemke (Ed.). *Mycology series* (pp. 281). New York: Marcel Dekker, Inc.

Kapoor, M., & Sveenivasan, G. M. (1988). The heat shock response of *Neurospora crassa*: Stress-induced thermotolerance in relation to peroxidase and superoxide dismutase levels. *Biochemical and Biophysical Research Communications, 156*, 1097–1102. https://doi.org/10.1016/s0006-291x(88)80745-9. S0006-291X(88)80745-9 [pii].

Karlsson, M., Atanasova, L., Jensen Dan, F., & Zeilinger, S. (2017). Necrotrophic mycoparasites and their genomes. *Microbiology Spectrum, 5*, 1128. https://doi.org/10.1128/microbiolspec.funk-0016-2016.

Krell, V., Jakobs-Schoenwandt, D., Persicke, M., & Patel, A. V. (2018). Endogenous arabitol and mannitol improve shelf life of encapsulated *Metarhizium brunneum*. *World Journal of Microbiology and Biotechnology, 34*, 108. https://doi.org/10.1007/s11274-018-2492-x.

Kthiri, A., Hidouri, S., Wiem, T., Jeridi, R., Sheehan, D., & Landouls, A. (2019). Biochemical and biomolecular effects induced by a static magnetic field in *Saccharomyces cerevisiae*: Evidence for oxidative stress. *PLoS One, 14*, e0209843. https://doi.org/10.1371/journal.pone.0209843.

Lewis, J. G., Learmonth, R. P., & Watson, K. (1995). Induction of heat, freezing and salt tolerance by heat and salt shock in *Saccharomyces cerevisiae*. *Microbiology (Reading), 141*(Pt 3), 687–694. https://doi.org/10.1099/13500872-141-3-687.

Leyer, G. J., & Johnson, E. A. (1993). Acid adaptation induces cross-protection against environmental stresses in *Salmonella typhimurium*. *Applied and Environmental Microbiology, 59*, 1842–1847. https://doi.org/10.1128/aem.59.6.1842-1847.1993.

Li, Z. Z., Alves, S. B., Roberts, D. W., Fan, M. Z., Delalibera, I., Tang, J., ... Rangel, D. E. N. (2010). Biological control of insects in Brazil and China: History, current programs and reasons for their successes using entomopathogenic fungi. *Biocontrol Science and Technology, 20*, 117–136. https://doi.org/10.1080/09583150903431665.

Licona-Juárez, K. C., Andrade, E. P., Medina, H. R., Oliveira, J. N. S., Sosa-Gómez, D. R., & Rangel, D. E. N. (2023a). Tolerance to UV-B radiation of the entomopathogenic fungus *Metarhizium rileyi*. *Fungal Biology-UK, 127*, 1250–1258. https://doi.org/10.1016/j.funbio.2023.04.004.

Licona-Juárez, K. C., Bezerra, A. V. S., Oliveira, I. T. C., Massingue, C. D., Medina, H. R., & Rangel, D. E. N. (2023b). Congo red induces trans-priming to UV-B radiation in *Metarhizium robertsii*. *Fungal Biology-UK, 127*, 1544–1550. https://doi.org/10.1016/j.funbio.2023.06.005.

Lima, D. M. C. G., Costa, T. P. C., Emri, T., Pócsi, I., Pupin, B., & Rangel, D. E. N. (2021). Fungal tolerance to Congo red, a cell wall integrity stress, as a promising indicator of ecological niche. *Fungal Biology-UK, 125*, 646–657. https://doi.org/10.1016/j.funbio.2021.03.007.

Lin, H. Y., Wang, J. J., Feng, M. G., & Ying, S. H. (2019). Autophagy-related gene ATG7 participates in the asexual development, stress response and virulence of filamentous insect pathogenic fungus *Beauveria bassiana*. *Current Genetics, 65*, 1015–1024. https://doi.org/10.1007/s00294-019-00955-1.

Lynch, S. V., Brodie, E. L., & Matin, A. (2004). Role and regulation of sigma S in general resistance conferred by low-shear simulated microgravity in *Escherichia coli*. *Journal of Bacteriology, 186*, 8207–8212. https://doi.org/10.1128/JB.186.24.8207-8212.2004.

Lynch, S. V., & Matin, A. (2005). Travails of microgravity: Man and microbes in space. *Biologist (Columbus, Ohio), 52*.

Machado, C. A. (2009). A Grande Aventura da Descoberta Científica. https://oscientistas.wordpress.com/kit-os-cientistas/albert-einstein, Os Cientistas. FUNBEC (Brazilian Foundation for the Development of Science Teaching) and Editora Abril Guarapuava, PR.

Magan, N. (2001). Physiological approaches to improving the ecological fitness of fungal biocontrol agents. In T. M. C. W. J. Butt, & N. Magan (Eds.). *Fungi as biocontrol agents*. Oxon: CAB Publishing.

Manoliu, A., Oprica, L., Olteanu, Z., Neacsu, I., Artenie, V., Creanga, D. E., ... Bodale, I. (2006). Peroxidase activity in magnetically exposed cellulolytic fungi. *Journal of Magnetism and Magnetic Materials, 300*, e323–e326. https://doi.org/10.1016/j.jmmm.2005.10.111.

Mateescu, C., Burunţea, N., & Stancu, N. (2011). Investigation of *Aspergillus niger* growth and activity in a static magnetic flux density field. *Romanian Biotechnological Letters, 16*, 6364–6368.

Matin, A., & Lynch, S. V. (2005). Investigating the threat of bacteria grown in space. *ASM News, 71*, 235–240.

McCann, M. P., Fraley, C. D., & Matin, A. (1993). The putative σ factor KatF is regulated posttranscriptionally during carbon starvation. *Journal of Bacteriology, 175*, 2143–2149.

Medina, E. Q. A., Oliveira, A. S., Medina, H. R., & Rangel, D. E. N. (2020). Serendipity in the wrestle between *Trichoderma* and *Metarhizium*. *Fungal Biology, 124*, 418–426. https://doi.org/10.1016/j.funbio.2020.01.002.

Minot, C. S. (1902). The problem of consciousness in its biological aspects. *Science (New York, N. Y.), 16*, 1–12. https://doi.org/10.1126/science.16.392.1. 16/392/1 [pii].

Mitchell, A., & Pilpel, Y. (2011). A mathematical model for adaptive prediction of environmental changes by microorganisms. *Proceedings of the National Academy of Sciences of the United States of America, 108*, 7271–7276. https://doi.org/10.1073/pnas.1019754108.

Mitchell, A., Romano, G. H., Groisman, B., Yona, A., Dekel, E., Kupiec, M., ... Pilpel, Y. (2009). Adaptive prediction of environmental changes by microorganisms. *Nature, 460*, 220–224. https://doi.org/10.1038/nature08112.

Money, N. P. (2021). Hyphal and mycelial consciousness: The concept of the fungal mind. *Fungal Biology, 125*, 257–259. https://doi.org/10.1016/j.funbio.2021.02.001. S1878-6146(21)00024-6 [pii].

Money, N. P. (2022). New theories expand cognition to fungi. *Research Outreach, 128.* https://doi.org/10.32907/RO-128-2230462618.

Morano, K. A., Grant, C. M., & Moye-Rowley, W. S. (2012). The response to heat shock and oxidative stress in *Saccharomyces cerevisiae. Genetics, 190,* 1157–1195. https://doi.org/10.1534/genetics.111.128033.

Nagy, P., & Fischl, G. (2004). Effect of static magnetic field on growth and sporulation of some plant pathogenic fungi. *Bioelectromagnetics, 25,* 316–318. https://doi.org/10.1002/bem.20015.

Nicholson, W. L., & Law, J. F. (1999). Method for purification of bacterial endospores from soils: UV resistance of natural Sonoran desert soil populations of *Bacillus* spp. with reference to *Bacillus subtilis* strain 168. *Journal of Microbiological Methods, 35,* 13–21.

Oliveira, A. S., Braga, G.Ú. L., & Rangel, D. E. N. (2018). *Metarhizium robertsii* illuminated during mycelial growth produces conidia with increased germination speed and virulence. *Fungal Biology, 122,* 555–562. https://doi.org/10.1016/j.funbio.2017.12.009.

Oliveira, A. S., & Rangel, D. E. N. (2018). Transient anoxia during *Metarhizium robertsii* growth increases conidial virulence to *Tenebrio molitor. Journal of Invertebrate Pathology, 153,* 130–133. https://doi.org/10.1016/j.jip.2018.03.007.

Panek, A. (1962). Synthesis of trehalose by baker's yeast (*Saccharomyces cerevisiae*). *Archives of Biochemistry and Biophysics, 98,* 349–355.

Panek, A. (1963). Function of trehalose in baker's yeast (*Saccharomyces cerevisiae*). *Archives of Biochemistry and Biophysics, 100,* 422–425.

Park, J.-I., Grant, C. M., Attfield, P. V., & Dawes, I. W. (1997). The freeze-thaw stress response of the yeast *Saccharomyces cerevisiae* is growth phase specific and is controlled by nutritional state via the RAS-cyclic AMP signal transduction pathway. *Applied and Environmental Microbiology, 63,* 3818–3824.

Potenza, L., Saltarelli, R., Polidori, E., Ceccaroli, P., Amicucci, A., Zeppa, S., ... Stocchi, V. (2012). Effect of 300 mT static and 50 Hz 0.1 mT extremely low frequency magnetic fields on *Tuber borchii* mycelium. *Canadian Journal of Microbiology, 58,* 1174–1182. https://doi.org/10.1139/w2012-093.

Rangel, D. E. N. (2011). Stress induced cross-protection against environmental challenges on prokaryotic and eukaryotic microbes. *World Journal of Microbiology and Biotechnology, 27,* 1281–1296. https://doi.org/10.1007/s11274-010-0584-3.

Rangel, D. E. N., & Alder-Rangel, A. (2020). History of the International Symposium on Fungal Stress—ISFUS, a dream come true!. *Fungal Biology-UK, 124,* 525–535. https://doi.org/10.1016/j.funbio.2020.02.004.

Rangel, D. E. N., Alder-Rangel, A., Dadachova, E., Finlay, R. D., Dijksterhuis, J., Braga, G.Ú. L., ... Hallsworth, J. E. (2015a). The International Symposium on Fungal Stress: ISFUS. *Current Genetics, 61,* 479–487. https://doi.org/10.1007/s00294-015-0501-2.

Rangel, D. E. N., Alder-Rangel, A., Dadachova, E., Finlay, R. D., Kupiec, M., Dijksterhuis, J., ... Hallsworth, J. E. (2015b). Fungal stress biology: A preface to the Fungal Stress Responses special edition. *Current Genetics, 61,* 231–238. https://doi.org/10.1007/s00294-015-0500-3.

Rangel, D. E. N., Alston, D. G., & Roberts, D. W. (2008a). Effects of physical and nutritional stress conditions during mycelial growth on conidial germination speed, adhesion to host cuticle, and virulence of *Metarhizium anisopliae*, an entomopathogenic fungus. *Mycological Research, 112,* 1355–1361.

Rangel, D. E. N., Anderson, A. J., & Roberts, D. W. (2006). Growth of *Metarhizium anisopliae* on non-preferred carbon sources yields conidia with increased UV-B tolerance. *Journal of Invertebrate Pathology, 93,* 127–134. https://doi.org/10.1016/j.jip.2006.05.011.

Rangel, D. E. N., Anderson, A. J., & Roberts, D. W. (2008b). Evaluating physical and nutritional stress during mycelial growth as inducers of tolerance to heat and UV-B radiation in *Metarhizium anisopliae* conidia. *Mycological Research, 112*, 1362–1372. https://doi.org/10.1016/j.mycres.2008.04.013.

Rangel, D. E. N., Bignayan, H. G., Golez, H. G., Keyser, C. A., Evans, E. W., & Roberts, D. W. (2022). Virulence of the insect-pathogenic fungi *Metarhizium* spp. to Mormon crickets, *Anabrus simplex* (Orthoptera: Tettigoniidae). *Bulletin of Entomological Research, 112*, 179–186. https://doi.org/10.1017/s0007485321000663.

Rangel, D. E. N., Braga, G. Ú. L., Anderson, A. J., & Roberts, D. W. (2005a). Influence of growth environment on tolerance to UV-B radiation, germination speed, and morphology of *Metarhizium anisopliae* var. *acridum* conidia. *Journal of Invertebrate Pathology, 90*, 55–58.

Rangel, D. E. N., Braga, G. Ú. L., Anderson, A. J., & Roberts, D. W. (2005b). Variability in conidial thermotolerance of *Metarhizium anisopliae* isolates from different geographic origins. *Journal of Invertebrate Pathology, 88*, 116–125. https://doi.org/10.1016/j.jip.2004.11.007.

Rangel, D. E. N., Braga, G. Ú. L., Fernandes, É. K. K., Keyser, C. A., Hallsworth, J. E., & Roberts, D. W. (2015c). Stress tolerance and virulence of insect-pathogenic fungi are determined by environmental conditions during conidial formation. *Current Genetics, 61*, 383–404. https://doi.org/10.1007/s00294-015-0477-y.

Rangel, D. E. N., Braga, G. Ú. L., Flint, S. D., Anderson, A. J., & Roberts, D. W. (2004). Variations in UV-B tolerance and germination speed of *Metarhizium anisopliae* conidia produced on artificial and natural substrates. *Journal of Invertebrate Pathology, 87*, 77–83. https://doi.org/10.1016/j.jip.2004.06.007.

Rangel, D. E. N., Dettenmaier, S. J., Fernandes, E. K. K., & Roberts, D. W. (2010). Susceptibility of *Metarhizium* spp. and other entomopathogenic fungi to dodine-based selective media. *Biocontrol Science and Technology, 20*, 375–389.

Rangel, D. E. N., Fernandes, E. K. K., Anderson, A. J., & Roberts, D. W. (2012). Culture of *Metarhizium robertsii* on salicylic-acid supplemented medium induces increased conidial thermotolerance. *Fungal Biology, 116*, 438–442.

Rangel, D. E. N., Fernandes, E. K. K., Braga, G. Ú. L., & Roberts, D. W. (2011). Visible light during mycelial growth and conidiation of *Metarhizium robertsii* produces conidia with increased stress tolerance. *FEMS Microbiology Letters, 315*, 81–86. https://doi.org/10.1111/j.1574-6968.2010.02168.x.

Rangel, D. E. N., Finlay, R. D., Hallsworth, J. E., Dadachova, E., & Gadd, G. M. (2018). Fungal strategies for dealing with environmental and agricultural stress. *Fungal Biology, 122*, 602–612. https://doi.org/10.1016/j.funbio.2018.02.002.

Ružič, R., Gogala, N., & Jerman, I. (1997). Sinusoidal magnetic fields: Effects on the growth and ergosterol content in mycorrhizal fungi. *Electro- and Magnetobiology, 16*, 129–142. https://doi.org/10.3109/15368379709009838.

Santos, M. P., Dias, L. P., Ferreira, P. C., Pasin, L. A., & Rangel, D. E. N. (2011). Cold activity and tolerance of the entomopathogenic fungus *Tolypocladium* spp. to UV-B irradiation and heat. *Journal of Invertebrate Pathology, 108*, 209–213. https://doi.org/10.1016/j.jip.2011.09.001.

Segundo-Ortin, M., & Calvo, P. (2022). Consciousness and cognition in plants. *Wiley Interdisciplinary Reviews—Cognitive Science, 13*, e1578. https://doi.org/10.1002/wcs.1578.

Sessa, L., Pedrini, N., Altier, N., & Abreo, E. (2022). Alkane-priming of *Beauveria bassiana* strains to improve biocontrol of the redbanded stink bug *Piezodorus guildinii* and the bronze bug *Thaumastocoris peregrinus*. *Journal of Invertebrate Pathology, 187*, 107700. https://doi.org/10.1016/j.jip.2021.107700. S0022-2011(21)00167-1 [pii].

Setiadi, E. R., Doedt, T., Cottier, F., Noffz, C., & Ernst, J. F. (2006). Transcriptional response of *Candida albicans* to hypoxia: Linkage of oxygen sensing and Efg1p-regulatory networks. *Journal of Molecular Biology, 361*, 399–411. https://doi.org/10.1016/j.jmb.2006.06.040.

Shama, G. (2021). Uninvited guests: A chronology of petri dish contaminations. *Advances in Applied Microbiology, 116*, 169–200. https://doi.org/10.1016/bs.aambs.2021.04.002.

Silva, A. M., Pedrini, N., Pupin, B., Roberts, D. W., & Rangel, D. E. N. (2023). Asphyxiation of Metarhizium robertsii during mycelial growth produces conidia with increased stress tolerance via increased expression of stress-related genes. *Fungal Biology-UK, 127*, 1209–1217. https://doi.org/10.1016/j.funbio.2023.01.005.

Souza, R. K. F., Azevedo, R. F. F., Lobo, A. O., & Rangel, D. E. N. (2014). Conidial water affinity is an important characteristic for thermotolerance in entomopathogenic fungi. *Biocontrol Science and Technology, 24*, 448–461. https://doi.org/10.1080/09583157.2013.871223.

Taylor, B. L., & Zhulin, I. B. (1999). PAS domains: Internal sensors of oxygen, redox potential, and light. *Microbiology and Molecular Biology Reviews, 63*, 479-+.

Thevelein, J. M. (1984). Regulation of trehalose mobilization in fungi. *Microbiological Reviews, 48*, 42–59.

Thevelein, J. M., & Hohmann, S. (1995). Trehalose synthase: Guard to the gate of glycolysis in yeast? *Trends in Biochemical Sciences, 20*, 3–10.

Trewavas, A., Baluska, F., Mancuso, S., & Calvo, P. (2020). Consciousness facilitates plant behavior. *Trends in Plant Science, 25*, 216–217. https://doi.org/10.1016/j.tplants.2019.12.015.

van Dijck, P., Colavizza, D., Smet, P., & Thevelein, J. M. (1995). Differential importance of trehalose in stress resistance in fermenting and nonfermenting *Saccharomyces cerevisiae* cells. *Applied and Environmental Microbiology, 61*, 109–115.

Verma, N. C., & Singh, R. K. (2001). Stress-inducible DNA repair in *Saccharomyces cerevisiae*. *Journal of Environmental Pathology, Toxicology and Oncology: Official Organ of the International Society for Environmental Toxicology and Cancer, 20*(1), 7.

Wainwright, M. (1993). Oligotrophic growth of fungi—Stress or natural state? In D. H. Jennings (Ed.). *Stress tolerance of fungi* (pp. 127–144). New York: Marcel Dekker, Inc.

Wittekindt, E., Broers, D., Kraepelin, G., & Lamprecht, I. (1990). Influence of nonthermic AC magnetic fields on spore germination in a dimorphic fungus. *Radiation and Environmental Biophysics, 29*, 143–152. https://doi.org/10.1007/bf01210559.

Ying, S.-H. (2024). Subcellular biochemistry and biology of filamentous entomopathogenic fungi. In D. E. N. Rangel (Ed.). *Advances in applied microbiology*. Academic Press.

Ying, S. H., & Feng, M. G. (2019). Insight into vital role of autophagy in sustaining biological control potential of fungal pathogens against pest insects and nematodes. *Virulence, 10*, 429–437. https://doi.org/10.1080/21505594.2018.1518089.

Zhang, Y., Zhao, J., Fang, W., Zhang, J., Luo, Z., Zhang, M., ... Pei, Y. (2009). Mitogen-activated protein kinase hog1 in the entomopathogenic fungus *Beauveria bassiana* regulates environmental stress responses and virulence to insects. *Applied and Environmental Microbiology, 75*, 3787–3795. https://doi.org/10.1128/AEM.01913-08.

CHAPTER TWO

Subcellular biochemistry and biology of filamentous entomopathogenic fungi

Sheng-Hua Ying*

Institute of Microbiology, College of Life Sciences, Zhejiang University, Hangzhou, P.R. China
*Corresponding author. e-mail address: yingsh@zju.edu.cn

Contents

Abstract

Filamentous entomopathogenic fungi (EPF) function as important biotic factors regulating the arthropod population in natural ecosystems and have great potential as biocontrol agents in modern agriculture. In the infection cycle, EPF undergo a plethora of physiological processes, including metabolism (e.g., cuticle hydrolysis and nutrient utilization), development (e.g., dimorphism and conidiation), stress response (e.g., oxidative and osmotic stresses), and immune evasion from the host. In-depth explorations of the mechanisms

Advances in Applied Microbiology, Volume 129
ISSN 0065-2164, https://doi.org/10.1016/bs.aambs.2024.04.002

involved in the lifecycle of EPF offer excellent opportunities to increase their virulence and stability, which increases the efficacy of EPF in biocontrol programs. This review discusses the current state of knowledge relating to the biological roles and regulatory mechanisms of organelles and subcellular structures in the physiology of EPF, as well as some suggestions for future investigation.

1. Introduction

Filamentous entomopathogenic fungi (EPF) naturally infect a wide range of arthropods via direct penetration of the host epicuticle and represent an important group of control factors for the host population (Meyling & Eilenberg, 2007). As archetypical insect pathogenic fungi, the species *Beauveria* and *Metarhizium* serve as a large source for the development of mycoinsecticides and mycoacaricides (De Faria & Wraight, 2007; Li et al., 2010; Wang & Feng, 2014). In general, the infection cycle of EPF consists of several steps, including conidial adhesion and germination, penetration through the host cuticle, proliferation in the host hemocoel, growth out of the cadaver, and conidiation (Huang, Shang, Chen, Gao, & Wang, 2015; Ortiz-Urquiza & Keyhani, 2013; Wang, He, Feng, & Ying, 2014). After maturation, conidia act as dormant cells that are essential for fungal dispersal and survival in the environment (Mei et al., 2020). Multiple environmental conditions, including high temperature, low humidity, and UV radiation, can have deleterious effects on conidial infectivity and the final control efficacy of EPF (Quesada-Moraga, González-Mas, Yousef-Yousef, Garrido-Jurado, & Fernández-Bravo, 2023). In the past two decades, numerous investigations have revealed the biological roles of a wide array of genes. Many reviews have summarized the molecular insights known about functional genes, signal transduction, and transcriptional regulation involved in various physiological processes associated with the EPF lifecycle (Ortiz-Urquiza & Keyhani, 2016; Wang & Wang, 2017). Here, we focus on recent advances in the biogenesis and functions of organelles and other subcellular structures, which provide a mechanistic overview on the lifecycle of EPF from a cell biology perspective.

2. Peroxisome

Peroxisome is a single membrane-bounded organelle observed in most eukaryotic cells and characterized by its catalytic roles in β-oxidation of fatty acids and detoxification of the reactive oxygen species (Pieuchot & Jedd, 2012). In filamentous EPF, peroxisomes perform multiple physiological functions, including metabolism, stress response, development, virulence, and biogenesis of Woronin bodies.

2.1 Peroxisome biogenesis factors

In fungi, more than 30 peroxins (Pex) have been genetically characterized as involved in peroxisomal biogenesis, proliferation, and matrix protein translocation (Kiel, Veenhuis, & van der Klei, 2006). Peroxisomal matrix proteins are synthesized in cytoplasm and then transported in organellar lumen. This process depends on the peroxisomal targeting signal (PTS) type 1 and type 2 at the carboxyl and amino terminus of protein, respectively. PTS1 and PTS2 proteins are recognized and translocated into peroxisome by Pex5 and Pex7, respectively (Baker, Lanyon-Hogg, & Warriner, 2016). A plethora of peroxins link peroxisomal functions to the physiological events of EPF. In *B. bassiana*, when fungal cells grow on alkane and develop into microsclerotia-like pellets, peroxisome proliferation is enhanced (Huarte-Bonnet et al., 2018, 2019). *B. bassiana* PTS1 pathway is more important than the PTS2 pathway in the protein translocation from cytosol into peroxisome. Both Pex5 and Pex7 contribute to the development, oxidation tolerance, and virulence of *B. bassiana*; in particular, Pex5 has additional roles in vegetative growth and cell-wall perturbing stress (Pang, Lin, Hou, Feng, & Ying, 2022). In yeast, receptor export from the peroxisomal matrix depends on the docking complex consisting of Pex1, Pex6, and Pex15. In filamentous fungi, Pex26 acts as the functional ortholog of yeast Pex15 (Pieuchot & Jedd, 2012). In *B. bassiana*, these three peroxins are absolutely necessary for the PTS1 pathway, and BbPex1is dispensable for the functionality of the PTS2 pathway. However, three peroxins play convergent roles in fungal nutritional utilization, development, and virulence. Notably, these peroxins contribute to fungal response to oxidative, osmotic, and cell-wall perturbing stresses as well as cytotoxicity caused by polyunsaturated fatty acids (Hou, Lin, Ding, Feng, & Ying, 2022). Pex14 is dispensable for peroxisome biogenesis but is necessary for peroxisomal targeting of proteins with PTS1 in *B. bassiana*. Its loss significantly impairs phenotypes in vegetative growth, stress response, differentiation, and virulence (Lin et al., 2023a). The RING-finger complex consists of peroxins Pex2, Pex10, and Pex12. Disruption of these three *PEX* genes does not block the peroxisomal biogenesis, and their mutants display different impaired phenotypes in growth on nutrients and under stress conditions but similar defects in acetyl-CoA biosynthesis, development, and virulence (Lei, Sun, Feng, & Ying, 2023). *B. bassiana* Pex14 contains two phosphorylation sites (S54 and T262), which are required for asexual development but dispensable for their roles in nutrient assimilation and oxidation tolerance (Lin, Feng, & Ying, 2023b). In *M. robertsii*, Pex33 is involved in fungal cell wall integrity and resistance to oxidative stress, with consequences in appressorial formation and pathogenicity (Wang et al., 2021a). During cuticle infection-related morphogenesis, histone lysine methyltransferase ASH1 activates Pex16 via H3K36 dimethylation (Wang et al., 2023a). In addition, peroxisome fission regulated by a dynamin-related protein contributes to vegetative growth, asexual development, and virulence of *M. robertsii* (Xie et al., 2020). Unlike *Metarhizium* species, *B. bassiana* infection does not involve appressorium formation (Zhang et al., 2013), which suggests that peroxisome contributes to fungal pathogenesis via different mechanisms.

2.2 Peroxisomal metabolic genes

Peroxisome harbors many proteins in the organellar lumen, which are involved in various metabolic activities (Wanders & Waterham, 2006). Peroxisomal roles in fungal utilization of fatty acids and tolerance to oxidative stress are attributed to the metabolic pathways for β-oxidation of fatty acids and detoxification of reactive oxygen species (Pieuchot & Jedd, 2012). *B. bassiana* adapts to growth on the insect-like hydrocarbons via enhancing the β-oxidation pathway during alkane degradation and catalase activities (Pedrini, Juárez, Crespo, & De Alaniz, 2006). Acetyl-coenzyme A (CoA) synthetase (Acs) catalyzes acetate and CoA into acetyl-CoA, which are an important metabolite in cellular metabolism and physiology (Starai & Escalante-Semerena, 2004). *B. bassiana* carries two Acs proteins that localize in cytoplasm (BbAcs1) and peroxisome (BbAcs2). BbAcs2 plays a more important role than BbAcs1, significantly contributing to lipid metabolism, blastospore formation, and virulence (Lei, Lin, Ding, Feng, & Ying, 2022). Sterol carrier protein 2 (Scp2) acts as a sterol-binding protein and exhibits affinity to a variety of fatty acids and their derivatives. In *B. bassiana*, a peroxisomal Scp2 displays different affinities to various lipids and contributes to fungal differentiation and virulence. This suggests this Scp2 mediates the trafficking process in peroxisomes during lipid degradation (Lin, Pang, Feng, & Ying, 2022a; Lin, Ding, Peng, Feng, & Ying, 2022b).

2.3 Woronin body

Woronin body is a peroxisome-derived organelle, in which Hex1 functions as a major protein in form of hexagonal crystal, and plays important roles in maintaining cellular integrity and heterogeneity in filamentous fungi (Bleichrodt et al., 2012). *B. bassiana* harbors a single Hex1, which is indispensable for the biogenesis of Woronin bodies. This organelle is associated with the whole lifecycle of *B. bassiana*, and its loss results in significant defects in its development and virulence, but it has no significant effects on fungal stress responses (Pang et al., 2022). Similar roles of Hex1 are also present in *M. robertsii*, although additional experiments demonstrated that Hex1 is functionally important in maintaining cell integrity and heterogeneity (Tang, Shang, Li, & Wang, 2020). These findings suggest that cell heterogeneity is involved in fungal development, which remains to be explored in detail.

3. Mitochondrion

Mitochondria are double membrane-bound organelles present in most eukaryotes and function as powerhouses to generate energy in the form of adenosine triphosphate (ATP) for cellular activities. From the point of physiological function, this organelle plays a key role in fungal virulence, pathogenicity, and drug resistance (Chatre & Ricchetti, 2014). The mitochondrial roles in fungal entomopathogens gradually become more and more attractive.

3.1 Electron transport chain proteins

The electron transport chain (ETC) is the essential machinery to generate energy in the form of ATP and involves a series of proteins and electron carriers within the mitochondrial membrane (Chatre & Ricchetti, 2014). Succinate dehydrogenase (SDH), also known as respiratory chain complex II, plays a critical role in mediating the flow of electrons through ETC (Cecchini, 2013). *B. bassiana* contains two SdhC domain-containing proteins (BbSdhC1 and BbSdhC2), in which BbSdhC1 is the functional homolog of SDH owing to its localization in mitochondria. BbSdhC1 significantly contributes to energy generation and maintaining intracellular levels of reactive oxygen species (ROS), which is essential for fungal growth, oxidation tolerance, and virulence (Ding, Li, Lei, Feng, & Ying, 2022). In *Saccharomyces cerevisiae*, Sdh3, the homolog of SdhC, is also involved in ROS production and oxidation resistance (Szeto, Reinke, Sykes, & Lemire, 2007). In the phytopathogen *Botrytis cinerea*, mutation of SdhC is detrimental to fungal resistance to succinate dehydrogenase inhibitors (Amiri, Zuniga, & Peres, 2020). Unlike the investigations in other fungi that are based on natural mutation of SdhC, the gene disruption mutant of *B. bassiana* SdhC is viable. This suggests *B. bassiana* as a model to elucidate the ETC roles in fungal physiology.

In complex III (bc_1 complex) of ETC, the Rieske iron-sulfur protein (Rip1) mediates electron transfer and proton transportation (Ndi, Marin-Buera, Salvatori, Singh, & Ott, 2018). Bcs1 is a mitochondrial AAA protein (ATPase associated with diverse cellular activities) required for the complex III assembly via mediating biosynthesis and translocation of Rip1 proteins (Cruciat, Hell, Folsch, Neupert, & Stuart, 1999; Nobrega, Nobrega, & Tzagoloff, 1992). *B. bassiana* has six BCS1-domain-containing proteins (BbBcs1a through f). Only Bbbcs1c is considered the homolog of yeast bcs1 owing to its localization in mitochondria. Although Bbbcs1c is dispensable for mitochondrial biogenesis, it significantly contributes to organellar functionality (e.g., ATP synthesis, mitochondrial targeting of proteins, and membrane potential). Its loss results in impaired phenotypes in vegetative growth, stress response, development, and virulence. In addition, BbBcs1c is involved in fungal lipid metabolism and resistance to linoleic acid stress (Hou, Ding, Peng, Feng, & Ying, 2023a). This suggests the conserved roles of ETC components in mitochondrial functionality by maintaining the homeostasis of the electron flow through the chain.

3.2 Mitochondrial membrane proteins

Mitochondria are surrounded by the inner mitochondrial membrane (IMM) and outer mitochondrial membrane (OMM). Mitofilin, an IMM protein, acts as an essential organizer in the mitochondrial inner membrane organization system that maintains the IMM architecture (Zerbes et al., 2012). In *S. cerevisiae*, mitofilin/Fcj1 localizes within crista junctions (CJs) and is indispensable for CJ formation (Harner et al., 2011). The role of mitofilin in CJ biogenesis is conserved in *B. bassiana* and is involved in energy generation, asexual development, and virulence (Wang, Peng, Feng, & Ying, 2019). Miro is a conserved mitochondrial Rho GTPase, belonging to the Ras superfamily. *B. bassiana* Miro is anchored on the OMM and has mild contributions to ATP formation, conidial germination, vegetative growth, thermotolerance, UV-B resistance, and virulence but has no significant roles in fungal responses to oxidative and osmotic stresses (Guan, Wang, Ying, & Feng, 2016). In *B. bassiana*, a mitochondrial membrane protein is called Ohmm, owing to its role in the oxidative homeostasis of the mitochondrial membrane. This protein increases under oxidative stress but has a negative effect on fungal tolerance to oxidative stress and virulence (He et al., 2015).

4. Lipid droplet

Lipid droplets (LDs) are monolayer membrane-cycled organelles that store neutral lipids (i.e., triacylglycerols and steryl esters) and are the center of lipid metabolism (Walther & Farese, 2012). Perilipins were first identified in mammalian and are localized on the surface of LDs (Marcinkiewicz, Gauthier, Garcia, & Brasaemle, 2006). In *M. anisopliae*, perlipins are also associated with LDs and engaged in accumulating lipids, which is required for appressorial turgor pressure generated by the accumulated solutes via lipid degradation. However, their loss results in a slightly decreased virulence (Wang & St Leger, 2007). The perilipin homolog displays the conserved localization and roles in lipid storage and has a slight contribution to virulence in *B. bassiana* (Wang et al., 2022). This reinforces that perilipin is an ideal marker for tracking the LDs (Li, Peng, Ding, Feng, & Ying, 2022; Li, Song, Wei, Tang, & Wang, 2022).

More and more studies have linked LDs to the biology of fungal entomopathogens. In *M. robertsii,* glycerol-3-phosphate acyltransferase (mrGAT) is localized to the endoplasmic reticulum and catalyzes the glycerolipid synthesis, which is essential for LD formation. Due to its role in fungal penetration through the host cuticle, mrGAT is indispensable for fungal pathogenesis (Gao, Shang, Huang, & Wang, 2013). DUF3129 proteins encoded by this fungus are localized to LDs. Their single loss does not affect conidial accumulation of LDs and appressorial formation; however, it impairs fungal virulence through cuticle infection (Huang, Hong, Tang, Lu, & Wang, 2019a). In *B. bassiana*, caleosin is embedded in the LD membrane and contributes moderately to LD generation and lipid storage but greatly to virulence via the cuticle infection route. Interestingly, caleosin has an additional role in determining conidial ability to disperse (Fan, Ortiz-Urquiza, Garrett, Pei, & Keyhani, 2015). In *S. cerevisiae*, steryl acetyl hydrolase 1 (Say1) catalyzes the reaction of sterol deacetylation and regulates the sterol homeostasis (Tiwari, Köffel, & Schneiter, 2007). *B. bassiana* Say1 is a cytoplasmic enzyme, and its loss results in the accumulation of enlarged LDs, which implies that BbSay1 plays a role in maintaining the degradation of lipids and indirectly affects LD formation. BbSay1 is involved in conidial germination, development, and virulence (Peng, Zhang, Feng, & Ying, 2022a). The linoleic acid (LA) response gene *LAR1* encodes a lipid-droplet protein, which plays an important role in the intracellular homeostasis of fatty acids in *B. bassiana* and fungal resistance to LA toxicity. BbLar1 mediates a comprehensive transcriptional response under LA stress. A repressed gene encoding a protein with the domain of common in the fungal extracellular membrane (CFEM) is named *BbCFEM8*. This gene also contributes to fungal resistance to LA stress and comparability with LA in bioassay (Hou, Zhang, Ding, Feng, & Ying, 2023b). In addition, BbCfem8 contributes to fungal survival under the iron-limited conditions (e.g., interspecies interaction) (Peng et al., 2022b), which suggests the interplay between iron acquisition and lipid metabolism. These reports indicate that LDs and their proteins contribute to the lipid metabolism in EPF, which is critical for the infection process, survival in the environment, and efficacy in practical application.

5. Vesicle and vacuole

In EPF, cell endomembranes form a variety of structures, including vesicles and vacuoles in various cell types (Lewis, Robalino, & Keyhani, 2009). Proteins are associated with vacuolar membrane or localized in the lumen. A small and domain-lacking protein (22.96 kDa) is localized in mycelial vacuoles and named vacuole-localized protein 4 (Vlp4). Its disruption mutant displays defective phenotypes in development, protease secretion, conidial hydrophobicity, and virulence (Chu, Sun, Zhu, Ying, & Feng, 2017). An integral membrane protein gene encoding a vacuolar membrane protein Imp has an important role in the blastospore formation but only a slight contribution to conidiation in *B. bassiana* (Ding, Lin, Feng, & Ying, 2020b). *B. bassiana* Wsc1 (cell-wall stress-responsive component) localizes in the vacuoles and cell wall/ membrane and is involved in fungal resistance to multiple stresses and virulence (Tong, Wang, Gao, Ying, & Feng, 2019). Na^+/H^+ antiporter (Nhx1) is enriched in endosomal and trans-Golgi network compartments in *B. bassiana*, and its absence leads to severe acidification and fusion defect in vacuoles. A loss-of-function mutant displays the compromised phenotypes in sporulation and virulence but enhanced tolerance to UV-B radiation (Zhu, Ying, & Feng, 2016). P4-ATPases are involved in vesicle formation and trafficking. In *B. bassiana*, cyclosporine A resistance-related P4-ATPase acts as a component involved in vesicle trafficking through the trans-Golgi-endosomes to vacuoles, which targets the drug compounds (e.g., cyclosporine A) into vacuoles for detoxification (Li et al., 2023). Enterotoxigenic bacteria commonly employ heat-labile enterotoxins (LT) as virulence factors that consist of one subunit A (LTA) and five B subunits (LTB). *B. bassiana* has 14 enterotoxin_A domain proteins, five of which display subcellular localizations at the vesicle and vacuole.

Loss-of-function mutation of five *LTA* genes significantly reduced fungal asexual development, stress resistance, and virulence (Ding, Wei, Feng, & Ying, 2023a). This indicates that the vesicle and its associated proteins play critical roles in maintaining the functionality of the endomembrane system.

6. Cell membrane

Like other living cells, fungal cells are surrounded by cell membranes, isolating a cell from its environment. Cell membranes are composed primarily of lipids and proteins (Douglas & Konopka, 2014). In filamentous fungi, membrane integrity is essential for conidial germination (Palma-Guerrero et al., 2009). In *B. bassiana*, the iron-limitation-responsive transcription factor HapX regulates conidial oleic acid homeostasis via transcriptional activation of the $\Delta 9$-fatty acid desaturase gene (*OLE1*), which is important for membrane functionality involved in the early stage of the infection process (Peng et al., 2020). Other physiological processes linked to the cytomembrane integrity of *B. bassiana* include lipid trafficking (e.g., BbScp2) (Lin et al., 2022a), thioesterification of fatty acid (e.g., BbFaa1) (Li et al., 2022), and the deacetylation of acetylated sterol (e.g., BbSay1) (Peng et al., 2022a). *M. robertsii* carries an insect-like sterol carrier gene NPC2a acquired through horizontal gene transfer. This gene is specifically expressed after fungus enters the insect hemolymph and is required to maintain membrane integrity, which consequentially contributes to fungal virulence (Zhao et al., 2014). *B. bassiana* Thm1 (transcription factor for heat and membrane integrity) belongs to the Zn (II)$_2$Cys$_6$ (Gal4–like) family. BbThm1 functions in the maintenance of cell membrane integrity, which results from its regulatory role in sterol synthesis (Huang, Keyhani, Zhao, & Zhang, 2019b). Current findings indicate that lipid metabolism is highly linked to membrane functionality in EPF, and the question about the involvement of other pathways is still open.

7. Cell wall

In filamentous fungi, the cell wall is a highly complex assembly of polysaccharides (e.g., chitin and β-glucan) and proteins. It provides mechanical strength to the cells, determines cell morphology, and mediates fungal interaction with biotic and abiotic environments (Chevalier et al., 2023; Gow & Lenardon, 2023). Generally speaking, entomopathogenic fungi are more sensitive to cell-wall stress than other filamentous fungi, including plant-pathogenic, mycoparasitic, and saprotrophic fungi (Lima et al., 2021). Numerous efforts have been devoted to characterizing genes that function in the cell-wall integrity of fungal entomopathogens.

7.1 Cell-wall carbohydrates

Chitin is a scaffold component in the cell wall of filamentous fungi. Seven chitin synthases function divergently in *M. acridum* growth, stress tolerance, cell-wall integrity, and virulence. MaChsIII, MaChsV, and MaChsVII have evident contributions to chitin organization and distribution in the cell wall, which suggests that the three genes are required for cell-wall integrity (Zhang et al., 2019). In *M. robertsii*, a bZIP-type transcription factor negatively regulates the expression of chitin synthase genes, and the increased chitin level impairs cell-wall integrity, conidial hydrophobic trait, and virulence via topical infection (Huang, Shang, Chen, Cen, & Wang, 2015a). β-glucan is another essential cell-wall component in most fungi. β-1,3-glucan synthase contributes to the synthesis of β-1,3-glucan in *M. acridum*, which is important to cell-wall integrity, tolerance to osmotic stress, and conidiation (Yang, Jin, & Xia, 2011). Calcineurin (CN), a Ca^{2+}-calmodulin-dependent serine-threonine protein phosphatase, is highly conserved in eukaryotes. Mutation of catalytic subunit A (CnA) greatly reduces chitin and β-1,3-glucan levels in the cell wall, with impaired phenotypes in conidiation, hyphal morphology, and virulence in *M. acridum* (Cao, Du, Luo, & Xia, 2014). In *B. bassiana*, mitogen-activated-protein kinases (MAPKs) play important roles in cell-wall integrity, development, and virulence. The Slt2 MAPK pathway becomes involved in regulating biosynthesis of chitin and glucan (Huang et al., 2015b; Luo et al., 2012). Under cell wall-perturbing conditions, Slt2 is required for CN phosphatase activity in *B. bassiana*, suggesting the interplay between these two pathways (Huang et al., 2015b). Ndt80-like transcription factor Ron1 and $Zn(II)_2Cys_6$ transcriptional regulator Tpc1 play important roles in chitin synthesis, conidiation, and virulence of *B. bassiana*, which is ascribed to their roles in the N-acetylglucosamine (GlcNAc) catabolism and transcriptional activation of chitin synthetase genes (Qiu et al., 2021, 2022). MADS-box transcription factor Mcm1 is involved in regulating the synthesis of chitin and other carbohydrates (Zhao et al., 2019a). These findings imply that a complex interplay and regulatory network is involved in chitin biosynthesis in EPF.

7.2 Cell-wall associated proteins

Filamentous EPFs generate three types of single cells, including conidium, in vitro blastospore, and in vivo hyphal body, which display distinct morphologies and surface components (Holder, Kirkland, Lewis, & Keyhani, 2007). Fungal cell-wall-associated proteins are important components of cell wall and have highly diverse functions in fungal physiology.

In the natural environment, conidia are produced on solid matrix or the host cadaver surface (Wang et al., 2014). Hydrophobins involve two classes (class I and II), dominate on the conidial surface, and confer the hydrophobic property. The coat protein Hyd1 mediates conidial attachment to the insect cuticle and competence in the rhizosphere when colonizing in plant roots (Moonjely, Keyhani, & Bidochka, 2018; Zhang, Xia, Kim, & Keyhani, 2011) as well as fungal ability to interfere with the host behavior (e.g., oviposition) (Falchi, Marche, Mura, & Ruiu, 2015). *M. acridum* (a specific pathogen of locusts and grasshoppers) employs Hyd3 to activate specifically the immune responses of its specific host, *Locusta migratoria manilensis* (Jiang, Ligoxygakis, & Xia, 2020). Therefore, the hydrophobin amount has been suggested as an index to evaluate the virulence of the EPF strains (Shahriari, Zibaee, Khodaparast, & Fazeli-Dinan, 2021). Hydrophobin synthesis and assembly are finely tuned during conidial maturation. Several genes have been linked to the expression of hydrophobin genes in EPF, including histone methyltransferase (Set2 and Ash1), the subunit of COP9 signalosome (Cns5), and adhesion (Adh2) (Mou, Ren, Tong, Ying, & Feng, 2021; Ren, Mou, Ying, & Feng, 2021; Zhou, Yu, Ying, & Feng, 2021). Asexual development activator WetA and a subunit of SWI/SNF complex Ssr4 are important regulators in conidial production and the formation of the rod-like structures of hydrophobin (Shao, Cai, Tong, Ying, & Feng, 2020; Zhang, Zhang, Xu, Ying, & Feng, 2023). *B. bassiana* Lec1, a lectin-like protein containing a Fruit Body_domain, binds to chitobiose and chitin in fungal cell wall.BbLec1 is essential for stabilizing the eisosome morphology, which contributes to conidial cell-wall biogenesis and hydrophobin assembly (Peng, Ding, Lin, Feng, & Ying, 2021). Acyl-CoA synthetase that catalyzes the synthesis of acyl-CoA is also known as fatty acid activation protein (Faa). In *B. bassiana*, an ortholog of yeast Faa1 is essential for cytomembrane integrity and the translocation of hydrophobin across cell membrane. The loss of BbFaa1 reduces conidial hydrophobicity and virulence (Li et al., 2022). More investigations are needed to elucidate the regulatory mechanisms involved in the hydrophobin functionality.

Hyphal bodies are generated by EPF via dimorphic transition in the host hemocoel and play essential roles in the fungus-host interaction (Wanchoo, Lewis, & Keyhani, 2009). To overcome the host immune defense, EPF employ a surface modification strategy. *M. anisopliae* utilizes a collagen-like protein MCL1 to block the recognition of β-1,3-glucan by the host immune system (Wang & St Leger, 2006). In *B. bassiana*, LysM effectors compromise insect immune responses, and the chitin-binding one protects fungal cells from chitinase hydrolysis (Cen, Li, Lu, Zhang, & Wang, 2017). Similarly, a cell-wall protein in *B. bassiana* (BbCwp) contributes to fungal development in host hemocoel and virulence. This protein maintains homeostasis of conidial lectin-binding traits and fungal resistance to osmotic and oxidative stresses when treated by hydrolytic enzymes present in the host hemolymph (Ding, Hou, Feng, & Ying, 2020a). *B. bassiana* endo-β-1,3-glucanase (BbEng1) is localized to the cell wall, where it remodels the pathogen-associated molecular patterns that initiate host defense responses, thus facilitating fungal colonization within the host hemocoel (Wang et al., 2023b). Many exclusive proteins have been characterized in hyphal bodies (Yang et al., 2017). This improves the understanding of the colonization mechanisms of EPF within the hosts.

7.3 Other proteins related to cell-wall integrity

Two-component signaling pathways generally include sensor histidine kinases and response regulators. *M. robertsii* Skn7 is required for cell-wall biosynthesis, sporulation, and virulence, but has no obvious contributions to fungal tolerance to oxidative and osmotic stresses (Shang, Chen, Chen, Lu, & Wang, 2015). Heterotrimeric G-proteins are crucial for fungal growth, differentiation, stress response, and virulence in *M. robertsii*. Two group II G alpha subunits (MrGPA2 and MrGPA4) are localized at the cytoplasm and required for cell-wall integrity (Tong et al., 2022).

Transmembrane proteins function as sensors mediating intracellular signal pathways in response to stress. *M. rileyi* Mid2 is a transmembrane protein and acts as a sensor protein in the cell-wall integrity pathway, and its absence results in phenotypic defects in conidiation, dimorphism, stress resistance, virulence (Xin et al., 2020). Glycosylphosphatidylinositol (GPI) anchoring is a kind of post-translation modification that targets proteins to the cytomembrane. Ecm33, a GPI-anchored protein, is involved in fungal cell wall integrity, with functions in conidiation, stress response, and virulence; however, its biological functions differ among fungal species (Chen, Zhu, Ying, & Feng, 2014).

Thiazole and pyrimidine are structural molecules of thiamine (vitamin B1). Thiamine-biosynthesis enzymes BbPyr and BbThi contribute to the integrity of conidial cell wall, and their losses result in significant reduction in conidial production (Jin et al., 2021).

Protein O-glycosylation is an essential post-translational modification process with O-mannosyltransferase mediating modification. *M. acridum* has three O-glycoside mannosyltransferase homologs (Pmt1, Pmt2, and Pmt4), which are required for cell-wall formation and fungal response to cell-wall perturbing stress, as well as fungal development and virulence (Wen, Tian, Xia, & Jin, 2021; Zhao, Tian, Xia, & Jin, 2019b).

These investigations link signal transduction, metabolism, and post-translation modification to the integrity and functionality of cell wall in EPF.

8. Autophagy

Autophagy is an intracellular mechanism that delivers cytoplasmic materials to vacuoles for degradation and is classified in three types: microautophagy, macroautophagy, and chaperone-mediated autophagy. Microautophagy and macroautophagy occur in a selectively or non-selectively manner. A plethora of autophagy-related genes (ATG) are organized into different functional groups, including ATG1 kinase complex (A1C), membrane recruiting system (MRS), phosphoinositide 3-kinasecomplex (PI3KC), ubiquitin-like conjugation system (ULCS), and degradation and transportation system (DTS) (Galluzzi et al., 2017; Zhang, 2022). In EPF, except for ATGs in A1C, other ATGs involved in different functional groups are not completely identical to those in yeast. Autophagic events are associated with the whole lifecycle of EPF, including pathogenic and saprotrophic growth. Autophagy plays important roles in various processes of the infection cycle of EPF, including nutritional starvation/shift, development, stress response, and virulence (Ying & Feng, 2019). In addition to findings primarily obtained from Atg1, 8, and 11, a series of forward genetic analyses must still be conducted to unravel the potential roles of autophagy and its related genes in EPF.

8.1 Biological roles of autophagy and its related genes

Macroautophagy involves autophagosome formation in which Atg8-phosphatidylethanolamine (PE) conjugate acts as a major structural component of the autophagosome membrane. Atg8-PE is generated by ULCS, in which the cysteine protease Atg4 exposes the glycine residue at the carboxyl terminus of Atg8 (Nakatogawa, 2020). *B. bassiana* Atg4 is required for the autophagic process, which plays important roles in mycelial growth, development, antioxidant response, and virulence (Ding, Wei, Feng, & Ying, 2023b). In *M. robertsii*, Atg4 is dispensable for appressorial formation but significantly contributes to lipid accumulation and virulence (Duan et al., 2013). Atg12 and Atg16 function in complex controlling Atg8–PE formation in yeast. Although their homologs in *B. bassiana* function divergently in autophagy, growth, and virulence, but have similar roles in fungal responses to oxidative stress (Hou, Wang, Lin, Feng, & Ying, 2020). In yeast, Atg22 is a permease that delivers the degradation products in the vacuole to the cytosol (Yang, Huang, Geng, Nair, & Klionsky, 2006). However, the number of Atg22 homologs differs significantly among filamentous fungi (Ying & Feng, 2019). The subcellular distributions of the four Atg22-like proteins (BbAtg22A through D) in *B. bassiana* significantly differ. BbAtg22A is localized in lipid droplets. BbAtg22B and BbAtg22C are associated with vacuoles, and BbAtg22D has dual localizations of cytomembrane and vacuole. Four Atg22-like proteins are involved in starvation response, development, and virulence with different contributions (Ding, Zhang, Feng, & Ying, 2023c). These findings indicate that the Atg homologs perform different roles in different fungi and even in the same species.

In addition to the well-known biological roles, new functions have been unraveled for autophagy in EPF. Conidial environmental persistence is critical for fungal dispersal in the environment and the higher biocontrol efficacy of EPF. Autophagy functions as a safeguard mechanism to maintain conidial lifespan during dormancy, wherein an Atg8-dependent vacuolar targeting of hydrolase is essential for conidial recovery from dormancy (Ding, Feng, & Ying, 2023d, 2023e). *M. robertsii* invades the host through cuticle penetration by appressoria in which autophagic flux is greatly mobilized. Microautophagy mediates the direct engulfment of the lipid droplets into vacuoles for degradation, which generates proper turgor pressure in appressoria (Li et al., 2023).

8.2 Vacuolar targeting pathway in autophagy

In non-selective macroautophagy, the cytoplasmic materials (also referred to as autophagic cargoes) are randomly engulfed in vacuoles. However, it has been well demonstrated that a plethora of cargoes (e.g., organelles and macromolecules) are targeted to vacuoles via a selective pathway (Farré & Subramani, 2016). In EPF, selective autophagy is associated with fungal stress response, development, and virulence (Ying & Feng, 2019). However, only pexophagy, the selective autophagy of peroxisome, has been clarified for its biological roles and targeting pathway in *B. bassiana*.

Pexophagy occurs when fungal cells are at the early stage of starvation and under oxidative stress, develop into spores, and propagate within the host (Ding, Peng, Chu, Feng, & Ying, 2018). This selective degradation of organelles involves a straightforward trafficking route. Peroxin 14 (Pex14) is a peroxisomal protein and mediates the organellar recognition by Nbr1 (neighbor of BRCA1 gene 1). Nbr1 sequentially interacts with Atg8 responsible for autophagosome biogenesis, which recruits the targeted peroxisomes to autophagosomes. Finally, the sequestrated cargo is trans-located into a vacuole for degradation (Lin et al., 2023a). In addition, two phosphorylation sites (S54 and T262) in BbPex14 play different roles in pexophagy, in which the site of T262 is indispensable (Lin et al., 2023b).

Autophagy has a role in the translocation of hydrolases into vacuoles. These enzymes include aminopeptidase I (Ape1), aspartyl aminopeptidase (Ape4), leucine aminopeptidase III (Lap3), and a–mannosidase (Ams1) (Kageyama, Suzuki, & Ohsumi, 2009; Lamark & Johansen, 2021; Yuga, Gomi, Klionsky, & Shintani, 2011). In S. cerevisiae, the cytoplasm-to-vacuole targeting (Cvt) pathway is considered a biosynthetic autophagy-related route for vacuolar targeting of cytoplasmic enzymes. Atg19 functions as the receptor for selective autophagy (SAR) to recognize the target enzymes and then targets the protein complex to autophagosomes to be fused with vacuole (Gatica, Lahiri, & Klionsky, 2018). As revealed in B. bassiana, there are two homologs of yeast aminopeptidase I (Ape1) (i.e., BbApe1A and BbApe1B). These two homologs contribute to fungal starvation tolerance, development, and virulence. Nbr1 acts as a SAR to target these two Ape1 homologs to Atg8 anchored on autopha-gosome during fungal development. As for Ape1A, Atg11 mediates Nbr1 interaction with Atg8; however, as for Ape1B, Nbr1 directly interacts with Atg8 (Ding, Wei, Feng, & Ying, 2023f). Filamentous fungi have no homolog of yeast Atg19 (Meijer, van der Klei, Veenhuis, & Kiel, 2007). Thus, Nbr1 functions as a functional homolog of yeast Atg19 in filamentous fungi, which is similar to the Nbr1-mediated vacuolar targeting (NVT) pathway in fission yeast (Wang et al., 2021). Conidial germination is critical for the pathogenicity of fungal insect pathogens (Sephton-Clark & Voelz, 2018). B. bassiana aspartyl aminopeptidase (Ape4) contributes to conidial persistence and germination after dormancy. Ape4 is translocated into vacuoles via its direct interaction with Atg8, which is dependent on the autophagic function of Atg8. This Ape4-Atg8 pathway is suggested as a simplified cytoplasm-to-vacuole pathway (sCvt) due to the absence of the autophagy receptor Atg19 in yeast Cvt pathway (Ding, Lin, Hou, Feng, & Ying, 2023e). These findings suggest the distinctiveness of a selective autophagy pathway in fungi.

8.3 Regulation of autophagy

As revealed in *B. bassiana*, repression or enhancement of autophagy inhibits conidial germination (Ding et al., 2023e). Therefore, the autophagic level is finely maintained in EPF. The target of rapamycin (TOR) pathway is a main regulator of autophagy and inhibits autophagy via phosphorylation of Atg13, a regulator of Atg1 complex. The inactivation of TOR kinase by initial signals results in dephosphorylation of Atg13, which activates Atg1 complex and autophagic flux (Russell, Yuan, & Guan, 2014). Atg1 mediates the comprehensive proteome and phosphoproteome in *B. bassiana* and directly phosphorylates the E2-like enzyme Atg3 of the ubiquitin–like conjugation system (ULCS). The phosphorylation of Atg3 is indispensable for the formation of Atg8-PE during fungal development and under starvation stress, which indicates that this phosphorylation process is critical for ULCS functionality. In addition to acting as a regulator of autophagy initiation, Atg1 orchestrates autophagosome biogenesis via its feedforward control on ULCS (Lin et al., 2022a, 2022b). As summarized in Ying and Feng (2019),

G-protein pathway and alternative splicing are involved in regulation of autophagy. However, the complicated regulatory network for autophagy remains largely unknown in EPF.

9. Future perspectives

In the past decade, great advances have been made in exploring functional genes of EPF, and this review shows a bird's eye view of subcellular biochemistry and cell biology. As a eukaryote, most organelles and subcellular structures are conserved in EPF, in which many subcellular structures remain to be functionally unraveled in future studies. Furthermore, different organelles and subcellular structures function on the same pathway. Thus, it is important to determine their regulatory network and interplay mechanism. EPF have a unique habit, so dynamic modifications are required for fungal cell biology to adapt to varying environments. A more complete understanding of cell biology in EPF will advance mechanistic breakthroughs involving fungal interaction with abiotic and biotic environments, facilitating application of this family of fungi in sustainable agriculture.

Acknowledgments

We gratefully acknowledge the funding by the National Natural Science Foundation of China (32170027). We are also thankful to Alene Alder Rangel for the English review.

References

Amiri, A., Zuniga, A. I., & Peres, N. A. (2020). Mutations in the membrane-anchored SdhC subunit affect fitness and sensitivity to succinate dehydrogenase inhibitors in *Botrytis cinerea* populations from multiple hosts. *Phytopathology, 110*(2), 327–335.

Baker, A., Lanyon-Hogg, T., & Warriner, S. L. (2016). Peroxisome protein import: A complex journey. *Biochemical Society Transactions, 44*(3), 783–789.

Bleichrodt, R. J., Van Veluw, G. J., Recter, B., Maruyama, J., Kitamoto, K., & Wösten, H. A. (2012). Hyphal heterogeneity in *Aspergillus oryzae* is the result of dynamic closure of septa by Woronin bodies. *Molecular Microbiology, 86*(6), 1334–1344.

Cao, Y., Du, M., Luo, S., & Xia, Y. (2014). Calcineurin modulates growth, stress tolerance, and virulence in *Metarhizium acridum* and its regulatory network. *Applied Microbiology and Biotechnology, 98*(19), 8253–8265.

Cecchini, G. (2013). Respiratory complex II: Role in cellular physiology and disease. *Biochimica et Biophysica Acta, 1827*(5), 541–542.

Cen, K., Li, B., Lu, Y., Zhang, S., & Wang, C. (2017). Divergent LysM effectors contribute to the virulence of *Beauveria bassiana* by evasion of insect immune defenses. *PLoS Pathogens, 13*(9), e1006604.

Chatre, L., & Ricchetti, M. (2014). Are mitochondria the Achilles' heel of the Kingdom Fungi? *Current Opinion in Microbiology, 20*, 49–54.

Chen, Y., Zhu, J., Ying, S. H., & Feng, M. G. (2014). The GPI-anchored protein Ecm33 is vital for conidiation, cell wall integrity, and multi-stress tolerance of two filamentous entomopathogens but not for virulence. *Applied Microbiology and Biotechnology, 98*(12), 5517–5529.

Chevalier, L., Pinar, M., Le Borgne, R., Durieu, C., Peñalva, M. A., Boudaoud, A., & Minc, N. (2023). Cell wall dynamics stabilize tip growth in a filamentous fungus. *PLoS Biology, 21*(1), e3001981.

Chu, Z. J., Sun, H. H., Zhu, X. G., Ying, S. H., & Feng, M. G. (2017). Discovery of a new intravacuolar protein required for the autophagy, development and virulence of *Beauveria bassiana*. *Environmental Microbiology, 19*(7), 2806–2818.

Ndi, M., Marin-Buera, L., Salvatori, R., Singh, A. P., & Ott, M. (2018). Biogenesis of the bc1 complex of the mitochondrial respiratory chain. *Journal of Molecular Biology, 430*(21), 3892–3905.

Cruciat, C. M., Hell, K., Folsch, H., Neupert, W., & Stuart, R. A. (1999). Bcs1p, an AAA familymember, is a chaperone for the assembly of the cytochrome bc(1) complex. *The EMBO Journal, 18*(19), 5226–5233.

De Faria, M. R., & Wraight, S. P. (2007). Mycoinsecticides and Mycoacaricides: A comprehensive list with worldwide coverage and international classification of formulation types. *Biological Control, 43*(3), 237256.

Ding, J. L., Wei, K., Feng, M. G., & Ying, S. H. (2023a). Homologs of bacterial heat-labile enterotoxin subunit A contribute to development, stress response, and virulence in filamentous entomopathogenic fungus *Beauveria bassiana*. *Frontiers in Immunology, 14*, 1264560.

Ding, J. L., Wei, K., Feng, M. G., & Ying, S. H. (2023b). Autophagy-related gene 4 participates in the asexual development, stress response and virulence of filamentous insect pathogenic fungus *Beauveria bassiana*. *Journal of Fungi (Basel), 9*(5), 543.

Ding, J. L., Zhang, H., Feng, M. G., & Ying, S. H. (2023c). Divergent physiological functions of four Atg22-like proteins in conidial germination, development, and virulence of the entomopathogenic fungus *Beauveria bassiana*. *Journal of Fungi (Basel), 9*(2), 262.

Ding, J. L., Feng, M. G., & Ying, S. H. (2023d). Autophagy safeguards conidial environmental persistence in filamentous fungi. *Autophagy Reports, 2*, 2205343.

Ding, J. L., Lin, H. Y., Hou, J., Feng, M. G., & Ying, S. H. (2023e). The entomopathogenic fungus *Beauveria bassiana* employs autophagy as a persistence and recovery mechanism during conidial dormancy. *mBio, 14*(2), e0304922.

Ding, J. L., Wei, K., Feng, M. G., & Ying, S. H. (2023f). Two aminopeptidase I homologs convergently contribute to pathobiology of fungal entomopathogen *Beauveria bassiana* via divergent physiology-dependent autophagy pathways for vacuolar targeting. *Journal of Advanced Research*. https://doi.org/10.1016/j.jare.2023.06.007.

Ding, J. L., Hou, J., Feng, M. G., & Ying, S. H. (2020a). Transcriptomic analyses reveal comprehensive responses of insect hemocytes to mycopathogen *Beauveria bassiana*, and fungal virulence-related cell wall protein assists pathogen to evade host cellular defense. *Virulence, 11*(1), 1352–1365.

Ding, J. L., Lin, H. Y., Feng, M. G., & Ying, S. H. (2020b). Mbp1, a component of the *MluI* cell cycle box-binding complex, contributes to morphological transition and virulence in the filamentous entomopathogenic fungus *Beauveria bassiana*. *Environmental Microbiology, 22*(2), 584–597.

Ding, J. L., Li, X. H., Lei, J. H., Feng, M. G., & Ying, S. H. (2022). Succinate dehydrogenase subunit C contributes to mycelial growth and development, stress response, and virulence in the insect parasitic fungus *Beauveria bassiana*. *Microbiology Spectrum, 10*(5) e0289122.

Ding, J. L., Peng, Y. J., Chu, X. L., Feng, M. G., & Ying, S. H. (2018). Autophagy-related gene *BbATG11* is indispensable for pexophagy and mitophagy, and contributes to stress response, conidiation and virulence in the insect mycopathogen *Beauveria bassiana*. *Environmental Microbiology, 20*(9), 3309–3324.

Douglas, L. M., & Konopka, J. B. (2014). Fungal membrane organization: The eisosome concept. *Annual Review of Microbiology, 68*, 377–393.

Duan, Z., Chen, Y., Huang, W., Shang, Y., Chen, P., & Wang, C. (2013). Linkage of autophagy to fungal development, lipid storage and virulence in *Metarhizium robertsii*. *Autophagy, 9*(4), 538–549.

Falchi, G., Marche, M. G., Mura, M. E., & Ruiu, L. (2015). Hydrophobins from aerial conidia of *Beauveria bassiana* interfere with *Ceratitis capitata* oviposition behavior. *Biological Control, 81*, 37–43.

Fan, Y., Ortiz-Urquiza, A., Garrett, T., Pei, Y., & Keyhani, N. O. (2015). Involvement of a caleosin in lipid storage, spore dispersal, and virulence in the entomopathogenic filamentous fungus, *Beauveria bassiana*. *Environmental Microbiology, 17*(11), 4600–4614.

Farré, J. C., & Subramani, S. (2016). Mechanistic insights into selective autophagy pathways: Lessons from yeast. *Nature Reviews. Molecular Cell Biology, 17*(9), 537–552.

Galluzzi, L., Baehrecke, E. H., Ballabio, A., Boya, P., Bravo-San Pedro, J. M., Cecconi, F., ... Kroemer, G. (2017). Molecular definitions of autophagy and related processes. *The EMBO Journal, 36*(13), 1811–1836.

Gao, Q., Shang, Y., Huang, W., & Wang, C. (2013). Glycerol-3-phosphate acyltransferase contributes to triacylglycerol biosynthesis, lipid droplet formation, and host invasion in *Metarhizium robertsii*. *Applied and Environmental Microbiology, 79*(24), 7646–7653.

Gatica, D., Lahiri, V., & Klionsky, D. J. (2018). Cargo recognition and degradation by selective autophagy. *Nature Cell Biology, 20*(3), 233–242.

Gow, N. A. R., & Lenardon, M. D. (2023). Architecture of the dynamic fungal cell wall. *Nature Reviews. Microbiology, 21*(4), 248–259.

Guan, Y., Wang, D. Y., Ying, S. H., & Feng, M. G. (2016). Miro GTPase controls mitochondrial behavior affecting stress tolerance and virulence of a fungal insect pathogen. *Fungal Genetics and Biology, 93*, 1–9.

He, Z., Zhang, S., Keyhani, N. O., Song, Y., Huang, S., Pei, Y., & Zhang, Y. (2015). A novel mitochondrial membrane protein, Ohmm, limits fungal oxidative stress resistance and virulence in the insect fungal pathogen *Beauveria bassiana*. *Environmental Microbiology, 17*(11), 4213–4238.

Hou, J., Ding, J. L., Peng, Y. J., Feng, M. G., & Ying, S. H. (2023a). Genome-wide identification of BCS1 domain-containing proteins reveals the mitochondrial bcs1 essential for growth, stress response, and virulence of the filamentous entomopathogenic fungus *Beauveria bassiana*. *Microbiological Research, 267*, 127262.

Hou, J., Lin, H. Y., Ding, J. L., Feng, M. G., & Ying, S. H. (2022). Peroxins in peroxisomal receptor export system contribute to development, stress response, and virulence of insect pathogenic fungus *Beauveria bassiana*. *Journal of Fungi (Basel), 8*(6), 622.

Hou, J., Wang, J. J., Lin, H. Y., Feng, M. G., & Ying, S. H. (2020). Roles of autophagy related genes in conidiogenesis and blastospore formation, virulence, and stress response of *Beauveria bassiana*. *Fungal Biology, 124*(12), 1052–1057.

Hou, J., Zhang, H., Ding, J. L., Feng, M. G., & Ying, S. H. (2023b). Transcriptomic investigation reveals a physiological mechanism for *Beauveria bassiana* to survive under linoleic acid stress. *iScience, 26*(4), 106551.

Huang, W., Shang, Y., Chen, P., Cen, K., & Wang, C. (2015a). Basic leucine zipper (bZIP) domain transcription factor MBZ1 regulates cell wall integrity, spore adherence, and virulence in *Metarhizium robertsii*. *The Journal of Biological Chemistry, 290*(13), 8218–8231.

Huang, S., He, Z., Zhang, S., Keyhani, N. O., Song, Y., Yang, Z., ... Zhang, Y. (2015b). Interplay between calcineurin and the Slt2 MAP-kinase in mediating cell wall integrity, conidiation and virulence in the insect fungal pathogen *Beauveria bassiana*. *Fungal Genetics and Biology, 83*, 78–91.

Huang, W., Hong, S., Tang, G., Lu, Y., & Wang, C. (2019a). Unveiling the function and regulation control of the DUF3129 family proteins in fungal infection of hosts. *Philosophical Transactions of the Royal Society B-Biological Sciences, 374*(1767), 20180321.

Huang, S., Keyhani, N. O., Zhao, X., & Zhang, Y. (2019b). The Thm1 $Zn(II)_2Cys_6$ transcription factor contributes to heat, membrane integrity and virulence in the insect pathogenic fungus *Beauveria bassiana*. *Environmental Microbiology, 21*(8), 3153–3171.

Huang, W., Shang, Y., Chen, P., Gao, Q., & Wang, C. (2015). MrpacC regulates sporulation, insect cuticle penetration and immune evasion in *Metarhizium robertsii*. *Environmental Microbiology, 17*(4), 994–1008.

Huarte-Bonnet, C., Paixão, F. R. S., Mascarin, G. M., Santana, M., Fernandes, É. K. K., & Pedrini, N. (2019). The entomopathogenic fungus *Beauveria bassiana* produces microsclerotia-like pellets mediated by oxidative stress and peroxisome biogenesis. *Environmental Microbiology Reports, 11*(4), 518–524.

Huarte-Bonnet, C., Paixão, F. R. S., Ponce, J. C., Santana, M., Prieto, E. D., & Pedrini, N. (2018). Alkane-grown *Beauveria bassiana* produce mycelial pellets displaying peroxisome proliferation, oxidative stress, and cell surface alterations. *Fungal Biology, 122*(6), 457–464.

Harner, M., Körner, C., Walther, D., Mokranjac, D., Kaesmacher, J., Welsch, U., ... Neupert, W. (2011). The mitochondrial contact site complex, a determinant of mitochondrial architecture. *The EMBO Journal, 30*(21), 4356–4370.

Jiang, Z. Y., Ligoxygakis, P., & Xia, Y. X. (2020). HYD3, a conidial hydrophobin of the fungal entomopathogen *Metarhizium acridum* induces the immunity of its specialist host locust. *International Journal of Biological Macromolecules, 165*(Pt A), 1303–1311.

Jin, D., Sun, B., Zhao, W., Ma, J., Zhou, Q., Han, X., ... Pei, Y. (2021). Thiamine-biosynthesis genes *Bbpyr* and *Bbthi* are required for conidial production and cell wall integrity of the entomopathogenic fungus *Beauveria bassiana*. *Journal of Invertebrate Pathology, 184*, 107639.

Kageyama, T., Suzuki, K., & Ohsumi, Y. (2009). Lap3 is a selective target of autophagy in yeast, *Saccharomyces cerevisiae*. *Biochemical and Biophysical Research Communications, 378*(3), 551–557.

Kiel, J. A., Veenhuis, M., & van der Klei, I. J. (2006). *PEX* genes in fungal genomes: Common, rare or redundant. *Traffic (Copenhagen, Denmark), 7*, 1291–1303.

Lamark, T., & Johansen, T. (2021). Mechanisms of selective autophagy. *Annual Review of Cell and Developmental Biology, 37*, 143–169.

Lei, J. H., Lin, H. Y., Ding, J. L., Feng, M. G., & Ying, S. H. (2022). Functional characterization of two homologs of yeast acetyl-coenzyme A synthetase in the entomopathogenic fungus *Beauveria bassiana*. *Archives of Microbiology, 204*(10), 653.

Lei, J. H., Sun, T. F., Feng, M. G., & Ying, S. H. (2023). Functional insights of three RING-finger peroxins in the life cycle of the insect pathogenic fungus *Beauveria bassiana*. *Current Genetics, 69*(4–6), 267–276.

Lewis, M. W., Robalino, I. V., & Keyhani, N. O. (2009). Uptake of the fluorescent probe FM4-64 by hyphae and haemolymph-derived in vivo hyphal bodies of the entomo-pathogenic fungus *Beauveria bassiana*. *Microbiology-SGM, 155*(Pt 9), 3110–3120.

Li, Z. Z., Alves, S. B., Roberts, D. W., Fan, M. Z., Delalibera, I., Tang, J., ... Rangel, D. E. N. (2010). Biological control of insects in Brazil and China: History, current programs and reasons for their successes using entomopathogenic fungi. *Biocontrol Science and Technology, 20*, 117–136.

Li, B., Song, S., Wei, X., Tang, G., & Wang, C. (2022). Activation of microlipophagy during early infection of insect hosts by *Metarhizium robertsii*. *Autophagy, 18*(3), 608–623.

Li, X. H., Peng, Y. J., Ding, J. L., Feng, M. G., & Ying, S. H. (2022). A homologue of yeast acyl-CoA synthetase Faa1 contributes to cytomembrane functionality involved in development and virulence in the insect pathogenic fungus *Beauveria bassiana*. *Microbial Pathogenesis, 164*, 105419.

Li, Y., Ren, H., Wang, F., Chen, J., Ma, L., Chen, Y., ... Pei, Y. (2023). Cell detoxification of secondary metabolites by P4-ATPase-mediated vesicle transport. *eLife, 12*, e79179.

Lima, D. M. C. G., Costa, T. P. C., Emri, T., Pócsi, I., Pupin, B., & Rangel, D. E. N. (2021). Fungal tolerance to Congo red, a cell wall integrity stress, as a promising indicator of ecological niche. *Fungal Biology, 125*(8), 646–657.

Lin, H. Y., Pang, M. Y., Feng, M. G., & Ying, S. H. (2022a). A peroxisomal sterol carrier protein 2 (Scp2) contributes to lipid trafficking in differentiation and virulence of the insect pathogenic fungus *Beauveria bassiana*. *Fungal Genetics and Biology, 158*, 103651.

Lin, H. Y., Ding, J. L., Peng, Y. J., Feng, M. G., & Ying, S. H. (2022b). Proteomic and phosphoryproteomic investigations reveal that autophagy-related protein 1, a protein kinase for autophagy initiation, synchronously deploys phosphoregulation on the ubiquitin-like conjugation system in the mycopathogen *Beauveria bassiana*. *mSystems, 7*(1), e0146321.

Lin, H. Y., Lei, J. H., Ding, J. L., Peng, Y. J., Zhang, H., Feng, M. G., & Ying, S. H. (2023a). Peroxin 14 tags peroxisomes and interacts with Nbr1 for pexophagy in the filamentous insect pathogenic fungus *Beauveria bassiana*. *Autophagy Reports, 2*, 2168337 2023.

Lin, H. Y., Feng, M. G., & Ying, S. H. (2023b). Phosphorylation modification orchestrates the functionalities of peroxin 14 in filamentous entomopathogenic fungus *Beauveria bassiana*. *Fungal Biology, 127*(9), 1284–1290.

Luo, X., Keyhani, N. O., Yu, X., He, Z., Luo, Z., Pei, Y., & Zhang, Y. (2012). The MAP kinase Bbslt2 controls growth, conidiation, cell wall integrity, and virulence in the insect pathogenic fungus *Beauveria bassiana*. *Fungal Genetics and Biology, 49*(7), 544–555.

Marcinkiewicz, A., Gauthier, D., Garcia, A., & Brasaemle, D. L. (2006). The phosphor-ylation of serine 492 of perilipin a directs lipid droplet fragmentation and dispersion. *The Journal of Biological Chemistry, 281*(17), 11901–11909.

Mei, L., Chen, M., Shang, Y., Tang, G., Tao, Y., Zeng, L., ... Wang, C. (2020). Population genomics and evolution of a fungal pathogen after releasing exotic strains to control insect pests for 20 years. *The ISME Journal, 14*(6), 1422–1434.

Meijer, W. H., van der Klei, I. J., Veenhuis, M., & Kiel, J. A. (2007). ATG genes involved in non-selective autophagy are conserved from yeast to man, but the selective Cvt and pexophagy pathways also require organism-specific genes. *Autophagy, 3*(2), 106–116.

Meyling, N. V., & Eilenberg, J. (2007). Ecology of the entomopathogenic fungi *Beauveria bassiana* and *Metarhizium anisopliae* in temperate agroecosystems: Potential for con-servation biological control. *Biological Control, 43*, 145–155.

Moonjely, S., Keyhani, N. O., & Bidochka, M. J. (2018). Hydrophobins contribute to root colonization and stress responses in the rhizosphere-competent insect pathogenic fungus *Beauveria bassiana*. *Microbiology-SGM, 164*(4), 517–528.

Mou, Y. N., Ren, K., Tong, S. M., Ying, S. H., & Feng, M. G. (2021). Essential role of COP9 signalosome subunit 5 (Csn5) in insect pathogenicity and asexual development of *Beauveria bassiana. Journal of Fungi (Basel), 7*(8), 642.

Nakatogawa, H. (2020). Mechanisms governing autophagosome biogenesis. *Nature Reviews. Molecular Cell Biology, 21*(8), 439–458.

Nobrega, F. G., Nobrega, M. P., & Tzagoloff, A. (1992). BCS1, a novel gene required for the expression of functional Rieske iron-sulfur protein in *Saccharomyces cerevisiae. The EMBO Journal, 11*(11), 3821–3829.

Holder, D. J., Kirkland, B. H., Lewis, M. W., & Keyhani, N. O. (2007). Surface characteristics of the entomopathogenic fungus *Beauveria (Cordyceps) bassiana. Microbiology-SGM, 153*(Pt 10), 3448–3457.

Ortiz-Urquiza, A., & Keyhani, N. O. (2013). Action on the surface: Entomopathogenic fungi versus the insect cuticle. *Insects, 4*(3), 357–374.

Ortiz-Urquiza, A., & Keyhani, N. O. (2016). Molecular genetics of *Beauveria bassiana* infection of insects. *Advances in Genetics, 94*, 165–249.

Palma-Guerrero, J., Huang, I. C., Jansson, H. B., Salinas, J., Lopez-Llorca, L. V., & Read, N. D. (2009). Chitosan permeabilizes the plasma membrane and kills cells of *Neurospora crassa* in an energy dependent manner. *Fungal Genetics and Biology, 46*(8), 585–594.

Pang, M. Y., Lin, H. Y., Hou, J., Feng, M. G., & Ying, S. H. (2022). Different contributions of the peroxisomal import protein Pex5 and Pex7 to development, stress response and virulence of insect fungal pathogen *Beauveria bassiana. Journal of Applied Microbiology, 132*(1), 509–519.

Pedrini, N., Juárez, M. P., Crespo, R., & De Alaniz, M. J. (2006). Clues on the role of *Beauveria bassiana* catalases in alkane degradation events. *Mycologia, 98*(4), 528–534.

Peng, Y. J., Ding, J. L., Lin, H. Y., Feng, M. G., & Ying, S. H. (2021). A virulence-related lectin traffics into eisosome and contributes to functionality of cytomembrane and cell-wall in the insect-pathogenic fungus *Beauveria bassiana. Fungal Biology, 125*(11), 914–922.

Peng, Y. J., Zhang, H., Feng, M. G., & Ying, S. H. (2022a). Steryl acetyl hydrolase 1 (BbSay1) links lipid homeostasis to conidiogenesis and virulence in the entomopathogenic fungus *Beauveria bassiana. Journal of Fungi (Basel), 8*(3), 292.

Peng, Y. J., Hou, J., Zhang, H., Lei, J. H., Lin, H. Y., Ding, J. L., ... Ying, S. H. (2022b). Systematic contributions of CFEM domain-containing proteins to iron acquisition are essential for interspecies interaction of the filamentous pathogenic fungus *Beauveria bassiana. Environmental Microbiology, 24*(8), 3693–3704.

Peng, Y. J., Wang, J. J., Lin, H. Y., Ding, J. L., Feng, M. G., & Ying, S. H. (2020). HapX, an indispensable bZIP transcription factor for iron acquisition, regulates infection initiation by orchestrating conidial oleic acid homeostasis and cytomembrane functionality in mycopathogen *Beauveria bassiana. mSystems, 5*(5) e00695-20.

Pieuchot, L., & Jedd, G. (2012). Peroxisome assembly and functional diversity in eukaryotic microorganisms. *Annual Review of Microbiology, 66*, 237–263.

Qiu, L., Song, J. Z., Li, J., Zhang, T. S., Li, Z., Hu, S. J., ... Wang, J. J. (2022). The transcription factor Ron1 is required for chitin metabolism, asexual development and pathogenicity in *Beauveria bassiana*, an entomopathogenic fungus. *International Journal of Biological Macromolecules, 206*, 875–885.

Qiu, L., Zhang, J., Song, J. Z., Hu, S. J., Zhang, T. S., Li, Z., ... Cheng, W. (2021). Involvement of BbTpc1, an important $Zn(II)_2Cys_6$ transcriptional regulator, in chitin biosynthesis, fungal development and virulence of an insect mycopathogen. *International Journalof Biological Macromolecules, 166*, 1162–1172.

Quesada-Moraga, E., González-Mas, N., Yousef-Yousef, M., Garrido-Jurado, I., & Fernández-Bravo, M. (2023). Key role of environmental competence in successful use of entomopathogenic fungi in microbial pest control. *Journal of Pest Science*. https://doi.org/10.1007/s10340-023-01622-8.

Ren, K., Mou, Y. N., Ying, S. H., & Feng, M. G. (2021). Conserved and noncanonical activities of two histone H3K36 methyltransferases required for insect-pathogenic lifestyle of *Beauveria bassiana*. *Journal of Fungi (Basel), 7*(11), 956.

Russell, R. C., Yuan, H. X., & Guan, K. L. (2014). Autophagy regulation by nutrient signaling. *Cell Research, 24*(1), 42–57.

Sephton-Clark, P. C. S., & Voelz, K. (2018). Spore germination of pathogenic filamentous fungi. *Advances in Applied Microbiology, 102*, 117–157.

Shahriari, M., Zibaee, A., Khodaparast, S. A., & Fazeli-Dinan, M. (2021). Screening and virulence of the entomopathogenic fungi associated with *Chilo suppressalis* Walker. *Journal of Fungi (Basel), 7*(1), 34.

Shang, Y., Chen, P., Chen, Y., Lu, Y., & Wang, C. (2015). MrSkn7 controls sporulation, cell wall integrity, autolysis, and virulence in *Metarhizium robertsii*. *Eukaryotic Cell, 14*(4), 396–405.

Shao, W., Cai, Q., Tong, S. M., Ying, S. H., & Feng, M. G. (2020). Nuclear Ssr4 is required for the in vitro and in vivo asexual cycles and global gene activity of *Beauveria bassiana*. *mSystems, 5*(2) e00677-19.

Starai, V. J., & Escalante-Semerena, J. C. (2004). Acetyl-coenzyme A synthetase (AMP forming). *Cellular and Molecular Life Sciences, 61*(16), 2020–2030.

Szeto, S. S. W., Reinke, S. N., Sykes, B. D., & Lemire, B. D. (2007). Ubiquinone-binding site mutations in the *Saccharomyces cerevisiae* succinate dehydrogenase generate superoxide and lead to the accumulation of succinate. *The Journal of Biological Chemistry, 282*(37), 27518–27526.

Tang, G., Shang, Y., Li, S., & Wang, C. (2020). MrHex1 is required for Woronin body formation, fungal development and virulence in *Metarhizium robertsii*. *Journal of Fungi (Basel), 6*, 172.

Tiwari, R., Köffel, R., & Schneiter, R. (2007). An acetylation/deacetylation cycle controls the export of sterols and steroids from *S. cerevisiae*. *The EMBO Journal, 26*(24), 5109–5119.

Tong, S. M., Wang, D. Y., Gao, B. J., Ying, S. H., & Feng, M. G. (2019). The DUF1996 and WSC domain-containing protein Wsc1I acts as a novel sensor of multiple stress cues in *Beauveria bassiana*. *Cellular Microbiology, 21*(12), e13100.

Tong, Y., Wu, H., He, L., Qu, J., Liu, Z., Wang, Y., ... Huang, B. (2022). Role of two G-protein α subunits in vegetative growth, cell wall integrity, and virulence of the entomopathogenic fungus *Metarhizium robertsii*. *Journal of Fungi (Basel), 8*(2), 132.

Walther, T. C., & Farese, R. V., Jr (2012). Lipid droplets and cellular lipid metabolism. *Annual Review of Biochemistry, 81*, 687–714.

Wanchoo, A., Lewis, M. W., & Keyhani, N. O. (2009). Lectin mapping reveals stage-specific display of surface carbohydrates in in vitro and haemolymph-derived cells of the entomopathogenic fungus *Beauveria bassiana*. *Microbiology-SGM, 155*(Pt 9), 3121–3133.

Wanders, R. J., & Waterham, H. R. (2006). Biochemistry of mammalian peroxisomes revisited. *Annual Review of Biochemistry, 75*, 295–332.

Wang, C., & St Leger, R. J. (2006). A collagenous protective coat enables *Metarhizium anisopliae* to evade insect immune responses. *Proceedings of the National Academy of Sciences of the United States of America, 103*(17), 6647–6652.

Wang, C., & St Leger, R. J. (2007). The *Metarhizium anisopliae* perilipin homolog MPL1 regulates lipid metabolism, appressorial turgor pressure, and virulence. *The Journal of Biological Chemistry, 282*(29), 21110–21115.

Wang, C., & Wang, S. (2017). Insect pathogenic fungi: Genomics, molecular interactions, and genetic improvements. *Annual Review of Entomology, 62*, 73–90.

Wang, C., & Feng, M. G. (2014). Advances in fundamental and applied studies in China of fungal biocontrol agents for use against arthropod pests. *Biological Control, 68*, 129–135.

Wang, L., Lai, Y., Chen, J., Cao, X., Zheng, W., Dong, L., ... Wang, S. (2023a). The ASH1-PEX16 regulatory pathway controls peroxisome biogenesis for appressorium-mediated insect infection by a fungal pathogen. *Proceedings of the National Academy of Sciences of the United States of America, 120*(4), e2217145120.

Wang, H., Lu, Z., Keyhani, N. O., Deng, J., Zhao, X., Huang, S., ... Zhang, Y. (2023b). Insect fungal pathogens secrete a cell wall-associated glucanase that acts to help avoid recognition by the host immune system. *PLoS Pathogens, 19*(8), e1011578.

Wang, J. J., Peng, Y. J., Feng, M. G., & Ying, S. H. (2019). Functional analysis of the mitochondrial gene mitofilin in the filamentous entomopathogenic fungus *Beauveria bassiana*. *Fungal Genetics and Biology, 132*, 103250.

Wang, X., Liu, Y., Keyhani, N. O., Zhu, S., Wang, J., Wang, J., ... Fan, Y. (2022). A perilipin affects lipid droplet homeostasis and aerial hyphal growth, but has only small effects on virulence in the insect pathogenic fungus *Beauveria bassiana*. *Journal of Fungi (Basel), 8*(6), 634.

Wang, X. X., He, P. H., Feng, M. G., & Ying, S. H. (2014). BbSNF1 contributes to cell differentiation, extracellular acidification, and virulence in *Beauveria bassiana*, a filamentous entomopathogenic fungus. *Applied Microbiology and Biotechnology, 98*(20), 8657–8673.

Wang, Z., Feng, J., Jiang, Y., Xu, X., Xu, L., Zhou, Q., & Huang, B. (2021a). MrPEX33 is involved in infection-related morphogenesis and pathogenicity of *Metarhizium robertsii*. *Applied Microbiology and Biotechnology, 105*(3), 1079–1090.

Wang, Y. Y., Zhang, J., Liu, X. M., Li, Y., Sui, J., Dong, M. Q., ... Du, L. L. (2021). Molecular and structural mechanisms of ZZ domain-mediated cargo selection by Nbr1. *The EMBO Journal, 40*(15), e107497.

Wen, Z., Tian, H., Xia, Y., & Jin, K. (2021). O-mannosyltransferase MaPmt2 contributes to stress tolerance, cell wall integrity and virulence in *Metarhizium acridum*. *Journal of Invertebrate Pathology, 184*, 107649.

Xie, X., Wang, Y., Yu, D., Xie, R., Liu, Z., & Huang, B. (2020). DNM1, a dynamin-related protein that contributes to endocytosis and peroxisome fission, is required for the vegetative growth, sporulation, and virulence of *Metarhizium robertsii*. *Applied and Environmental Microbiology, 86*(17), e01217–e01220.

Xin, C., Xing, X., Wang, F., Liu, J., Ran, Z., Chen, W., ... Song, Z. (2020). MrMid2, encoding a cell wall stress sensor protein, is required for conidium production, stress tolerance, microsclerotium formation and virulence in the entomopathogenic fungus *Metarhizium rileyi*. *Fungal Genetics and Biology, 134*, 103278.

Yang, M., Jin, K., & Xia, Y. (2011). MaFKS, a β-1,3-glucan synthase, is involved in cell wall integrity, hyperosmotic pressure tolerance and conidiation in *Metarhizium acridum*. *Current Genetics, 57*(4), 253–260.

Yang, Z., Jiang, H., Zhao, X., Lu, Z., Luo, Z., Li, X., ... Zhang, Y. (2017). Correlation of cell surface proteins of distinct *Beauveria bassiana* cell types and adaption to varied environment and interaction with the host insect. *Fungal Genetics and Biology, 99*, 13–25.

Yang, Z., Huang, J., Geng, J., Nair, U., & Klionsky, D. J. (2006). Atg22 recycles amino acids to link the degradative and recycling functions of autophagy. *Molecular Biology of the Cell, 17*(12), 5094–5104.

Ying, S. H., & Feng, M. G. (2019). Insight into vital role of autophagy in sustaining biological control potential of fungal pathogens against pest insects and nematodes. *Virulence, 10*(1), 429–437.

Yuga, M., Gomi, K., Klionsky, D. J., & Shintani, T. (2011). Aspartyl aminopeptidase is imported from the cytoplasm to the vacuole by selective autophagy in *Saccharomyces cerevisiae*. *The Journal of Biological Chemistry, 286*(15), 13704–13713.

Zerbes, R. M., Van Der Klei, I. J., Veenhuis, M., Pfanner, N., van der Laan, M., & Bohnert, M. (2012). Mitofilin complexes: Conserved organizers of mitochondrial membrane architecture. *Biological Chemistry, 393*(11), 1247–1261.

Zhang, H. (2022). The genetics of autophagy in multicellular organisms. *Annual Review of Genetics, 56*, 17–39.

Zhang, J., Jiang, H., Du, Y., Keyhani, N. O., Xia, Y., & Jin, K. (2019). Members of chitin synthase family in *Metarhizium acridum* differentially affect fungal growth, stress tolerances, cell wall integrity and virulence. *PLoS Pathogens, 15*(8), e1007964.

Zhang, J. G., Zhang, K., Xu, S. Y., Ying, S. H., & Feng, M. G. (2023). Essential role of WetA, but no role of VosA, in asexual development, conidial maturation and insect pathogenicity of *Metarhizium robertsii*. *Microbiology Spectrum, 11*(2), e0007023.

Zhang, L., Wang, J., Xie, X. Q., Keyhani, N. O., Feng, M. G., & Ying, S. H. (2013). The autophagy gene *BbATG5*, involved in the formation of the autophagosome, contributes to cell differentiation and growth but is dispensable for pathogenesis in the entomopathogenic fungus *Beauveria bassiana*. *Microbiology-SGM, 159*(Pt 2), 243–252.

Zhang, S., Xia, Y. X., Kim, B., & Keyhani, N. O. (2011). Two hydrophobins are involved in fungal spore coat rodlet layer assembly and each play distinct roles in surface interactions, development and pathogenesis in the entomopathogenic fungus, *Beauveria bassiana*. *Molecular Microbiology, 80*(3), 811–826.

Zhao, H., Xu, C., Lu, H. L., Chen, X., St Leger, R. J., & Fang, W. (2014). Host-to-pathogen gene transfer facilitated infection of insects by a pathogenic fungus. *PLoS Pathogens, 10*(4) e1004009.

Zhao, X., Yang, X., Lu, Z., Wang, H., He, Z., Zhou, G., ... Zhang, Y. (2019a). MADS-box transcription factor Mcm1 controls cell cycle, fungal development, cell integrity and virulence in the filamentous insect pathogenic fungus *Beauveria bassiana*. *Environmental Microbiology, 21*(9), 3392–3416.

Zhao, T., Tian, H., Xia, Y., & Jin, K. (2019b). MaPmt4, a protein O-mannosyltransferase, contributes to cell wall integrity, stress tolerance and virulence in *Metarhizium acridum*. *Current Genetics, 65*(4), 1025–1040.

Zhou, Q., Yu, L., Ying, S. H., & Feng, M. G. (2021). Comparative roles of three adhesin genes (adh1-3) in insect-pathogenic lifecycle of *Beauveria bassiana*. *Applied Microbiology and Biotechnology, 105*(13), 5491–5502.

Zhu, J., Ying, S. H., & Feng, M. G. (2016). The Na^+/H^+ antiporter Nhx1 controls vacuolar fusion indispensible for life cycles in vitro and in vivo in a fungal insect pathogen. *Environmental Microbiology, 18*(11), 3884–3895.

CHAPTER THREE

Recovery of insect-pathogenic fungi from solar UV damage: Molecular mechanisms and prospects

Ming-Guang Feng*
Institute of Microbiology, College of Life Sciences, Zhejiang University, Hangzhou, P.R. China
*Corresponding author. e-mail address: mgfeng@zju.edu.cn

Contents

Abstract

Molecular mechanisms underlying insect-pathogenic fungal tolerance to solar ultraviolet (UV) damage have been increasingly understood. This chapter reviews the methodology established to quantify fungal response to solar UV radiation, which consists of UVB and UVA, and characterize a pattern of the solar UV dose (damage) accumulated from sunrise to sunset on sunny summer days. An emphasis is placed on anti-UV mechanisms of fungal insect pathogens in comparison to those well documented in model yeast. Principles are discussed for properly timing the application of a fungal pesticide to improve pest control during summer months. Fungal UV tolerance depends on either nucleotide excision repair (NER) or photorepair of UV-induced DNA lesions to recover UV-impaired cells in the darkness or the light. NER is a slow process independent of light and depends on a large family of anti-UV radiation (RAD) proteins studied intensively in model yeast but rarely in non-yeast fungi. Photorepair is a rapid process that had long been considered to depend on only one or two photolyases in filamentous fungi. However, recent studies have greatly expanded a genetic/molecular basis for photorepair-dependent photoreactivation

Advances in Applied Microbiology, Volume 129
ISSN 0065-2164, https://doi.org/10.1016/bs.aambs.2024.04.003

that serves as a primary anti-UV mechanism in insect-pathogenic fungi, in which photolyase regulators required for photorepair and multiple RAD homologs have higher or much higher photoreactivation activities than do photolyases. The NER activities of those homologs in dark reactivation cannot recover the severe UV damage recovered by their activities in photoreactivation. Future studies are expected to further expand the genetic/molecular basis of photoreactivation and enrich principles for the recovery of insect-pathogenic fungi from solar UV damage.

1. Introduction

Fungal insecticides and acaricides are developed from qualified candidate strains of insect-pathogenic fungi, such as *Beauveria* and *Metarhizium* (Arthurs & Dara, 2019; de Faria & Wraight, 2007), and increasingly employed as alternatives to fight against arthropod pests (Mascarin et al., 2019; Peng, Guo, Tong, Ying, & Feng, 2021; Peng, Xie, et al., 2021). Active agents of such pesticides are aerial conidia, aggregated pellets, or microsclerotia formulated for easy application (Feng et al., 1994; Corval et al., 2021; Roberts & Leger, 2004). After field application, formulated conidia or the conidia produced by hydrated pellets or microsclerotia germinate for hyphal invasion into insect hemocoel. The hyphae become unicellular hyphal bodies for yeast-like budding proliferation to accelerate mycosis development. Upon host death from mummification, the hyphal bodies turn back into hyphae to penetrate through host cuticle for outgrowth and conidiation on insect cadavers (Guo et al., 2022; Guo, Peng, Tong, Ying, & Feng, 2021; Lewis, Robalino, & Keyhani, 2009). This type of killing action makes fungal pesticides environmentally friendly.

Despite the advantage of fungal pesticidal activity, conidia applied directly or produced by the hydrated pellets or microsclerotia are highly vulnerable to solar ultraviolet (UV) radiation (Braga, Rangel, Fernandes, Flint, & Roberts, 2015; Tong & Feng, 2022). Solar UV on the Earth's surface comprises wavelengths of UVA (320–400 nm) and UVB (290–320 nm) but not UVC (<290 nm), which is most harmful to microbial cells but completely scavenged by the atmospheric ozone layer (Madronich, 1993). In general, fungal conidia are far more vulnerable to shorter wavelengths of UVB than of UVA (Braga, Rangel, Flint, Anderson, & Roberts, 2006; Fernandes, Rangel, Moraes, Bittencourt, & Roberts, 2007; Yao, Ying, Feng, & Hatting, 2010), making UVB a major source of solar UV damage to conidia applied for pest control during summer months. UVB causes direct damage to intracellular macromolecules, such as DNA, ribosomes, and biomembranes, unlike UVA to induce the generation of reactive oxygen species as a source of oxidative stress on exposed cells (Griffiths, Mistry, Herbert, & Lunec, 1998; Sancar, 2003).

UV induces covalent linkages of adjacent bases in DNA duplex, forming cyclobutane pyrimidine dimer (CPD) and (6-4)-pyrimidine-pyrimidine (6-4PP) photoproducts, which are detrimental or lethal to eukaryotic cells (Sancar, 2003). Such CPD and 6–4PP DNA lesions can be reversed through nucleotide excision repair (NER) and photorepair, two distinctive mechanisms underlying cellular resistance to UV damage (Boiteux & Jinks-Robertson, 2013; de Laat, Jaspers, & Hoeijmakers, 1999; Sancar, 1996, 2003; Suter, Wellinger, & Thoma, 2000; Yasui et al., 1994). However, the understanding of such mechanisms is still limited in filamentous fungi, including insect-pathogenic fungi.

Pioneer researchers devoted to the application of entomopathogens for biological control of insect pests realized the importance of fungal UV tolerance for successful biocontrol in the last century (Alder-Rangel, 2021; Ignoffo, Hostetter, Sikorowski, Sutter, & Brooks, 1977; Roberts & Campbel, 1977). However, entomopathogenic fungal resistance to solar UV damage was not effectively studied until this century. This chapter reviews the progresses of anti-UV roles and mechanisms elucidated in fungal entomopathogens in the past two decades. Emphasis is placed upon their anti-UV mechanisms in comparison to those documented in model yeast and principles for properly timing the application of a fungal pesticide to improve pest control during summer months. Other reviews are available about the physiological effects of UV on conidia (Braga et al., 2015), screening of anti-UV strains and formulations (Fernandes, Rangel, Braga, & Roberts, 2015), contributions of non-anti-UV genes to fungal UV resistance (Tong & Feng, 2022), *Metarhizium* photobiology (Brancini, Hallsworth, Corrochano, & Braga, 2022), fungal stress tolerance in general sense (Rangel et al., 2015; Tong & Feng, 2019, 2020; Zhang & Feng, 2018), and agricultural photo-antimicrobials (Braga, Silva-Junior, Brancini, Hallsworth, & Wainwright, 2022).

2. Fungal tolerance to ultraviolet damage

2.1 Indices of fungal ultraviolet tolerance

The first assay system of UV damage to fungal conidia was effectively established early in this century. UVB damage to *Metarhizium anisopliae* was quantified as medial lethal time (LT_{50}) to cause 50% viability loss of conidia exposed for 1–8 h to a fixed UVB irradiation of 920 or 1200 mW/m^2 (Braga, Flint, Messias, Anderson, & Roberts, 2001; 2001). LT_{50} and relative viability (culturability) of conidia exposed to a given period of irradiation from a fixed UV source have since been used to reveal inter- and intraspecific variations in conidial UVB resistance among the strains of *Metarhizium* spp. and *Beauveria* spp. with different geographic and host origins (Braga, Flint, Messias, Anderson, & Roberts, 2001; Fernandes et al., 2007; Rangel, Butler, et al., 2006), conidial tolerance to UVA and UVB (Braga et al., 2002; Braga, Flint, Miller, Anderson, & Roberts, 2001), and the effects of culture conditions on fungal UV resistance (Rangel, Anderson, & Roberts, 2006, 2008; Rangel, Braga, Anderson, & Roberts, 2005; Rangel, Braga, Flint, Anderson, & Roberts, 2004; Rangel, Fernandes, Braga, & Roberts, 2011). This assay system has also been modified to estimate LT_{10} and LT_{90} as alternative parameters of conidial resistance to sunlight-simulated full-spectrum UV irradiation (Dias et al., 2020).

Either UVB or UVA damage to fungal cells depends on the irradiation dose, which is a function of irradiation intensity and time. To assess dose-dependent fungal response, a more robust assay system has been introduced to quantify conidial response to irradiation of weighted UVB wavelength of 312 nm at the doses (d) of 0.1–1.6 J/cm^2 or of weighted UVA wavelengths of 365 nm at the doses (d) of 1–18 J/cm^2 in a Bio-Sun^{++} UV Chamber (Huang & Feng, 2009; Yao et al., 2010). The employed device is integrated with a microprocessor that can automatically adjust both the intensity and the wavelength of irradiation four times per second for an error control of less than 1 µJ/cm^2 (10^{-6}) at a given UVB or UVA dose (J/cm^2) released in minutes. The irradiated conidia (on agar plates) are incubated at the optimum 25 °C for 24 h in darkness to assess a maximal germination percentage after exposure to each dose, followed by computing conidial survival index (I_s) as a ratio of irradiated versus non-irradiated (control) conidia. The resulting $I_s - d$ trends fit the modified logistic (survival) equation $I_s = 1/[1 - \exp(a + rd)]$, in which the parameter r denotes a declining rate of conidial viability with increasing UVB or UVA doses (Huang & Feng, 2009). Solving the fitted equation at $I_s = 0.5$ leads to estimation of median lethal dose ($LD_{50} = -a/r$), which serves as an absolute index of conidial resistance to UVB or UVA. The LD_{50} estimates reveal a wide variation of conidial UVB resistance from 0.15 to 0.93 J/cm^2 among 20 *Beauveria bassiana* strains with different host and geographic origins (Huang & Feng, 2009), ~10-fold higher sensitivities of fungal strains to UVB than to UVA, and large variations in conidial UVB or UVA resistance among 60 strains of *B. bassiana* and several *Metarhizium* species (Yao et al., 2010).

2.2 Exploitability of fungal ultraviolet tolerance

Recent studies have revealed that conidial UVB resistance depends not only on the irradiated doses but also on the time of post-radiation incubation and the presence or absence of light during the incubation due to the great difference between NER and photorepair activities in recovering UVB-impaired or inactivated conidia of *B. bassiana* (Wang, Fu, Tong, Ying, & Feng, 2019; Wang, Mou, Tong, Ying, & Feng, 2020; Xu, Yu, Luo, Ying, & Feng, 2023; Yu, Xu, Luo, Ying, & Feng, 2022; Yu, Xu, Luo, Ying, & Feng, 2023) and *Metarhizium robertsii*, as (Peng, Guo, Tong, Ying, & Feng, 2021; Peng, Zhang, Ying, & Feng, 2024; Zhang, Peng, Ying, & Feng, 2023; Zhang, Peng, Zhang, Ying, & Feng, 2023) discussed later. Survival index of *B. bassiana* conidia impaired by UVB to different degrees declines much more rapidly with increasing UVB doses during the full-dark incubation (dark reactivation) period of 12 h than of 24 or 40 h. The declining trend is greatly slowed down by an incubation of 3 or 5 h under visible light (photoreactivation) plus 9 or 7 h in the dark or plus 21 or 19 h in the dark, as illustrated in Fig. 1A. Consequently, the LD_{50} values estimated from the fitted survival trends increase significantly with the time length of post-radiation incubation and differ greatly between the same periods of dark incubation and light plus dark incubation (Fig. 1B). The LD_{50} values of irradiated conidia in the 12-h dark treatment mimicking nighttime on the Earth's surface average only 0.116 J/cm^2, which is far below 0.281 and 0.294 J/cm^2 in the 12-h treatments of 3-h light plus 9-h dark and 5-h light plus 7-h dark, respectively. The UVB LD_{50} is further increased to 0.456 and 0.539 J/cm^2 in the 24-h treatments of 3-h light plus 21-h dark and 5-h light plus 19-h dark, respectively. These differences demonstrate that *B. bassiana* has much greater activity in photoreactivation than in dark reactivation. A UVB damage of 0.2–0.3 J/cm^2 to conidia can be reversed by 3 or 5 h of light exposure (photoreactivation) following ~10 h of dark incubation. Therefore, fungal photoreactivation activity is exploitable for properly timing the application of a fungal formulation for improved pest control according to an accumulation pattern of solar UV damage on sunny summer days, as discussed below.

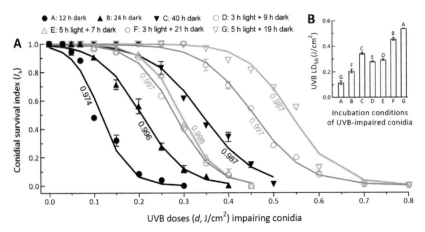

Fig. 1 View of fungal activity in NER-dependent dark reactivation and photorepair-dependent photoreactivation. (A) Trends of dark reactivation rates (black symbols) and photoreactivation rates (color symbols) for fungal conidia impaired at gradient UVB doses of 0.03–0.8 J/cm^2. (B) LD$_{50}$ values estimated from the fitted trends. A decimal value on each curve fitted to observations (symbols) denotes the coefficient of determination (r^2) for the fitted curve ($P < 0.001$ in F test for fitness). Different uppercase letters marked on bars (B) differ significantly from one another ($P < 0.01$ in Tukey's test). Error bars denote standard deviations from three independent replicates. Reanalyzed from the experimental data of a wild-type *Beauveria bassiana* strain in Yu et al. (2024).

3. Daily accumulation pattern of solar ultraviolet damage

Solar UV irradiation is a ubiquitous stress to filamentous fungi associated with insects and plants on the Earth's surface. Assessment of solar UV accumulation on sunny summer days is a prerequisite to understand the biological significance of fungal UV tolerance for rational application of fungal formulations against arthropod pests.

Solar UV intensity varies greatly over the daytime period of a sunny day from sunrise to sunset at a given geographic site and can be instantly quantified using photoelectronic sensors. For example, the solar UVB intensities (μW/cm^2) hourly recorded at the center (30°17′57″ N × 120°5′7″ E) of a large shade-free lawn from 5:00 a.m. to 7:00 p.m. in five cloud-free sunny summer days can be converted to time-specific UVB doses (J/cm^2; mean intensity × 3600 × 10^{-6} due to 1 μW = 10^{-6} W and 1 J = 1 W·s) assuming that continuous solar irradiation intensity is similar within a given hour and roughly divided into 3600 s-based points per hour (Yu, Xu, Tong, Ying, & Feng, 2022). As illustrated in Fig. 2A, a bell-shaped curve of time-specific UVB doses (d) can fit the symmetrical equation

$$d = a. \exp[b(t - c)^2]$$

in which parameter c denotes a symmetrical point of the fitted curve, namely, a point of daytime (t) at which the UVB intensity in sunlight is strongest; and a and b are two other parameters to be fitted. The fitted parameter c is equal to 11.8722, a time point of 11:52 a.m. approaching noon. Solar UVB dose (d_c) accumulated daily from 5:00 a.m. to 7:00 p.m. is then computed as definite integrals of the fitted equation in the following form

$$d_C = \int_5^j a. \exp[b(t - c)^2]\,dt)\,(j = 6, \quad 7, \quad ..., \quad 19).$$

Based on the computed integrals and associated 95% confidential limits (Fig. 2B), solar UVB damage (dose) accumulated in sunny daytime is 2.331 (2.055−2.655) J/cm^2, which is far above the upper limit of fungal activity for recovery of conidia from UVB damage through photoreactivation (Fig. 1). Notably, 90.2% of the accumulation occurs prior to 3:00 p.m. This is in contrast with 8.2% accumulated between 3:00 and 5:00 p.m. and only 1.6% accumulated after 5:00 p.m. The daytime accumulation pattern of solar UVB damage suggests the need to optimize application strategy of fungal insecticides on sunny summer days, namely, either low-risk application between 3:00 and 5:00 p.m. or no-risk application after 5:00 p.m. (Yu, Xu, Tong, et al., 2022). This is because conidia impaired at the UVB doses of 0.2−0.3 J/cm^2 can be readily photoreactivated by 3 or 5 h of pre-evening light exposure and subsequent nighttime. Fungal insecticides applied in the late afternoon to avoid solar UV exposure have proved more efficacious against rice planthoppers and leaf rollers than those applied in the morning for increased exposure to solar UV damage (Qi et al., 2023; Xu, Wen, et al., 2023).

Importantly, an accumulation pattern of solar UV damage may vary with altitudes and latitudes that are relevant to the daytime length of summer. Therefore, the above pattern is applicable to those regions adjacent to the geographic site where solar UV intensity was monitored. In other regions at different altitudes and latitudes, the same method can be used to characterize local patterns of solar UV accumulation to optimize application strategies of fungal pesticides.

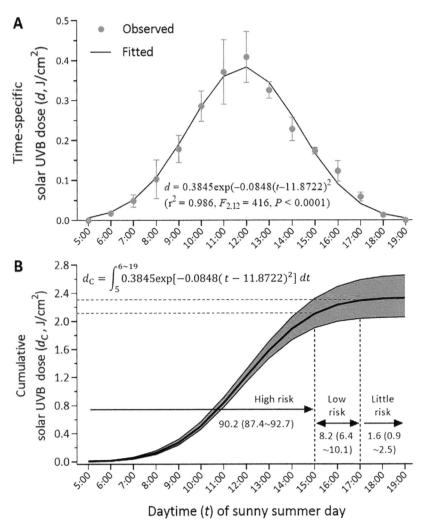

Fig. 2 Accumulation pattern of solar UVB damage (dose) on sunny summer days. (A) Distribution of time-specific UVB doses (d) over the points of daytime (t) based on the hourly records of solar UVB intensity taken at the center of a large shade-free lawn from 5:00 a.m. to 7:00 p.m. The bell-shaped trends fit the symmetrical equation with high coefficient of determination ($r^2 \geq 0.986$, $P < 0.0001$ in F test for fitness). (B) The definite integral curve (with 95% confidential limits) of the fitted equation for an accumulation of solar UVB dose (d_C) from 5:00 a.m. to 7:00 p.m. The three daytime intervals recognized to apply fungal pesticides are before 3:00 p.m. at high risk; between 3:00 and 5:00 p.m. at low risk; and after 5:00 p.m. at little risk in sunny summer. The values below the arrow (in blue) are a percentage and 95% confidential limits of d_C in the corresponding daylight interval. The red dash lines indicate a light UVB damage to be readily recovered by photoreactivation in the pre-evening daytime (light exposure). Error bars (A): standard deviations from hourly records in five cloud-free sunny summer days at 30°17′57″ N × 120°5′7″ E. Reanalyzed *from data in Yu, Xu, Tong, et al. (2022).*

4. Mechanisms underlying dark reactivation and photoreactivation

4.1 Dark reactivation depending on nucleotide excision repair

In general, NER is a slow process independent of light. A large family of anti-UV radiation (RAD) proteins and their partners have been intensively studied in the model yeast *Saccharomyces cerevisiae* and revealed to function in the NER pathway, which is divided into several subpathways, such as global-genome NER (GG-NER), transcription-coupled NER (TC-NER), and error-free post-replication repair (PRR) to bypass DNA damage (Boiteux & Jinks-Robertson, 2013; Prakash, Sung, & Prakash, 1993). However, homologs of those yeast RAD proteins are functionally unknown in non-yeast fungi. It was long unclear whether NER serves as a main anti-UV mechanism in filamentous fungi until recent reports about the anti-UV activities of those homologs elucidated in fungal insect pathogens (Peng et al., 2024; Wang et al., 2020; Yu et al., 2023; Yu, Xu, Luo, et al., 2022; Zhang, Peng, Ying, et al., 2023; Zhang, Peng, Zhang, et al., 2023).

In *S. cerevisiae*, RAD proteins and partners interact with one another, forming manifold RAD-RAD and RAD-containing protein complexes, which act as endonucleases, helicases, ligases, and polymerases in the processes of GG-NER, TC-NER, and/or error-free PRR (Godderz, Giovannucci, Lalakova, Menendez-Benito, & Dantuma, 2017; Gong, Fahy, & Smerdon, 2006; Smerdon & Thoma, 1990; Suter et al., 2000; Tsuchiya et al., 2017), as well reviewed by Boiteux and Jinks-Robertson (2013). Among those complexes, the Rad4-Rad23-Rad33 complex can initiate GG-NER by recognition of impaired DNA (Den Dulk, Sun, De Ruijter, Brandsma, & Brouwer, 2006; Gong et al., 2006; Sarkar, Kiely, & McHugh, 2010). The Rad1-Rad10 complex serves as an endonuclease essential for GG-NER (Bailly, Sommers, Sung, Prakash, & Prakash, 1992; Bardwell, Bardwell, Tomkinson, & Friedberg, 1994; Davies, Friedberg, Tomkinson, Wood, & West, 1995; Rodriguez, Wang, Friedberg, & Tomkinson, 1996; Tomkinson, Bardwell, Bardwell, Tappe, & Friedberg, 1993). Rad2 and Rad1-Rad10 needed for 5′ and 3′ incisions of DNA lesions (Evans, Moggs, Hwang, Egly, & Wood, 1997; Staresincic et al., 2009) are also the components of a huge complex comprising Rad14, the transcription factor TFIIH with eight subunits needed for NER, and replication protein A (RPA1), a subunit of the single-strand heterodimeric DNA-binding protein RPA that binds to Rad14 and TFIIH (Huang, Feaver, Tomkinson, & Friedberg, 1998; Lafrance-Vanasse et al., 2012). Rad7 and

Rad16 enable removal of CPD DNA lesions from non-transcribed strand of active genes (Verhage et al., 1994). They interact with each other to form a stable complex, which binds specifically to DNA lesions (Guzder, Sung, Prakash, & Prakash, 1997), stimulates dual cleavage of UV-irradiated DNA in vitro (Guzder, Sung, Prakash, & Prakash, 1999), and play an essential role in GG-NER (Reed, Akiyama, Stillman, & Friedberg, 1999; Reed, You, & Friedberg, 1998). Rad16 triggers histone H3 acetylation under UV light and induces an effect of histone H3 methylation on NER at silenced loci (Chaudhuri, Wyrick, & Smerdon, 2009; Teng et al., 2008). A combination of Rad4-Rad23 and Rad7-Rad16 increases a binding activity to UV-damaged DNA (Ramsey et al., 2004). Rad7-Rad16 interacts with Rad4 and Elc1-Cul3, forming a cullin-based E3 ubiquitin ligase complex that facilitates UV-reliant ubiquitination of Rad4 and proteins involved in chromatin remodeling (Gillette et al., 2006; Guzder, Habraken, Sung, Prakash, & Prakash, 1996). Rad26 is tied to RNA polymerase II and serves as a DNA-dependent ATPase needed for TC-NER and transcriptional bypass of lesions (Malik et al., 2010; Taschner et al., 2010). Phosphorylated Rad26 may enhance the TC-NER rate of DNA damage (Ulrich & Jentsch, 2000). The E3 ubiquitin ligase Rad5 interacts with Rad18 and Ubc13 to form a large protein complex composed of Rad5, Rad6-Rad18, and Ubc13-Mms2 needed for error-free PRR (Gangavarapu et al., 2006; Pagès et al., 2008; Ulrich & Jentsch, 2000). The Rad6-Rad18 complex catalyzes mono-ubiquitination (Ub1) of PCNA (proliferating cell nuclear antigen) on K164 (Hoege, Pfander, Moldovan, Pyrowolakis, & Jentsch, 2002) and competes with DNA polymerases for PCNA enriching DNA (Li, Larsen, & Hedglin, 2020). The complex also serves as a key player in the process of DNA damage bypass for the synthesis of DNA across lesions and the generation of a complementary, undamaged strand as a template in subsequent NER or base excision repair (Boiteux & Jinks-Robertson, 2013; Hedglin & Benkovic, 2015).

Anti-UVB roles of several proteins homologous to the yeast RAD proteins mentioned above have been elucidated in *B. bassiana* and *M. robertsii* in the past few years, including Rad1, Rad2, two Rad4 paralogs (Rad4A and Rad4B), Rad5, Rad6, Rad10, Rad14, two Rad16 paralogs (Rad16A and Rad16B), Rad23, and Rad26. Dark reactivation and photoreactivation rates are assessed as the respective indices of NER and photorepair (discussed later) activities in the null mutants and complemented strains of their coding genes versus a parental wild-type strain, as shown in Fig. 1. All except Rad4B exhibit extant NER activities, which are demonstrated by changes of dark reactivation rates between the null

Table 1 Anti-UVB roles for enzymes and proteins characterized in *Beauveria bassiana* (Bb) and *Metarhizium robertsii* (Mr).

Anti-UV factors	Subcellular feature[a]	% decrease in LD_{50}[b]	% change in PR[c]	Interacting factors	References
Bb Phr1	Nucleus	38	28 vs 80	Unknown	Wang et al. (2019, 2020)
Bb Phr2	Nucleus	19	38 vs 80	Rad23	
Bb CryD	Cytoplasm	19	80 vs 80	Unknown	
Bb WC1	Not shown	63	4 vs 83	Rad1, Rad10, Rad2	Yu, Xu, Tong, et al. (2022); Yu et al. (2024)
Bb WC2	Not shown	68	4 vs 83	Rad1, Rad10, Rad2, Rad14, Rad23, Rad6	Xu, Yu, et al. (2023); Xu, Wen, et al. (2023); Luo et al. (2024a); Luo et al. (2024b)
Bb Rad1	Nucleus	79	3 vs 87	Rad10, WC1, WC2, Rad6	Yu, Xu, Tong, et al. (2022); Yu et al. (2024)
Bb Rad10	Nucleus	80	4 vs 87	Rad1, WC1, WC2, Rad6, Rad14	Luo et al. (2024a)
Bb Rad4A	Nucleus	83	43 vs 88	Rad23	Yu et al. (2023)
Bb Rad4B	Nucleus	0	88 vs 88	Rad23	
Bb Rad23	N/C shuttling	84	5 vs 82	Phr2, WC2, Rad4A, Rad4B, Rad16B	Wang et al. (2020); Yu et al. (2023, 2024)

(continued)

Table 1 Anti-UVB roles for enzymes and proteins characterized in *Beauveria bassiana* (Bb) and *Metarhizium robertsii* (Mr). (*cont'd*)

Anti-UV factors	Subcellular feature[a]	% decrease in LD_{50}[b]	% change in PR[c]	Interacting factors	References
Bb Rad2	Nucleus	76	4 vs 71	WC1, WC2, Rad14	
Bb Rad14	Nucleus	88	1 vs 71	WC2, Rad2, Rad10, RFA1	
Bb Rad6	N/C shuttling	74	0 vs 75	WC2, Rad1, Rad10, Rad18,	Luo et al. (2024a)
Bb Rad5	N/C shuttling	40	10 vs 76	Mms2 (E2 ubiquitin conjugator)	Luo et al. (2024b)
Bb Rad16A	Nucleus	29	31 vs 76	Elc1-like E3 ubiquitin ligase	
Bb Rad16B	N/C shuttling	50	9 vs 76	Rad23, Elc1-like E3 ubiquitin ligase	
Mr WC1	N/C shuttling	54	30 vs 80	WC2, Phr1, Phr2, Rad2	Peng, Xie, et al. (2021); Peng, Guo, et al. (2021); Peng, Zhang, Ying, & Feng, (2024)
Mr WC2	N/C shuttling	67	9 vs 80	WC1, Phr1, Phr2, Rad1, Rad26	Zhang, Peng, Zhang, et al. (2023)
Mr Rad1	N/C shuttling	92	0 vs 63	Rad10, Phr1, WC2, Rad26	Zhang, Peng, Zhang, et al. (2023); Zhang, Peng, Ying, et al. (2023)

Mr Rad10	Nucleus	94	0 vs 63	Rad1, Rad4A, Rad26	Peng, Zhang, Ying, & Feng, (2024)
Mr Rad4A	Nucleus	93	0 vs 68	Rad10, Rad23	Zhang, Peng, Ying, et al. (2023)
Mr Rad4B	Nucleus	0	0 vs 68	Unknown	
Mr Rad23	N/C shuttling	66	70 vs 68	Rad4A	
Mr Rad2	Nucleus	94	2 vs 70	WC1, Rad1	Peng, Zhang, Ying, & Feng, (2024)
Mr Rad14	Nucleus	94	2 vs 70	Rad1	
Mr Rad26	Nucleus	89	2 vs 70	WC2, Rad1, Rad10	

[a]N/C shuttling, nucleocytoplasmic shuttling.

[b]Percent decrease of conidial UVB LD_{50} in the absence versus presence of each factor denotes its contribution to UVB resistance, which is assessed as an index of NER activity in 24-h dark reactivation of conidia impaired by UVB to different degrees.

[c]Percent change in photoreactivation rate (PR; 5-h light plus 19-h dark) of UVB-inactivated conidia in the absence versus presence of each factor. The UVB dose used in the referred studies is lethal (0.4 or 0.5 J/cm^2) of UVB-inactivated conidia in the absence versus presence of each factor. The UVB dose used in the referred studies is lethal (0.4 or 0.5 J/cm^2). *Metarhizium robertsii* Phr1 and Phr2 characterized in a different manner (Fang & St. Leger, 2012) are not listed because their LD_{50} values are not available.

mutants and the wild-type strain (Peng et al., 2024; Wang et al., 2020; Yu et al., 2023; Yu, Xu, Luo, et al., 2022; Zhang, Peng, Ying, et al., 2023; Zhang, Peng, Zhang, et al., 2023). However, their NER activities are limited to recovering light UVB damage in the dark and significantly lower than their activities in photoreactivation of lethal UVB dose-irradiated conidia due to their direct/indirect links to photolyases and photolyase regulators through multiple protein-protein interactions (Wang et al., 2019; Xu, Yu, et al., 2023; Yu, Xu, Luo, et al., 2022), as discussed below. Their NER and photorepair activities are summarized in Table 1 for comparison.

4.2 Photorepair-dependent photoreactivation

Photorepair is a light-dependent process, in which shorter UV-induced DNA lesions are rapidly repaired under longer UV or visible light as a supply of energy for direct transfer of electrons to CPD or 6–4PP lesions to break down the covalent linkages under the actions of photolyases (Chaves et al., 2011; Jans et al., 2005; Sancar, 1996; Sancar, 2003). Consequently, eukaryotic cells impaired by UV are photoreactivated through photorepair. Photolyases are classified to the cryptochrome-photolyase family (CPF) composed of one to four members in filamentous fungi, namely, one or two photolyases (Phr1 and/or Phr2) or one or two DASH-type crypto-chromes (Cry-DASHs) (Yu & Fischer, 2019). This is in contrast with the large family of RAD proteins and partners involved in the yeast NER pathway (Boiteux & Jinks-Robertson, 2013).

All fungal CPF members share a DNA_Photolyase domain needed for photorepair activity. Intriguingly, fungal photorepair is dependent on the CPD-specific photolyase Phr1 and/or the 6–4PP-specific photolyase Phr2 but independent of one or two Cry-DASHs (Berrocal-Tito, Esquivel-Naranjo, Horwitz, & Herrera-Estrella, 2007; Brych et al., 2016; Cohrs & Schumacher, 2017; Garcia-Esquivel, Esquivel-Naranjo, Hernandez-Onate, Ibarra-Laclette, & Herrera-Estrella, 2016). Exposure of *Metarhizium acridum* mycelium to light can upregulate photolyase expression to increase the fungal UVB resistance and photoreactivation activity (Brancini, Bachmann, & Braga, 2021; Brancini, Bachmann, Ferreira, Rangel, & Braga, 2018; Brancini, Rangel, & Braga, 2016). Why fungal photorepair depends on the actions of Phr1 and/or Phr2 instead of one or two Cry-DASHs was unclear until the nucleus-specific localization of a CPF member was revealed to be a prerequisite of its photorepair activity. In *B. bassiana*, Phr1 and Phr2 are intranuclear enzymes showing strong photorepair activities to CPD and

6–4PP DNA lesions, respectively, while the unique Cry-DASH is a cytosolic protein not exhibiting any photorepair activity (Wang et al., 2019). The nucleus-specific localization of either Phr1 or Phr2 is associated with a nuclear localization signal (NLS) motif predicted from its amino acid sequence at high probability, contrasting without any NLS motif predictable from the sequences of true Cry-DASHs in a survey of fungal genomes (Tong & Feng, 2022). Some Mucoromycetes have one CPF member, that is, Cry-DASH/CryA that repairs CPD lesions in vitro or exhibits photoreactivation activity (Navarro et al., 2020; Tagua et al., 2015). This CPF member is virtually identical to photolyases in either domain architecture or high-probability NLS motif (Tong & Feng, 2022). However, it was unknown why fungal photorepair depends on only one or two photolyases while the yeast NER relies upon dozens of RAD proteins and partners. An answer to the question has been revealed in recent studies, as summarized in Table 1 and discussed below.

WC1 and WC2 are two white-collar proteins that interact with each other to form a white-collar complex (WCC), which acts as a master regulator of light-responsive genes and circadian clock in filamentous fungi (Baker, Loros, & Dunlap, 2012; Hurley, Loros, & Dunlap, 2016; Yu & Fischer, 2019). Although no domain is needed for photorepair activity, WC1 and WC2 exhibit higher photorepair activities against CPD and 6–4PP DNA lesions and much higher activities in photoreactivation of UVB-impaired conidia than do Phr1 and Phr2 in *B. bassiana* because each regulates the expression of both photolyases by binding to the promoter regions of *phr1* and *phr2* (Xu, Yu, Luo, Ying, & Feng, 2023; Yu, Xu, Tong, Ying, & Feng, 2022). In *M. robertsii*, either WC1 or WC2 interacts with both Phr1 and Phr2 and displays as high activity in photorepair of 6–4PP or CPD lesions as does Phr2 or Phr1 and similar photoreactivation activity as seen in *B. bassiana* due to the *wc1* or *wc2* knockout mutation leading to abolished expression of *phr2* or *phr1* and downregulation of many other anti-UV genes (Peng et al., 2021). These studies unveil that the WCC formed by WC1 and WC2 acts as a core regulator of photolyases and many other ant-UV factors in insect-pathogenic fungi.

In *B. bassiana*, the ortholog of the yeast Rad23 is a nucleocytoplasmic shuttling protein interacting with Phr2; hence, it can recover the viability of lethal UVB dose-inactivated conidia through an incubation of 3 h under visible light plus 21 h in darkness but cannot do so through a 24-h dark incubation (Wang et al., 2020). This finding indicates that the Rad23-Phr2 interaction endows Rad23 with high photoreactivation activity. Unlike the tight link of Rad23 to Rad4 and Rad33 orthologs in the yeast, Rad23 lacks the partner Rad33 and interacts with Rad4A and Rad4B in *B. bassiana* or with Rad4A rather than with Rad4B or any photolyase in *M. robertsii*, leading to a marked difference in photoreactivation activity of either Rad23 or Rad4A between the two insect pathogens, in which Rad4B has no activity in either dark reactivation or photoreactivation (Yu et al., 2023; Zhang, Peng, Zhang, et al., 2023).

Rad1 and Rad10 interact with each other and exhibit high activities in photoreactivation due to their interactions with WC2 and Phr1 in *M. robertsii* (Zhang, Peng, Ying, et al., 2023), respectively, and with both WC1 and WC2 in *B. bassiana* (Yu, Xu, Luo, et al., 2022). Interactions of Rad2 with WC1, Rad14 with Rad1, and Rad26 with WC2, Rad1 and Rad10 confer very high photoreactivation activities on Rad2, Rad14, and Rad26 in *M. robertsii* (Peng, Zhang, Ying, & Feng, 2024). In *B. bassiana*, Rad2 and Rad14 also exhibit high photoreactivation activities due to the interactions of Rad2 with WC1, WC2 and Rad14 and of Rad14 with WC2, Rad2, Rad10 and RFA1 (Yu, Xu, Luo, Ying, & Feng, 2024). Similarly, interactions of Rad6 with WC2, Rad1, Rad10, and Rad18 provides Rad6 with extraordinarily high activity in photoreactivation (Luo, Yu, Xu, Ying, & Feng, 2024a). Rad5 interacting with Mms2 (E2 ubiquitin conjugator), Rad16A interacting with Elc1-like E3 ubiquitin ligase, and Rad16B interacting with Rad23 and the Elc1-like ligase also show differential photoreactivation activities in *B. bassiana* (Luo, Yu, Xu, Ying, & Feng, 2024b).

Most of the protein-protein interactions detected in *B. bassiana* and *M. robertsii* in recent studies (Table 1) are also documented in the yeast (Boiteux & Jinks-Robertson, 2013). A big difference lies in direct/indirect links of insect-pathogenic fungal RAD proteins to photolyases and photolyase regulators required for photorepair, leading to the formation of a complicated protein-protein interaction network dominated by WCC (Peng et al., 2024; Yu et al., 2024; Zhang, Peng, Ying, et al., 2023; Zhang, Peng, Zhang, et al., 2023). In contrast, neither photolyases nor photolyase regulators are present in the yeast RAD-RAD or RAD-containing protein complexes needed for GG-NER, TC-NER, or error-free PRR. Most of those anti-UV RAD proteins characterized in the insect pathogens have evolutionarily acquired high activities in photoreactivation of sublethal/lethal UVB dose-impaired conidia but contribute little or limitedly to the fungal lifecycle in vitro or in vivo. As an exception, Rad23 has pleiotropic effect in radial growth, conidiation, conidial quality, virulence, non-UV stress tolerance, and transcriptional expression of phenotype-related genes (Wang et al., 2020; Zhang, Peng, Ying, et al., 2023). Rad26 is required for not only solar UV damage repair but also the asexual and insect-pathogenic lifecycles and the genomic expression of *B. bassiana* (Luo et al., 2024a). The NER activities of those RAD proteins recently studied are extant but limited to recovering low UVB dose-impaired conidia during the night-time period available on the Earth's surface. Therefore, their anti-UV roles depend primarily on photoreactivation rather than on NER-dependent dark reactivation. These findings support a hypothesis that molecular machinery operating fungal photorepair and photoreactivation is far more intricate than what was long considered to depend on only one or two photolyases (Tong & Feng, 2022).

5. Conclusion and prospects

In conclusion, the tolerance of insect-pathogenic fungi to solar UV damage is far below daily UV damage accumulated on sunny summer days and depends primarily on photoreactivation. This tolerance is especially important for properly timing the application of fungal formulations to enhance field control efficacy against arthropod pests. Solar UV damage to conidia applied in the late afternoon can be recovered by fungal activity in photoreactivation on the Earth's surface. The activity of a candidate strain to be registered as a fungal pesticide must be well elucidated for proper timing of its application in combination with local pattern of solar UV accumulation.

The marked advances in recent studies have uncovered a much more robust mechanism behind photoreactivation of UVB-impaired/inactivated conidia than a mechanism considered to depend on only one or two photolyases in filamentous fungi, including insect pathogens. Anti-UV roles of the yeast RAD proteins and partners are well known to depend on the NER activities of small or large complexes, which act as endonucleases, helicases, ligases, and polymerases formed through interactions with one another (Boiteux & Jinks-Robertson, 2013). In contrast, their homologs characterized in *B. bassiana* and *M. robertsii* to date have been directly or indirectly linked to photolyases and photolyases regulators through multiple protein-protein interactions and evolutionarily acquired high photo-reactivation activities (Table 1). The NER activities of those homologs in dark reactivation cannot recover severe UV damage as do their photorepair activities in photoreactivation. The recent advances have greatly expanded the genetic/molecular basis for the recovery of filamentous fungi from solar UV damage and revealed that main anti-UV mechanisms are different between the yeast and the non-yeast fungi. Such a difference is due to anti-UV RAD proteins being directly or indirectly linked to photolyases and photolyase regulators in the insect pathogens but not in the yeast. These links suggest a core role of the regulator WCC in the protein-protein interaction network to orchestrate photorepair-dependent photoreactivation. Nonetheless, the number of RAD homologs characterized in fungal insect pathogens is still very limited compared to the large family of anti-UV RAD proteins and partners well studied in yeast. Future research is needed to elucidate functions of many more functionally unknown RAD proteins and their links to the WCC-cored network in the insect pathogens. Special attention should be paid to the anti-UV roles and mechanisms of RAD homologs that remain unexplored in other fungi with various lifestyles. Such efforts will further expand a genetic/molecular basis for fungal photoreactivation and enrich the principles for fungal adaptation and resistance to solar UV damage.

Acknowledgments

Funding of this study was provided by the National Natural Science Foundation of China (Grant No. 32372613). Alene Alder-Rangel reviewed the English in the manuscript.

References

Alder-Rangel, Alene. (2021) *The Adventures of Donald W. Roberts, International insect pathologist* (459 pp.), São José dos Campos: Inbioter.

Arthurs, S., & Dara, S. K. (2019). Microbial biopesticides for invertebrate pests and their markets in the United States. *Journal of Invertebrate Pathology, 165*, 13–21.

Bailly, V., Sommers, C. H., Sung, P., Prakash, L., & Prakash, S. (1992). Specific complex formation between proteins encoded by the yeast DNA repair and recombination genes RAD1 and RAD10. *Proceedings of the National Academy of Sciences of the United States of America, 89*, 8273–8277.

Baker, C. L., Loros, J. J., & Dunlap, J. C. (2012). The circadian clock of *Neurospora crassa*. *FEMS Microbiology Reviews, 36*, 95–110.

Bardwell, A. J., Bardwell, L., Tomkinson, A. E., & Friedberg, E. C. (1994). Specific cleavage of model recombination and repair intermediates by the yeast Rad1-Rad10 DNA endonuclease. *Science (New York), 265*, 2082–2085.

Berrocal-Tito, G. M., Esquivel-Naranjo, E. U., Horwitz, B. A., & Herrera-Estrella, A. (2007). *Trichoderma atroviride* PHR1, a fungal photolyase responsible for DNA repair, autoregulates its own photoinduction. *Eukaryotic Cell, 6*, 1682–1692.

Boiteux, S., & Jinks-Robertson, S. (2013). DNA repair mechanisms and the bypass of DNA damage in *Saccharomyces cerevisiae*. *Genetics, 193*, 1025–1064.

Braga, G. U. L., Flint, S. D., Messias, C. L., Anderson, A. J., & Roberts, D. W. (2001). Effect of UV-B on conidia and germlings of the entomopathogenic hyphomycete *Metarhizium anisopliae*. *Mycological Research, 105*, 874–882.

Braga, G. U. L., Flint, S. D., Messias, C. L., Anderson, A. J., & Roberts, D. W. (2001). Variability in response to UV-B among species and strains of *Metarhizium* isolated from sites at latitudes from 61°N to 54°S. *Journal of Invertebrate Pathology, 78*, 98–108.

Braga, G. U. L., Flint, S. D., Messias, C. L., Anderson, A. J., & Roberts, D. W. (2001). Effects of UVB irradiance on conidia and germinants of the entomopathogenic hyphomycete *Metarhizium anisopliae*: A study of reciprocity and recovery. *Photochemistry and Photobiology, 73*, 140–146.

Braga, G. U. L., Flint, S. D., Miller, C. D., Anderson, A. J., & Roberts, D. W. (2001). Both solar UVA and UVB radiation impair conidial culturability and delay germination in the entomopathogenic fungus *Metarhizium anisopliae*. *Photochemistray and Photobiology, 74*, 734–739.

Braga, G. U. L., Rangel, D. E. N., Fernandes, E. K. K., Flint, S. D., & Roberts, D. W. (2015). Molecular and physiological effects of environmental UV radiation on fungal conidia. *Current Genetics, 61*, 405–425.

Braga, G. U. L., Rangel, D. E. N., Flint, S. D., Miller, C. D., Anderson, A. J., & Roberts, D. W. (2002). Damage and recovery from UV-B exposure in conidia of the entomopathogens *Verticillium lecanii* and *Aphanocladium album*. *Mycologia, 94*, 912–920.

Braga, G. U. L., Rangel, D. E. N., Flint, S. D., Anderson, A. J., & Roberts, D. W. (2006). Conidial pigmentation is important to tolerance against solar-simulated radiation in the entomopathogenic fungus *Metarhizium anisopliae*. *Photochemistray and Photobiology, 82*, 418–422.

Braga, G. U. L., Silva-Junior, G. J., Brancini, G. T. P., Hallsworth, J. E., & Wainwright, M. (2022). Photoantimicrobials in agriculture. *Journal of Photochemistry and Photobiology B-Biology, 235*, 112548.

Brancini, G. T. P., Bachmann, L., & Braga, G. U. L. (2021). Timing and duration of light exposure during conidia development determine tolerance to ultraviolet radiation. *FEMS Microbiology Letters, 368*, fnab133.

Brancini, G. T. P., Bachmann, L., Ferreira, M. E. D., Rangel, D. E. N., & Braga, G. U. L. (2018). Exposing *Metarhizium acridum* mycelium to visible light up-regulates a photolyase gene and increases photoreactivating ability. *Journal of Invertebrate Pathology, 152*, 35–37.

Brancini, G. T. P., Hallsworth, J. E., Corrochano, L. M., & Braga, G. U. L. (2022). Photobiology of the keystone genus *Metarhizium*. *Journal of Photochemistry and Photobiology B-Biology, 226*, 112374.

Brancini, G. T. P., Rangel, D. E. N., & Braga, G. U. L. (2016). Exposure of *Metarhizium acridum* mycelium to light induces tolerance to UV-B radiation. *FEMS Microbiology Letters, 363*, fnw036.

Brych, A., Mascarenhas, J., Jaeger, E., Charkiewicz, E., Pokorny, R., Boelker, M., et al. (2016). White collar 1-induced photolyase expression contributes to UV-tolerance of *Ustilago maydis*. *MicrobiologyOpen, 5*, 224–243.

Chaudhuri, S., Wyrick, J. J., & Smerdon, M. J. (2009). Histone H3 Lys79 methylation is required for efficient nucleotide excision repair in a silenced locus of Saccharomyces cerevisiae. *Nucleic Acids Research, 37*, 1690–1700.

Chaves, I., Pokorny, R., Byrdin, M., Hoang, N., Ritz, T., Brettel, K., et al. (2011). The cryptochromes: Blue light photoreceptors in plants and animals. *Annual Review of Plant Biology, 62*, 335–364.

Cohrs, K. C., & Schumacher, J. (2017). The two cryptochrome/photolyase family proteins fulfill distinct roles in DNA photorepair and regulation of conidiation in the gray mold fungus *Botrytis cinerea*. *Applied and Environmental Microbiology, 83*, e00812.

Corval, A. R. C., Mesquita, E., Correa, T. A., Silva, C. D. R., Bitencourt, R. D. B., Fernandes, E. K. K., et al. (2021). UV-B tolerances of conidia, blastospores, and microsclerotia of *Metarhizium* spp. entomopathogenic fungi. *Journal of Basic Microbiology, 61*, 15–26.

Davies, A. A., Friedberg, E. C., Tomkinson, A. E., Wood, R. D., & West, S. C. (1995). Role of the Rad1 and Rad10 proteins in nucleotide excision repair and recombination. *Journal of Biological Chemistry, 270*, 24638–24641.

de Faria, M., & Wraight, S. P. (2007). Mycoinsecticides and mycoacaricides: A comprehensive list with worldwide coverage and international classification of formulation types. *Biollogical Control, 43*, 237–256.

de Laat, W. L., Jaspers, N. G. J., & Hoeijmakers, J. H. J. (1999). Molecular mechanism of nucleotide excision repair. *Genes & Development, 13*(7), 768–785.

Den Dulk, B., Sun, S. M., De Ruijter, M., Brandsma, J. A., & Brouwer, J. (2006). Rad33, a new factor involved in nucleotide excision repair in *Saccharomyces cerevisiae*. *DNA Repair, 5*, 683–692.

Dias, L. P., Pedrini, N., Braga, G. U. L., Ferreira, P. C., Pupin, B., Araujo, C. A. S., et al. (2020). Outcome of blue, green, red, and white light on *Metarhizium robertsii* during mycelial growth on conidial stress tolerance and gene expression. *Fungal Biology, 124*, 263–272.

Evans, E. J., Moggs, G., Hwang, J. R., Egly, J. M., & Wood, R. D. (1997). Mechanism of open complex and dual incision formation by human nucleotide excision repair factors. *EMBO Journal, 16*, 6559–6573.

Fang, W. G., & St. Leger, R. J. (2012). Enhanced UV resistance and improved killing of malaria mosquitoes by photolyase transgenic entomopathogenic fungi. *PLoS One, 7*, e43069.

Feng, M. G., Poprawski, T. J., & Khachatourians, G. G. (1994). Production, formulation and application of the entomopathogenic fungus *Beauveria bassiana* for insect control: Current status. *Biocontrol Science and Technology, 4*, 3–34.

Fernandes, E. K. K., Rangel, D. E. N., Braga, G. U. L., & Roberts, D. W. (2015). Tolerance of entomopathogenic fungi to ultraviolet radiation: A review on screening of strains and their formulation. *Current Genetics, 61*, 427–440.

Fernandes, E. K. K., Rangel, D. E. N., Moraes, A. M. L., Bittencourt, V. R. E. P., & Roberts, D. W. (2007). Variability in tolerance to UV-B radiation among *Beauveria* spp. isolates. *Journal of Invertebrate. Pathololology, 96*, 237–243.

Gangavarapu, V., Haracska, L., Unk, I., Johnson, R. E., Prakash, S., & Prakash, L. (2006). Mms2-Ubc13-dependent and -independent roles of Rad5 ligase in postreplication repair and translesion DNA synthesis in *Saccharomyces cerevisiae*. *Molecular and Cellular Biology, 26*, 7783–7790.

Garcia-Esquivel, M., Esquivel-Naranjo, E. U., Hernandez-Onate, M. A., Ibarra-Laclette, E., & Herrera-Estrella, A. (2016). The *Trichoderma atroviride* cryptochrome/photolyase genes regulate the expression of *blr1*-independent genes both in red and blue light. *Fungal Biology, 120*, 500–512.

Gillette, T. G., Yu, S., Zhou, Z., Waters, R., Johnston, S. A., & Reed, S. H. (2006). Distinct functions of the ubiquitin-proteasome pathway influence nucleotide excision repair. *EMBO Journal, 25*, 2529–2538.

Godderz, D., Giovannucci, T. A., Lalakova, J., Menendez-Benito, V., & Dantuma, N. P. (2017). The deubiquitylating enzyme Ubp12 regulates Rad23-dependent proteasomal degradation. *Journal of Cell Science, 130*, 3336–3346.

Gong, F., Fahy, D., & Smerdon, M. J. (2006). Rad4-Rad23 interaction with SWI/SNF links ATP-dependent chromatin remodeling with nucleotide excision repair. *Nature Structural and Molecular Biology, 13*, 902–907.

Griffiths, H. R., Mistry, P., Herbert, E., & Lunec, J. (1998). Molecular and cellular effects of ultraviolet light-induced genotoxicity. *Critical Reviews in Clinical Laboratory Sciences, 35*, 189–237.

Guo, C. T., Luo, X. C., Tong, S. M., Zhou, Y., Ying, S. H., & Feng, M. G. (2022). FluG and FluG-like FrlA coregulate manifold gene sets vital for fungal insect-pathogenic lifestyle but not involved in asexual development. *mSystems, 7*, e00318–e00322.

Guo, C. T., Peng, H., Tong, S. M., Ying, S. H., & Feng, M. G. (2021). Distinctive role of *fluG* in the adaptation of *Beauveria bassiana* to insect-pathogenic lifecycle and environmental stresses. *Environmental Microbiology, 23*, 5184–5199.

Guzder, S. N., Habraken, Y., Sung, P., Prakash, L., & Prakash, S. (1996). RAD26, the yeast homolog of human Cockayne's syndrome group B gene, encodes a DNA-dependent ATPase. *Journal of Biological Chemistry, 271*, 18314–18317.

Guzder, S. N., Sung, P., Prakash, L., & Prakash, S. (1997). Yeast Rad7-Rad16 complex, specific for the nucleotide excision repair of the nontranscribed DNA strand, is an ATP-dependent DNA damage sensor. *Journal of Biological Chemistry, 272*, 21665–21668.

Guzder, S. N., Sung, P., Prakash, L., & Prakash, S. (1999). Synergistic interaction between yeast nucleotide excision repair factors NEF2 and NEF4 in the binding of ultraviolet-damaged DNA. *Journal of Biological Chemistry, 274*, 24257–24262.

Hedglin, M., & Benkovic, S. J. (2015). Regulation of Rad6/Rad18 activity during DNA damage tolerance. *Annual Review of Biophysics, 44*, 207–228.

Hoege, C., Pfander, B., Moldovan, G. L., Pyrowolakis, G., & Jentsch, S. (2002). RAD6-dependent DNA repair is linked to modification of PCNA by ubiquitin and SUMO. *Nature, 419*, 135–141.

Huang, W., Feaver, W. J., Tomkinson, A. E., & Friedberg, E. C. (1998). The N-degron protein degradation strategy for investigating the function of essential genes: Requirement for replication protein A and proliferating cell nuclear antigen proteins for nucleotide excision repair in yeast extracts. *Mutation Research, 408*, 183–194.

Huang, B. F., & Feng, M. G. (2009). Comparative tolerances of various *Beauveria bassiana* isolates to UV-B irradiation with a description of a modeling method to assess lethal dose. *Mycopathologia, 168*, 145–152.

Hurley, J. M., Loros, J. J., & Dunlap, J. C. (2016). Circadian oscillators: Around the transcription- translation feedback loop and on to output. *Trends in Biochemical Sciences, 41*, 834–846.

Ignoffo, C. M., Hostetter, D. L., Sikorowski, P. P., Sutter, G., & Brooks, W. M. (1977). Inactivation of representative species of entomopathogenic viruses, a bacterium, fungus, and protozoan by an UV light-source. *Environmental Entomology, 6*, 411–415.

Jans, J., Schul, W., Sert, Y. G., Rijksen, Y., Rebel, H., Eker, A. P. M., et al. (2005). Powerful skin cancer protection by a CPD-photolyase transgene. *Current Biology, 15*, 105–115.

Lafrance-Vanasse, J., Arseneault, G., Cappadocia, L., Chen, H. T., Legault, P., & Omichinski, J. G. (2012). Structural and functional characterization of interactions involving the Tfb1 subunit of TFIIH and the NER factor Rad2. *Nucleic Acids Research, 40*, 5739–5750.

Lewis, M. W., Robalino, I. V., & Keyhani, N. O. (2009). Uptake of the fluorescent probe FM4-64 by hyphae and haemolymph-derived in vivo hyphal bodies of the entomopathogenic fungus *Beauveria bassiana*. *Microbiology-UK, 155*, 3110–3120.

Li, M. J., Larsen, L., & Hedglin, M. (2020). Rad6/Rad18 competes with DNA polymerases eta and delta for PCNA encircling DNA. *Biochemistry, 59*, 407–416.

Luo, X. C., Yu, L., Xu, S. Y., Ying, S. H., & Feng, M. G. (2024a). Rad6, a ubiquitin conjugator required for insect-pathogenic lifestyle, UV damage repair, and genomic expression of *Beauveria bassiana*. *Microbiological Research, 281*, 127622.

Luo, X. C., Yu, L., Xu, S. Y., Ying, S. H., & Feng, M. G. (2024b). Photoreactivation activities of Rad5, Rad16A and Rad16B help *Beauveria bassiana* to recover from solar ultraviolet damage. *Journal of Fungi, 10*, 420.

Madronich, S. (1993). UV radiation in the natural and perturbed atmosphere. In M. Tevini (Ed.). *UV-B radiation and ozone depletion* (pp. 17−69)Boca Raton: Lewis.

Malik, S., Chaurasia, P., Lahudkar, S., Durairaj, G., Shukla, A., & Bhaumik, S. R. (2010). Rad26p, a transcription-coupled repair factor, is recruited to the site of DNA lesion in an elongating RNA polymerase II-dependent manner in vivo. *Nucleic Acids Research, 38*, 1461–1477.

Mascarin, G. M., Lopes, R. B., Delalibera, I., Fernandes, E.K, K., Luz, C., & Faria, M. (2019). Current status and perspectives of fungal entomopathogens used for microbial control of arthropod pests in Brazil. *Journal of Invertebrate Pathology, 165*, 46–53.

Navarro, E., Niemann, N., Kock, D., Dadaeva, T., Gutierrez, G., Engelsdor, T., et al. (2020). The DASH-type cryptochrome from the fungus *Mucor circinelloides* is a canonical CPD-photolyase. *Current Biology, 30*, 4483–4490.

Pagès, V., Bresson, A., Acharya, N., Prakash, S., Fuchs, R. P., & Prakash, L. (2008). Requirement of Rad5 for DNA polymerase ζ-dependent translesion synthesis in *Saccharomyces cerevisiae*. *Genetics, 180*, 73–82.

Peng, H., Guo, C. T., Tong, S. M., Ying, S. H., & Feng, M. G. (2021). Two white collar proteins protect fungal cells from solar UV damage by their interactions with two photolyases in *Metarhizium robertsii*. *Environmental Microbiology, 23*, 4925–4938.

Peng, G. X., Xie, J. Q., Guo, R., Keyhani, N. O., Zeng, D. Y., Yang, P. Y., et al. (2021). Long-term field evaluation and large-scale application of a *Metarhizium anisopliae* strain for controlling major rice pests. *Journal of Pest Science, 94*, 969–980.

Peng, H., Zhang, Y. L., Ying, S. H., & Feng, M. G. (2024). Rad2, Rad14 and Rad26 recover solar UV damage by their links to photoreactivation-depending factors in *Metarhizium robertsii*. *Microbiological Research, 280*, 127589.

Prakash, S., Sung, P., & Prakash, L. (1993). DNA repair genes and proteins of *Saccharomyces cerevisiae*. *Annual Review of Genetics, 27*, 33–70.

Qi, D. Y., Xu, W. Y., Shao, Y. Z., Feng, J. R., Feng, M. G., & Tong, S. M. (2023). Mycoinsecticides applied in late afternoon are more efficacious against rice leaf-rolling insect pests than those in the morning. *Biological Control, 186*, 105352.

Ramsey, K. L., Smith, J. J., Dasgupta, A., Maqani, N., Grant, P., & Auble, D. T. (2004). The NEF4 complex regulates Rad4 levels and utilizes Snf2/Swi2-related ATPase activity for nucleotide excision repair. *Molecular and Cellular Biology, 24*, 6362–6378.

Rangel, D. E. N., Anderson, A. J., & Roberts, D. W. (2006). Growth of *Metarhizium anisopliae* on non-preferred carbon sources yields conidia with increased UV-B tolerance. *Journal of Invertebrate Pathology, 93*, 127–134. https://doi.org/10.1016/j.jip.2006.05.011.

Rangel, D. E. N., Anderson, A. J., & Roberts, D. W. (2008). Evaluating physical and nutritional stress during mycelial growth as inducers of tolerance to heat and UV-B radiation in *Metarhizium anisopliae* conidia. *Mycological Research, 112*, 1362–1372.

Rangel, D. E. N., Braga, G. U. L., Anderson, A. J., & Roberts, D. W. (2005). Influence of growth environment on tolerance to UV-B radiation, germination speed, and morphology of *Metarhizium anisopliae* var. *acridum* conidia. *Journal of Invertebrate Pathology, 90*, 55–58.

Rangel, D. E. N., Braga, G. U. L., Fernandes, E. K. K., Keyser, C. A., Hallsworth, J. E., & Roberts, D. W. (2015). Stress tolerance and virulence of insect-pathogenic fungi are determined by environmental conditions during conidial formation. *Current Genetics, 61*, 383–404.

Rangel, D. E. N., Braga, G. U. L., Flint, S. D., Anderson, A. J., & Roberts, D. W. (2004). Variations in UV-B tolerance and germination speed of *Metarhizium anisopliae* conidia produced on insects and artificial substrates. *Journal of Invertebrate Pathology, 87*, 77–83.

Rangel, D. E. N., Butler, M. J., Torabinejad, J., Anderson, A. J., Braga, G. U. L., Day, A. W., et al. (2006). Mutants and isolates of *Metarhizium anisopliae* are diverse in their relationships between conidial pigmentation and stress tolerance. *Journal of Invertebrate Pathology, 93*, 170–182.

Rangel, D. E. N., Fernandes, É. K. K., Braga, G. U. L., & Roberts, D. W. (2011). Visible light during mycelial growth and conidiation of *Metarhizium robertsii* produces conidia with increased stress tolerance. *FEMS Microbiology Letters, 315*, 81–86.

Reed, S. H., Akiyama, M., Stillman, B., & Friedberg, E. C. (1999). Yeast autonomously replicating sequence binding factor is involved in nucleotide excision repair. *Genes & Development, 13*, 3052–3058.

Reed, S. H., You, Z., & Friedberg, E. C. (1998). The yeast *RAD7* and *RAD16* genes are required for postincision events during nucleotide excision repair. In vitro and in vivo studies with *rad7* and *rad16* mutants and purification of a Rad7/Rad16-containing protein complex. *Journal of Biological Chemistry, 273,* 29481–29488.

Roberts, D. W., & Campbel, A. S. (1977). Stability of entomopathogenic fungi. *Miscellaneous Publications of the Entomological Society of America, 10,* 19–76.

Roberts, D. W., & Leger, R. J., St. (2004). *Metarhizium* spp., cosmopolitan insect pathogenic fungi: Mycological aspects. *Advances in Applied Microbiology, 54,* 1–70.

Rodriguez, K., Wang, Z., Friedberg, E. C., & Tomkinson, A. E. (1996). Identification of functional domains within the RAD1•RAD10 repair and recombination endonuclease of *Saccharomyces cerevisiae. Journal of Biological Chemistry, 271,* 20551–20558.

Sancar, A. (1996). No "end of history" for photolyases. *Science (New York), 272*(5258), 48–49.

Sancar, A. (2003). Structure and function of DNA photolyase and cryptochrome blue-light photoreceptors. *Chemical Reviews, 103,* 2203–2237.

Sarkar, S., Kiely, R., & McHugh, P. J. (2010). The Ino80 chromatin-remodeling complex restores chromatin structure during UV DNA damage repair. *Journal of Cell Biology, 191,* 1061–1068.

Smerdon, M. J., & Thoma, F. (1990). Site-specific DNA-repair at the nucleosome level in a yeast minichromosome. *Cell, 61,* 675–684.

Staresincic, L., Fagbemi, A. F., Enzlin, J. H., Gourdin, A. M., Wijgers, N., Dunand-Sauthier, I., et al. (2009). Coordination of dual incision and repair synthesis in human nucleotide excision repair. *EMBO Journal, 28,* 1111–1120.

Suter, B., Wellinger, R. E., & Thoma, F. (2000). DNA repair in a yeast origin of replication: Contributions of photolyase and nucleotide excision repair. *Nucleic Acids Research, 28,* 2060–2068.

Tagua, V. G., Pausch, M., Eckel, M., Gutiérrez, G., Miralles-Duran, A., Sanz, C., et al. (2015). Fungal cryptochrome with DNA repair activity reveals an early stage in cryptochrome evolution. *Proceedings of the National Academy of Sciences of the United States of America, 112,* 15130–15135.

Taschner, M., Harreman, M., Teng, Y., Gill, H., Anindya, R., Maslen, S. L., et al. (2010). A role for checkpoint kinase-dependent Rad26 phosphorylation in transcription-coupled DNA repair in *Saccharomyces cerevisiae. Molecular and Cellular Biology, 30,* 436–446.

Teng, Y., Liu, H., Gill, H. W., Yu, Y., Waters, R., & Reed, S. H. (2008). *Saccharomyces cerevisiae* Rad16 mediates ultraviolet-dependent histone H3 acetylation required for efficient global genome nucleotide-excision repair. *EMBO Reports, 9,* 97–102.

Tomkinson, A. E., Bardwell, A. J., Bardwell, L., Tappe, N. J., & Friedberg, E. C. (1993). Yeast DNA repair and recombination proteins Rad1 and Rad10 constitute a single-stranded-DNA endonuclease. *Nature, 362,* 860–862.

Tong, S. M., & Feng, M. G. (2019). Insights into regulatory roles of MAPK-cascaded pathways in multiple stress responses and life cycles of insect and nematode myco-pathogens. *Applied Microbiology and Biotechnology, 103,* 577–587.

Tong, S. M., & Feng, M. G. (2020). Phenotypic and molecular insights into heat tolerance of formulated cells as active ingredients of fungal insecticides. *Applied Microbiology and Biotechnology, 104*(13), 5711–5724.

Tong, S. M., & Feng, M. G. (2022). Molecular basis and regulatory mechanisms underlying fungal insecticides' resistance to solar ultraviolet irradiation. *Pest Management Science, 78,* 30–42.

Tsuchiya, H., Ohtake, F., Arai, N., Kaiho, A., Yasuda, S., Tanaka, K., et al. (2017). *In vivo* ubiquitin linkage-type analysis reveals that the Cdc48-Rad23/Dsk2 axis contributes to K48-linked chain specificity of the proteasome. *Molecular Cell, 66,* 488–502.

Ulrich, H. D., & Jentsch, S. (2000). Two RING finger proteins mediate cooperation between ubiquitin-conjugating enzymes in DNA repair. *EMBO Journal, 19,* 3388–3397.

Verhage, R., Zeeman, A. M., De Groot, N., Gleig, F., Bang, D. D., Vandeputte, P., et al. (1994). The RAD7 and RAD16 genes, which are essential for pyrimidine dimer removal from the silent mating type loci, are also required for repair of the non-transcribed strand of an active gene in *Saccharomyces cerevisiae*. *Molecular and Cellular Biology, 14*, 6135–6142.

Wang, D. Y., Fu, B., Tong, S. M., Ying, S. H., & Feng, M. G. (2019). Two photolyases repair distinct DNA lesions and reactivate UVB-inactivated conidia of an insect myco-pathogen under visible light. *Applied and Environmental Microbiology, 85*, e02459–18.

Wang, D. Y., Mou, Y. N., Tong, S. M., Ying, S. H., & Feng, M. G. (2020). Photoprotective role of photolyase-interacting RAD23 and its pleiotropic effect on the insect-pathogenic fungus *Beauveria bassiana*. *Applied and Environmental Microbiology, 86*, e00287–20.

Xu, W. Y., Wen, Z. X., Li, X. J., Hu, E. Z., Qi, D. Y., Feng, M. G., & Tong, S. M. (2023). Timing of fungal insecticide application to avoid solar ultraviolet irradiation enhances field control of rice planthoppers. *Insects, 14*, 307.

Xu, S. Y., Yu, L., Luo, X. C., Ying, S. H., & Feng, M. G. (2023). Co-regulatory roles of WC1 and WC2 in asexual development and photoreactivation of *Beauveria bassiana*. *Journal of Fungi, 9*, 290.

Yao, S. L., Ying, S. H., Feng, M. G., & Hatting, J. L. (2010). In vitro and in vivo responses of fungal biocontrol agents to the gradient doses of UV-B and UV-A irradiation. *BioControl, 55*, 413–422.

Yasui, A., Eker, A. P. M., Yasuhira, S., Yajima, H., Kobayashi, T., Takao, M., et al. (1994). A new class of DNA photolyases present in various organisms including aplacental mammals. *EMBO Journal, 13*, 6143–6151.

Yu, Z. Z., & Fischer, R. (2019). Light sensing and responses in fungi. *Nature Reviews. Microbiology, 17*, 25–36.

Yu, L., Xu, S. Y., Luo, X. C., Ying, S. H., & Feng, M. G. (2022). Rad1 and Rad10 tied to photolyase regulators protect insecticidal fungal cells from solar UV damage by pho-toreactivation. *Journal of Fungi, 8*, 1124.

Yu, L., Xu, S. Y., Luo, X. C., Ying, S. H., & Feng, M. G. (2023). Comparative roles of Rad4A and Rad4B in photoprotection of *Beauveria bassiana* from solar ultraviolet damage. *Journal of Fungi, 9*, 154.

Yu, L., Xu, S. Y., Luo, X. C., Ying, S. H., & Feng, M. G. (2024). High photoreactivation activities of Rad2 and Rad14 in recovering insecticidal *Beauveria bassiana* from solar UV damage. *Journal of Photochemistry and Photobiology B-Biology, 251*, 112849.

Yu, L., Xu, S. Y., Tong, S. M., Ying, S. H., & Feng, M. G. (2022). Optional strategies for low-risk and non-risk applications of fungal pesticides to avoid solar ultraviolet damage. *Pest Managements Science, 78*, 4660–4667.

Zhang, L. B., & Feng, M. G. (2018). Antioxidant enzymes and their contributions to biological control potential of fungal insect pathogens. *Applied Microbiology and Biotechnology, 102*, 4995–5004.

Zhang, Y. L., Peng, H., Ying, S. H., & Feng, M. G. (2023). Efficient photoreactivation of solar UV-injured *Metarhizium robertsii* by Rad1 and Rad10 linked to DNA photorepair-required proteins. *Photochemistry and Photobiology, 99*, 1122–1130.

Zhang, Y. L., Peng, H., Zhang, K., Ying, S. H., & Feng, M. G. (2023). Divergent roles of Rad4 and Rad23 homologs in *Metarhizium robertsii*'s resistance to solar ultraviolet damage. *Applied and Environmental Microbiology, 89*, e0099423.

CHAPTER FOUR

Basidiomycetes to the rescue: Mycoremediation of metal-organics co-contaminated soils

Lea Traxler, Katrin Krause, and Erika Kothe[*,1]

Friedrich Schiller University Jena, Institute of Microbiology, Jena, Germany
*Corresponding author. e-mail address: erika.kothe@uni-jena.de

Contents

Abstract

The increasing need for metals leads to contaminated post-mining landscapes. At the same time, the contamination with organic, recalcitrant contamination increases. This poses a problem of reuse of large areas, often co-contaminated with both, metals, and organic pollutants. For the remediation of areas contaminated with multiple contaminants and combining many stress factors, technical solutions including groundwater treatment, where necessary, have been devised. However, this is applied to highly contaminated, small sites. The reuse of larger, co-contaminated landscapes remains a major challenge. Mycoremediation with fungi offers a good option for such areas. Fungi cope particularly well with heterogeneous conditions due to their adaptability and their large hyphal network. This chapter summarizes the advantages of basidiomycetes with a focus on wood rot fungi in terms of their ability to tolerate metals, radionuclides, and organic contaminants

[1] www.mikrobiologie.uni-jena.de

Advances in Applied Microbiology, Volume 129
ISSN 0065-2164, https://doi.org/10.1016/bs.aambs.2024.06.001

such as polycyclic aromatic hydrocarbons. It also shows how these fungi can reduce toxicity of contaminants to other organisms including plants to allow for restored land-use. The processes based on diverse molecular mechanisms are introduced and their use for mycoremediation is discussed.

1. Introduction

Microbes are omnipresent, and bacteria and fungi in soil are essential for element cycling and soil formation processes (Fierer, 2017). Hence, their impact on agriculture, human nutrition, health, and renewable energy production cannot be overestimated and land-use changes will, in turn, impact the fungal and bacterial diversity (see, e.g., Tian et al., 2019). However, only a tiny fraction of the 2–11 million species estimated are known (Phukhamsakda et al., 2022). These estimates are thus in the same range as insect biodiversity, where the current estimates converge at approximately 6 million species (Stork, 2018). With soil harboring the highest biodiversity on Earth, and 90% of fungi living in soil, the soil ecosystem is of specific importance to biological balance (Anthony, Bender, & Van Der Heijden, 2023). The decomposition of plant material depends on fungi that can degrade the lignin present in plant matter (Morgan, Lee, Lewis, Sheppard, & Watkinson, 1993). Hence, their contribution to mineralization processes in soil is essential for soil fertility.

Toxic metals and xenobiotic compounds, such as persistent organic pollutants, pesticides, and polycyclic aromatic hydrocarbons (PAHs) are recognized as major chemical groups that are responsible for soil and water pollution worldwide (Morillo et al., 2007; Tobiszewski & Namieśnik, 2012). PAHs are composed of two or more fused aromatic rings of carbon and hydrogen atoms and are derived mainly from the incomplete combustion or pyrolysis of organic material. PAHs are found in biotic and abiotic components on Earth (the atmosphere, soil, and sediments, the water column) and in both terrestrial and aquatic organisms. In the co-contamination of metals with PAHs, the remediation using biological agents is specifically challenging.

The high prevalence of fungi warrants a closer look into their roles in pedogenesis and will also lead to more directional bioremediation of degraded soils, including mycoremediation approaches. Fungi, along with bacteria, are the most important microorganisms for reclamation, immobilization, or detoxification of metallic, semi-metallic, and radionuclide pollutants in terrestrial environments (Gadd, 2004, 2010; Gadd, Rhee, Stephenson, & Wei, 2012), and they may even structure bacterial microbiomes (Bogdanova, 2022). In addition, fungi can absorb large amounts of metals and radionuclides, and survive under adverse conditions (Árvay et al., 2014; Guillén & Baeza, 2014; Zhdanova et al., 2003).

White-rot fungi have proven to be particularly suitable for remediation, as they have a large variety of ligninolytic enzymes that help them to be metabolically flexible but also to break down a broad range of contaminations. One of these fungi is the model organism *Schizophyllum commune*. It is easily cultivated in haploid as well as mated, fruiting body-forming dikaryotic stages, and can tolerate appreciable metal concentrations to provide a good model system (Gabriel, Mokrejš, Bílý, & Rychlovský, 1994; Günther et al., 2014; Kirtzel, Scherwietes, Merten, Krause, & Kothe, 2019; Traxler et al., 2022). The amount of data on cellular and molecular biology, bioweathering activities, gene expression analyses, and proteome studies makes this fungus an excellent target for studying its potential in mycoremediation (Brunsch et al., 2015; Erdmann et al., 2012; Freihorst et al., 2018; Knabe et al., 2013; Ohm et al., 2010). The use of less well-known model fungi, which may have an even higher wood-degradation potential with more of the versatile ligninolytic enzymes, may profit from fundamental findings obtained with *S. commune*.

Among the basidiomycetes, the ectomycorrhizal species may attract additional interest for their association with host trees. Mycorrhizal fungi acting at the plant root-soil interfaces have a good potential for remediation, because they additionally protect plants from pollutants to allow for growth promotion in stressful conditions, and thy are known to improve their hosts' nutrient and water supply (Gadd & Raven, 2010; Smith & Read, 2008). Fruiting bodies of ectomycorrhizal species, e.g., *Paxillus involutus* or *Pisolithus tinctorius*, are frequently found in mining areas with deciduous and coniferous trees. These fungi show faster growth compared to other ectomycorrhizal fungi, but still exhibit high metal tolerance, are known to degrade polysaccharides and to degrade polyphenols (Rineau et al., 2012; Wallander & Söderström, 1999). Ectomycorrhizal fungi, as well as the arbuscular mycorrhiza forming *Rhizophagus irregularis* (Kokkoris et al., 2023), have been used in ecological and molecular biology studies showing their high potential to contribute to remediation and land reclamation (Abdulsalam et al., 2020; Bizo et al., 2017; Bogdanova, Kothe, & Krause, 2023; Kothe & Reinicke, 2017; Sammer, Krause, Gube, Wagner, & Kothe, 2016; Schlunk, Krause, Wirth, & Kothe, 2015).

2. The potential of mycoremediation

Metals and other pollutants like xenobiotic organic compounds are omnipresent in the environment and are mostly of anthropogenic origin. Nearly all habitats in soil, air, and water are affected by pollution.

Some metals, such as Cd, Cu, Pb, and Cr are leading to anthropogenic contamination of rather large areas of soil. They find applications as fertilizers, pigments, or lubricants. For example, Hg, As, and several PAHs are used as pesticides (Nriagu & Pacyna, 1988). Especially, PAHs and organochlorine pesticides are contaminants of concern due to their highly toxic, persistent, and bioaccumulative properties (Zhonghua, Zhang, & Wu, 2016). Nevertheless, the two main sources of metal contamination in the soil are mining and disposal of ash residues from coal combustion or commercial products (Nriagu & Pacyna, 1988).

While atmospheric pollution with volatile and particulate contaminants is of concern, we will concentrate on soil remediation. Metals tend to contribute to fine particulate pollution whereas most PAHs are semi-volatile (Gundel, Lee, Mahanama, Stevens, & Daisey, 1995; Sitaras, Bakeas, & Siskos, 2004). The particulate pollution decreased during the past decades, mostly due to particle filters and the decreased use of leaded gasoline (Maciejczyk, Chen, & Thurston, 2021). In aquatic ecosystems, PAHs and organochlorine pesticides are tightly connected to sediments because of their hydrophobic properties (Warren, Allan, Carter, House, & Parker, 2003). PAHs often originate from pyrolytic processes such as marine fuel or oily sewage (He et al., 2014). The main source of metal contamination in aquatic environments is due to the cooling water introduction of coal-burning power plants, iron, and steel plants, dumping of sewage sludge, and domestic wastewater effluents (Nriagu & Pacyna, 1988). While specific remediation approaches avoiding dust emission or water treatment have been devised, soil remediation still largely depends on excavation and safe deposition. However, this costly process is not feasible at landscape scale, and hence, more gentle (and cheaper) approaches involving bioremediation are called for.

One way of making all these contaminated habitats usable again is biotic soil remediation using plants, bacteria, or fungi. Fungi are known to cope with a wide range of contaminants, e.g., by degrading recalcitrant organic pollutants or changing metal mobility in soil (Gray, 1998). Hence, mycoremediation can be seen with mobilizing and immobilizing effects. While organic pollutants can be degraded by fungi, especially white-rot, and ectomycorrhizal species (Treu & Falandysz, 2017), metals that are

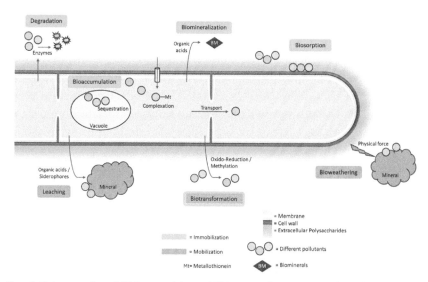

Fig. 1 Scheme of mobilizing and immobilizing mechanisms employed by basidio-mycete fungi that can be harnassed for mycoremediation.

mobilized can be taken up into plant biomass, and subsequently be safely deposited in ashes. Here, mycoremediation is only aiding phytoremediation. However, fungi are also famous for the uptake of metals and radionuclides and deposition within their fruiting bodies, as it has been seen after the Chernobyl accident (Kalač, 2001). Here, sampling of fruiting bodies would be a more direct way to extract the contaminants from soil.

In contrast to using the potential of mobilizing activities, fungi also can immobilize metals and radionuclides in soil, which lowers bioavailability and hence the threat to other life forms and food chains (Fig. 1). The immobilization of metals, e.g., by bioaccumulation or biomineral formation, can directly reduce the bioavailability and toxicity.

Finally, fungi are not only very well adapted to life in soil but also have acquired high resilience against environmental stresses.

The most common genera of filamentous fungi found at metal contaminated soils are the ascomycete genera *Aspergillus*, *Penicillium* and *Fusarium* (Torres-Cruz, Hesse, Kuske, & Porras-Alfaro, 2018). For a *Penicillium* sp. strain isolated from contaminated soil, a minimal inhibitory concentration of 4000 mg/L Cd has been reported (Zafar, Aqil, & Ahmad, 2007), whereas the basidiomycete *S. commune* grows up to a concentration of 65 mg/L Cd (Traxler et al., 2022), and the ectomycorrhizal *Tricholoma terreum* and *Tricholoma vaccinum* grew on media with

236 mg/L Cd (Krause, 2004), while *Amanita muscaria* was reported with 421 mg/L (Hartley, Cairney, & Meharg, 1997).

Especially with the latter, in addition to the direct use for removal of organic and metal pollutants, fungi can directly promote plant growth on contaminated land leading to new potentials for both, land-use and phytoremediation. The plant growth-promoting feature is specifically valid for mycorrhizal fungi, as they can act as biofilters determining plant uptake of toxic metals. White-rot fungi as well as brown- rot fungi offer additional advantages for mycoremediation by being able to degrade a wide range of organic pollutants (Chen, Zhang, Zhang, Zhu, & Zhuo, 2022; Peuke & Rennenberg, 2005; Zhuo & Fan, 2021). Since co-contamination of organic with metal contaminants is a spreading problem, this feature should be considered with special care.

2.1 Mycoremediation through immobilization

An important mechanism of immobilization is biosorption (compare to Fig. 1). Here, contaminants, mostly metals or radionuclides, but potentially also organic substances, are bound to functional groups of the cell wall (Dhankhar & Hooda, 2011; Gadd, 2000). The biosorbent does not even have to be alive for this, as it is a passive process that is not dependent on metabolism. The mechanisms of biosorption are generally based on physicochemical interactions between the metallic ions and functional groups of the cell wall, such as electrostatic interactions, ion exchange, and chelation or complex formation of metal ions. The functional groups most frequently involved in such interactions are carboxylate, hydroxyl, amine, and phosphoryl groups present in the components of the cell wall, such as polysaccharides, lipids, and proteins of fungi (Michalak, Chojnacka, & Witek-Krowiak, 2013). Therefore, the amount of potentially absorbed pollutants is dependent on the ionic state of its functional groups and thus depends on the pH value. Fungi are a powerful tool for biosorption because the cell wall makes up 30% or more of the dry weight of the fungus and can therefore bind large quantities (Dhankhar & Hooda, 2011). Algae, waste products from the food industry, but also ascomycetes such as *Aspergillus niger* are frequently examined as biosorbents (Boczonádi et al., 2021; Michalak et al., 2013). Basidiomycetes are rarely used for such studies, although the sorption capacity of the basidiomycete *Auricularia polytricha* is comparable to that of *A. niger* for e.g., copper (Mukhopadhyay, Noronha, & Suraishkumar, 2007). *S. commune* offers the advantage of forming the extracellular polysaccharide schizophyllan when cultivated in liquid culture. Schizophyllan belongs to the water-soluble polysaccharides and has many binding sites for ionic pollutants at its functional groups (Zhang, Kong, Fang, Nishinari, & Phillips, 2013). Thus, metals including radionuclides as well as organic pollutants can be accumulated. The physicochemical process would also allow applying fungal pre-cultivated biomass to treat wastewater, since the biomass does not have to be alive.

Another feasible way of mycoremediation is immobilization through bioaccumulation. Here, the pollutants are accumulated by the living hyphae, reducing again the bioavailability and hence toxicity for other biota. This also implies that the harmful substances can no longer be spread through wind or water erosion or with the soil water phase into surface and groundwater. The difference to biosorption is that the pollutants are metabolically actively taken up. The disadvantages of this method are that metabolically active accumulation is slower than passive sorption, the binding capacity is lower, or in remarkably high concentrations, toxicity may prevail (Dhankhar & Hooda, 2011). A major advantage is that the accumulation can be applied directly in soil in situ over an extended period of time.

Another immobilizing pathway of mycoremediation is biomineral formation (Fig. 2). This can be achieved by either changing physico-chemical conditions, or via metabolite induced biomineralization (Kirtzel et al., 2020). Fungi can change their microenvironment by the secretion of metabolites like organic acids such as oxalate, but also carbonates, sulfates or phosphates, which results in over-saturation and hence precipitation of biominerals (Gadd, 2010). In addition, the cell wall components can act as nucleation point and thus lead to formation of biominerals around the hyphae.

2.2 Mycoremediation through mobilization

Fungi can leach metals out of stable bindings. In most fungi, leaching is mediated by the production of organic acids like oxalate, which provide a source of protons and metal-complexing organic acid anions (Gadd et al., 2014). *S. commune* can leach metals like Ni, Mn, and Fe out of black slate, a rock rich in organic carbon (Wengel, Kothe, Schmidt, Heide, & Gleixner, 2006). The secretion of oxalic acid and siderophores is increased in the presence of black slate, which indicates participation in this process (Kirtzel, Siegel, Krause, & Kothe, 2017). Siderophores are small oligopeptides that are secreted to bind as chelators to Fe as well as Cd, Cu, Pb, and Zn (e.g., Dimkpa, Merten, Svatoš, Büchel, & Kothe, 2009; Kraemer et al., 1999; Neubauer, Furrer, Kayser, & Schulin, 2000). Metal removal into chelate can promote the dissolution of minerals, with oxalate biominerals being the

Fig. 2 Immobilization through biomineralization. The contaminants are removed from the water-soluble phase in the environment with concomitant protection of cells from toxic effects. Biominerals formed by *Serpula lacrymans* in interaction with co-occurring *Bacillus sp.* bacteria are shown. The blue color is caused by the crystals' own fluorescence and not by staining (excitation and emission like Calcofluor).

most prominent examples (Gadd et al., 2014). Therefore, the metal-binding capabilities of siderophores can also contribute to the biological weathering and metal solubilization of soil (Williamson et al., 2021).

One feature that makes filamentous fungi particularly useful for remediation is their ability to transport contaminants along their hyphae (Colpaert, Wevers, Krznaric, & Adriaensen, 2011). Fungi can spread in soil for very large distances—the largest known organism is a wood decay basidiomycete (Sipos, Anderson, & Nagy, 2018). Long hyphae, which are composed of hundreds of cellular compartments, are found in basidiomycetes with a length of 100–500 μm (Greening & Moore, 1996; Jung, Kothe, & Raudaskoski, 2018)—they can transport nutrients, water, and ions over large distances (Krause et al., 2020). Their important ecosystem functions influence biogeochemical cycles, and they support and control microbial communities in soil (Schlunk et al., 2015; Wagner et al., 2019). The accumulation and transport of pollutants within hyphae (Fig. 3) could be useful for heterogeneously contaminated areas to protect co-occurring organisms, like plants.

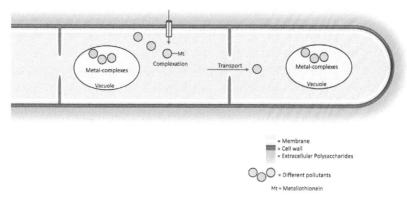

Fig. 3 Uptake and transport of metals along the basidiomycete hyphae. Sequestration in vacuoles leads to long-distance transport along the cytoskeleton.

A famous example of such an area is the Chernobyl Exclusion Zone where in 1986 a severe nuclear power plant accident happened. The released radiation was mostly due to ^{134}Cs and ^{137}Cs, as well as to a lesser extent to ^{90}Sr. It is estimated that 4×10^{16} Bq ^{137}Cs have been released worldwide. In a radius of 30 km, which is the Chernobyl Exclusion Zone, where no taxable stay is allowed, over 1.1×10^{5} Bq m^{-2} ^{90}Sr were found. In addition, highly radioactive particles, so-called hot particles, were released. A highly dose-relevant hot particle is ^{241}Am, which was distributed heterogeneously on the ground immediately after the reactor accident in the Chernobyl Exclusion Zone (Steinhauser, 2018; Walther & Denecke, 2013). *S. commune* can transport Cs and Sr along its hyphae, survive the harsh conditions of the Chernobyl Exclusion Zone, and spread in soil (Traxler et al., 2021, 2022). A study demonstrated that the introduction of *S. commune* in proximity to crops that are grown in radionuclide-contaminated soil results in a higher yield of harvested potatoes (Vuković et al., 2020). As the presence of *S. commune* increases the rate of ^{137}Cs transfer into the green of winter rye and the potato tuber, it is feasible to extract these plant materials and safely remove contaminants. This approach could potentially provide a land management scheme for the reuse of contaminated soils, such as the Chernobyl Exclusion Zone soil, in the future. Consequently, this fungus can contribute to mycoremediation for this area with its mycelial network even without building fruiting bodies. Metal and radionuclide uptake into fruiting bodies has been well documented for different basidiomycetes (Table 1), for example, the white-rot fungus *Pleurotus ostreatus* (Golian et al., 2021; Saba, Falandysz, &

Table 1 Fruiting body metal content mean metal contents [ppm] found at a former uranium mining site near Ronneburg, Germany.

species	Cd	Cs	Ni	Pb	Sr	U	Zn	Cu	Life style
Coprinus comatus	7	0.1	0.4	0.6	1	15	75	158	Saprotrophic
Leucoagaricus leucothites	6	0.3	0.6	8	4	0.1	100	53	Saprotrophic
Lycoperdon perlatum	0.9	0.1	0.8	2	0.6	0.1	122	170	Ectomycorrhizal
Boletus badius	0.2	0.5	5	0.2	0.7	0.1	105	53	Ectomycorrhizal
Paxillus involutus	0.4	7	2	0.2	7	1	180	95	Ectomycorrhizal
Thelephora terrestris	1	11	111	12	24	5	106	84	Ectomycorrhizal
Mean bioavailable soil contents	0.2	0.1	8	0.3	3	2	3	0.5	
Mean total soil contents	0.5	7	46	20	115	7	62	30	

The relevant bioavailable and the total soil contents of the respective metal concentrations at the site are given to show (hyper)accumulation in the fruiting bodies.

Loganathan, 2020). A potential mycoremediation approach is to continuously harvest fruiting bodies to decrease the level of contamination over time.

2.3 Mycoremediation using wood rot fungi ligninolytic enzymes

Another big advantage of white-rot but also brown-rot and mycorrhizal fungi for mycoremediation is their ability to produce ligninolytic enzymes (Chen, Taylor, Burke, & Cairney, 2001). White-rot fungi can degrade lignin, lignocellulose, and cellulose simultaneously through a non-specific, multi-enzyme machinery in which the enzymes can work separately or cooperatively (Harms, Wick, & Schlosser, 2017). They have a broad extracellular enzyme spectrum to degrade the complex polymer lignin, for example, lignin peroxidase, manganese-dependent peroxidases, laccases, and dioxygenases. Not all wood rot fungi are able to produce all these enzymes, but only a proportion of this multi-enzyme machinery is sufficient to degrade organic polymers (Almási et al., 2019).

Because these enzymes have a broad substrate spectrum, they can also break down other persistent compounds like polyaromatic hydrocarbons, aromatic diamines, polyphenols, and polyamines (Baldrian, 2006; Rhodes, 2015; Singh, 2006). For example, a laccase of the white-rot fungus *P. ostreatus* was shown to degrade 95% of the PAH anthracene and 14% of the higher molecular weight fluoranthene after 2 days (Pozdnyakova, Rodakiewicz-Nowak, & Turkovskaya, 2004). Both substances are mutagenic and are used in pharmaceuticals and pesticides. *Trametes versicolor* was reported to oxidize a large number of different PAHs via laccase activity (Majcherczyk, Johannes, & Hüttermann, 1998). Fungi can use these organic pollutants as a carbon source (Blasi et al., 2016). Fungi, that connect ligninolytic enzymes and thus degradation abilities towards PAHs or other xenobiotics with metal tolerance are especially potent candidates for mycoremediation.

However, the large substrate spectrum of laccases and many other ligninolytic enzymes can not only help with the breakdown of complex molecules, but oxidation may also change the composition of minerals, which could lead to an altered bioavailability of metals. For *S. commune*, secreted laccases lead to oxidation of the carboniferous rock black slate, which results in realease and subsequent sequestration of metals (Kirtzel et al., 2019).

3. Tolerance mechanisms against metals and organic pollutants of white-rot fungi

Prerequisite for application in mycoremediation is tolerance against increased levels of pollutants and in cases of co-contamination of both pollutants. Here, fungi often outperform bacterial bioremediation agents. Heavy metals can be essential as micronutrients and harmful at elevated concentrations (e.g., Zn, Cu, Ni, Fe, Mn, Mo), while others are non-essential and merely harmful (e.g., Cd, Pb, Hg, Ag, S, Cs). The toxic effect is caused by the inhibition of functional groups of crucial molecules, such as conformational changes of enzymes triggered by the binding of metals, or the oxidative deterioration of DNA by the binding to polynucleotides (Ochiai, 1987). For example, metals can change the conformation of DNA and proteins and can also attack these molecules by forming radicals. The process of radical formation involves the interaction with reactive oxygen species, such as hydrogen peroxide, based on the type of the Fenton reaction (Lloyd & Phillips, 1999).

Metal toxicity differs with different abiotic factors and the environmental conditions change significantly in different habitats. Metals are better tolerated near the optimal growth conditions for the respective fungus (Priyadarshini, Priyadarshini, Cousins, & Pradhan, 2021). Thus, the temperature and availability of nutrients play a major role. E.g., *S. commune* tolerates higher concentrations of Cs, Zn, and Cd at its optimal growth temperature of 28 °C than at lower temperatures (Traxler et al., 2022). The pH has an even more direct influence on the mobility of metals (Recer & Boylen, 1986). An acidic pH increases the chemical bioavailability of metals, but it tends to lower the toxicity for fungi (Babich & Stotzky, 1977; Gadd & Griffiths, 1980). The better tolerance at low pH values could be due to enhanced biosorption rate as decreasing pH provides a large number of free binding sites at functional groups and hence increases metal uptake (Dwivedi, 2012). At really low pH values, metals like Pb will precipitate (Babich & Stotzky, 1979). Furthermore, the environmental pH influences the synthesis of enzymes, which are important for tolerance mechanism of white-rot fungi as seen, e.g., for peroxidases (Elisashvili, 1993). The growth substrate also makes a difference to metal sensitivity. For filamentous fungi, inhibition occurred in liquid media at lower Cd concentrations than on solid media (Babich & Stotzky, 1977).

Radionuclides share the heavy metal toxicity with stable isotopes. However, radiation causes additional DNA damage, combined with their high uptake capacity poses an additional risk regarding the toxic effects of their stable isotopes (Borio et al., 1991 and citations therein). Although mycorrhizal fungi seem to accumulate the highest amount of radiocesium, saprotrophic basidiomycetes such as *S. commune* take up significant amounts (Smith, Taylor, & Sharma, 1993; Wasser, Grodzinska, & Lyugin, 1991). Nevertheless, the heavy metal toxicity of radiocesium exceeds its radio-toxicity (Schindler, Gube, & Kothe, 2012).

3.1 Tolerance mechanisms against metal pollution

Fungi, including white rots, can grow in metal-polluted soils because of a wide variety of tolerance mechanisms that lower metal stress. As already mentioned, these mechanisms can be divided into mobilization or immobilization of potentially toxic elements (see Fig. 1; Gadd, 1994, 2010). Immobilizing tolerance is mostly based on extracellular reduction of bioavailability including: (1) change in redox potential by a change of the oxidation state resulting in a less toxic, immobile species, e.g., in iron (hydro)oxides. This can result in precipitation and the formation of bio-minerals (Gadd et al., 2014; Gadd, 2007); (2) biosorption to cell wall compartments, pigments, polysaccharides, and extracellular polymeric substances (Volesky, 1990); (3) the formation of secondary biominerals via the excretion of metabolites like oxalate, or biomineralization of carbonates and phosphates (Gadd et al., 2014).

Mechanisms to reduce the mobility of contaminations by uptake into the hyphae include: (1) Intracellular compartmentalization in vacuoles that may then be transported along the intercellular microtubular network (Brunsch et al., 2015). (2) Intracellular complexation, e.g., by metal-lothioneins or phytochelatins, or precipitation, can be followed again by compartmentalization (Schlunk et al., 2015; Xu et al., 2014). (3) Change in the efflux or influx of metals can reduce the intracellular metal stress (Zafar et al. 2007). In addition, fungi can emit metals by producing guttation droplets (Fig. 4). Guttation droplets containing organic as well as inorganic constituents were screened from ectomycorrhizal and saprotrophic basi-diomycete fungi. Secondary metabolites, nutrients (sugar and amino acids), and metals (lead) were identified to be exuded with the guttation. Detoxification of lead could be seen with up to 18 mg/mL with guttation in axenically grown ectomycorrhizal fungi (Formann, 2016). (4) Through enzymatic detoxification, metals in the cell that cause oxidative can be counteracted with enzymes like superoxide dismutase or catalase (Ott, Fritz, Polle, & Schützendübel, 2002).

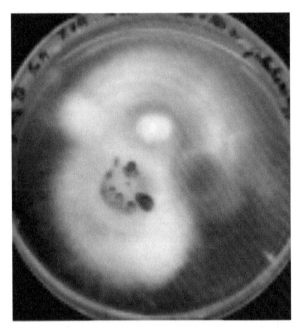

Fig. 4 Contaminant metal exudation with guttation droplets on aerial mycelium of *Pisolithus tinctorius* in co-culture with *Paxillus* sp. and *Tricholoma vaccinum*.

On the other hand, some tolerance mechanisms in fungi can result in the mobilization of metals, for example: (1) The change in redox potential dissolving formerly stable metal complexes through, e.g., Mn (IV) to Mn (II) reduction; (2) the complexation of metals, e.g., the chelation of iron by siderophores; (3) the chemoorganotrophic leaching by other excreted metabolites with metal-complexing properties, such as organic acids, amino acids, and phenolic compounds; (4) the methylation of metals, which can result in volatilization; and finally, (5) bioweathering and bio-corrosion through physical (mechanical), chemical, and biological processes are possible ways to mobilize metals (Bentley & Chasteen, 2002; Gadd et al., 2012; Gadd, 2007).

Fungi usually employ not just one of these mechanisms but a mixture of them (compare Fig. 1), or they use different mechanisms for different metals. *S. commune* tolerates Cd with a specific set of traits that differs from Sr, Cs, and Zn tolerance (Kirtzel et al., 2019; Traxler et al., 2021, 2022). Transcriptome analysis showed differential regulation of transporters under metal stress, with decreased influx and enhanced efflux being part in metal tolerance (Traxler et al., 2022). Furthermore, the accumulation of large amounts of U on the cell wall of *S. commune* has been found, thus biosorption is also employed (Wollenberg et al., 2021).

Other white-rot fungi also use a variety of ways to tolerate metals. For example, *Phanerochaete chrysosporium* can bind large amounts of Pb on its cell wall due to biosorption but is also able to precipitate Pb to $Pb_5(PO_4)_3OH$ or accumulate it intracellularly (Huang et al., 2017; Noormohamadi, Fat'hi, Ghaedi, & Ghezelbash, 2019). In the presence of Cd, *P. chrysosporium* excretes oxalate which leads to precipitation of this metal, but it can tolerate Cd also by biosorption (Zhao, Zhang, Zeng, Cheng, & Liu, 2016).

3.2 Tolerance mechanisms against organic pollutants

Persistent organic compounds are often volatile and thus move long distances in the atmosphere. As a result, this pollution is found all over the world, even in virtually untouched places like the Arctic (El-Shahawi, Hamza, Bashammakh, & Al-Saggaf, 2010). In aquatic habitats, this kind of pollutant is also widely distributed, because they bind easily to small particles (Van den Berg et al., 1998). These compounds have a very low solubility in water but a high one in fat and are consequently accumulated in tissue. They are primarily toxic in that they cause cancer in eukaryotes and bacteria (Baderna et al., 2011; Kim, Ahn, Woo, Jung, & Park, 2009), but they also have effects on membrane functions and enzyme systems (Mazzeo, Fernandes, & Marin-Morales, 2011).

The concurrent presence of several of these toxicants poses challenging environmental conditions, as intricate chemical interactions and synergies may arise, resulting in severe and toxic effects on organisms. Soils and wastewater containing organic and inorganic toxic substances, including metals, pesticides and herbicides, mineral oils, and petroleum derivatives are not rare (Alisi et al., 2009; Coulibaly, Gourene, & Agathos, 2003). Natural populations that thrive in toxic metal or poly-contaminated ecosystems are frequently subjected to selective pressures, resulting in heightened resistance. Mostly it has been reported that the occurrence of metals negatively effects the degradation ability of filamentous fungi (Bhattacharya, Das, Prashanthi, Palaniswamy, & Angayarkanni, 2014; Markowicz, Cycon, & Piotrowska, 2016).

The complete degradation of complex organic matter, including lignin, depends on the activity of basidiomycete fungi, specifically those possessing laccases (Fig. 5; Madhavan, Krause, Jung, & Kothe, 2014 and citations therein). The enzyme of the multicopper oxidases family of proteins catalyzes the oxidation of a variety of substrates in addition to lignin in radical

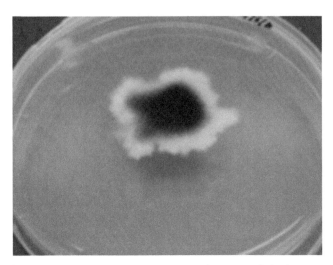

Fig. 5 Production of ligninolytic enzymes, shown by extracellular laccase activity on *S. commune* mycelium using ABTS (2,2′-azinobis (3-ethylbenz-thiazoline-6-sulfonic acid)) staining.

reactions, including the possible role of mediators. This opened interest in using white-rot, laccase-forming fungi for complex settings in bioremediation, including (co-)contamination with PAHs and similar contaminants.

White-rot fungi have significant potential in the biodegradation of organic pollutants because of the extensive array of ligninolytic enzymes that participate in crucial steps of aromatic hydrocarbon biodegradation and other synthetic organic pollutions (Baldrian & Gabriel, 2002; Gadd, 2010). The ligninolytic enzymes that are mainly involved in these unspecific degradations are lignin peroxidase, Mn peroxidase, dye-decolorizing peroxidase, versatile peroxidase, and laccases (Zhuo & Fan, 2021). With this array and with the help of the cytochrome P450 system *Trametes versicolor* can degrade several pharmaceutical active compounds (Rodríguez-Rodríguez, Marco-Urrea, & Caminal, 2010; Tran, Urase, & Kusakabe, 2010). Furthermore, the enzymatic breakdown of other substances like phenol, resorcinol, p-cresol, and o-nitrophenol was reported for *T. versicolor* (Yemendzhiev, Terziyska, Manasiev, & Alexieva, 2011). For *P. ostreatus,* the degradation of a wide range of PAHs was reported (Covino et al., 2010; Pozdnyakova et al., 2010). However, synthetic dyes such as "golden yellow R" can also be broken down by this fungus (Jilani, Asgher, Bhatti, & Mushtaq, 2011). These are just two examples of the many white-rot fungi that use the enzymes available to them to break down complex toxic compounds.

3.3 Harnessing fungal resilience

Among microorganisms, fungi exhibit a high degree of adaptability and versatility responding quickly to stress. Some fungi are more resistant to metal stress or the stress of other pollutants than others. It is assumed that the environmental presence of heavy metals can lead to the acquisition of tolerance mechanisms or metal resistance in fungi by selective pressure. Some fungi are naturally less susceptible to metal stress, with survival depending largely on the biochemical and structural capabilities, while others achieve their tolerance through genetic and physiological adaptations (Gadd & Griffiths, 1978). Possible properties that increase metal tolerance include impermeable and/or pigmented cell walls, increased binding capacity of metals to the cell wall or extracellular polysaccharides, and the excretion of metabolites, which bind or precipitate metals, e.g., oxalates (Gadd et al., 2014; Gadd, 1992). The tolerance against organic pollutants is partly based on the activity of extracellular enzymes. The induction of these enzymes can depend on available nutrition, the mixture of pollutants, and other factors (Harms et al., 2017). An increased production of such enzymes, e.g., ligninolytic enzymes or superoxide dismutase, will help increased tolerance.

Adaptation may be achieved through mutation (the genetic mechanism of adaptation) or a modified expression pattern, resulting in altered gene expression profiles (e.g., Chen, Ng, Cheow, & Ting, 2017). This shift in metabolic flow and enzymatic activities would be visible in transcriptome analyses. For *S. commune,* gene expression analysis of a metal-adapted linage showed higher expression of a major facilitator superfamily transporter indicating that efflux plays a role in enhanced metal tolerance (Traxler et al., 2022). This adapted linage was able to grow 4–6 times better than its non-adapted precursor strain on Cd, Cs, and Sr. *Panus tigrinus* can adapt to polychlorinated phenols and then survive the 4-fold amount of 2,4,6-trichlorophenol as well as degrade higher amounts than the unadapted strain (Leontievsky, Myasoedova, Golovleva, Sedarati, & Evans, 2002). The adaptation led to a higher activity of a Mn peroxidase, which led increased degradation of the chlorophenol, and an increased formation of extracellular polysaccharides, increasing the tolerance through biosorption.

The tolerance mechanisms employed by fungal species can be either specific or non-specific. Some fungi are not specific in their ability to biodegrade xenobiotics or cope with toxic metals, and this is due to a coincidence of mechanisms that evolved in some taxa that populate niches.

Natural populations that thrive in toxic metal or poly-contaminated eco-systems are frequently subjected to selective pressures, resulting in heightened resistance (Colpaert et al., 2011). It is not only important to adapt to the pollution, but also to the natural conditions of the region to be remediated. Here, natural populations have a major advantage over artificially added organisms. Many wood rot fungi can tolerate high concentrations of PAHs, such as phenanthrene, anthracene, fluoranthene, and pyrene, under laboratory conditions, but cannot survive in contaminated soils without an adaptation period (Lee et al., 2014). This still does not prevent use of basidiomycetes for mycoremediation, as they are exceptionally adaptive, including their broad saprotrophic nutrition range (Choi et al., 2009).

Furthermore, the fungal mycelial network is a perfect help for the adaptation to heterogeneous contaminated areas like soil. Since the mycelium is large and distributed, it is found in locations with varying degrees of contamination and has time to adapt. Even if the contamination is too high for the unadapted organism to survive in some places, it can survive and adapt in others. Furthermore, the hyphae can serve bacteria as a kind of highway, promoting bacterial access to the soil habitat and increasing the bioavailability of xenobiotics and potentially toxic elements for microbial communities (Harms et al., 2017).

4. White-rot fungi in soil

White-rot fungi occur naturally in rotting wood and live saprotrophically, which means that soil is not their natural habitat. However, some studies have already investigated the survival of white-rot fungi including *P. chrysosporium*, *T. versicolor*, *P. ostreatus* in soil (Canet, Birnstingl, Malcolm, Lopez-Real, & Beck, 2001; Morgan et al., 1993). These fungi require an added carbon source to survive in soil. The best carbon source turned out to be hemicellulose-containing substrates such as straw (Hultgren, Pizzul, Castillo, & Granhall, 2009).

In addition, the survival of *S. commune* in soil was tested without any additional carbon source and in natural and contaminated soil at a test field in the Chernobyl Exclusion Zone (Fig. 6). The survival was analyzed by DNA extraction from the soil and quantitative PCR with species-specific primers. Even 12 months after inoculation, significantly increased DNA values were still detected in the soil; thus, it was assumed that the fungus

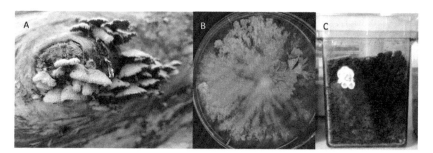

Fig. 6 Growth of the white-rot fungus *S. commune* with fruiting body formation. (A) In native stands on wood; (B) in laboratory conditions on complex yeast medium; and (C) growth in microcosms on sterilized soil.

had survived (Traxler et al., 2021). If the fungus had died, the dead biomass would certainly have been digested by other soil organisms. The survival of this unadapted *S. commune* strain on Chernobyl soil without additional nutrients was also demonstrated under laboratory conditions (Traxler et al., 2021). Furthermore, a spread in soil could be detected. as *S. commune* covered a distance of approximately one meter within 6 months. Thus, a growth of approx. 8 mm per day was calculated, corresponding well with the growth of the fungus on artificial media, which is 5–10 mm per day (Brunsch et al., 2015). According to Camel et al. (1991), the mycorrhizal fungus *Glomus mosseae* has the ability to traverse a distance of 2.3 cm per week when grown in sandy soil like that in the Chernobyl Exclusion Zone. This would result in a total of 60 cm in 6 months; therefore, this strictly biotrophic fungus is slower in soil than *S. commune*. This may be related to the reported observation that fungi exhibit faster growth rates and longer, unbranched hyphae, commonly referred to as runner hyphae when culti-vated in soil under stressful conditions, such as nutrient depletion (Agerer, 1999; Allen, 2007; Muthukumar, Priyadharsini, Uma, Jaison, & Pandey, 2014). The growth rates of runner hyphae are higher than those observed in well-fed hyphae, as well as in other fungi that thrive in soil (Camel et al., 1991; Smith, Bruhn, & Anderson, 1992). It can be inferred that the detection of *S. commune* DNA in the soil is based on the natural occurrence or spread by spores, as a monokaryotic strain was inoculated. The soil of the Chernobyl Exclusion Zone has several adverse properties. On one hand the fungus had to cope with increased radiation values of 2–6 μGy/h at ground level (Traxler et al., 2021). This radiation is heterogeneously dis-tributed due to hot particles. They were distributed immediately after the reactor accident in the Chernobyl Exclusion Zone (Steinhauser, 2018; Walther & Denecke, 2013). Consequently, *S. commune* must be able to survive such doses even without a long period of adaptation. Due to the inhomogeneous distribution of the radiation, the fungus might have survived by avoiding hot spots. This would offer a longer adjustment time for the resulting hyphae and thus the possibility of adaptation. Some studies have already been conducted on the adaptation to radiation over a longer period in the Chernobyl Exclusion Zone, but none with basidiomycetes (Galván et al., 2014; Klubicova et al., 2012; Kovalchuk, Abramov, Pogribny, & Kovalchuk, 2004). It was reported that several fungi can survive doses up to 1000 Gy/h (Aziz, El-Fouly, Abu-Shady, & Moussa, 1997). However, these tests took place under laboratory conditions. *S. commune* had to survive several adverse conditions at the same time in the dry, sandy Chernobyl Exclusion Zone soil.

Furthermore, *Phanerochaete sordida* and *P. chrysosporium* survived in contaminated soil in a field trial for at least 50 days. Here, the climatic conditions were also suboptimal for the fungus, and it had to cope with the contamination, in this case with the wood preservative pentachlorophenol. However, peat was added to the soil as an organic carbon source for the inoculated fungi and they did not have to assert themselves against the existing microbial community, because the soil was fumigated (Lamar & Dietrich, 1990).

This artificial setup of adding white-rot fungi to soil is of interest as these fungi displayed under laboratory conditions great potential for bioremediation of contaminated soils. The ligninolytic lifestyle includes laccase formation, which may induce changes even in mineral composition (Kirtzel et al., 2017), and hence bioavailability of radionuclides and other heavy metals bound in the minerals (Günther et al., 2014). The diminished bioavailability, in turn, may account for survival even in the presence of high levels of contamination. Furthermore, it was reported that laccase, which is excreted by white-rot fungi, can immobilize phenolic compounds by coupling to soil humic substances naturally occurring in soil (Bollag & Myers, 1992). This immobilization lowers the biological availability of the xenobiotic pollutants and thus their toxicity. On the other hand, these humic acids can inhibit the enzymatic activity of laccases (Claus & Filip, 1988). Soil is a very complex substrate in which there are many different interactions with the environment and many different ecological niches, which are populated by a diverse community.

4.1 Effects on microbial communities

Since fungi can still survive in particularly highly contaminated environments, they can have an impact on the remediation of pollution in many habitats with this variety of possibilities (Árvay et al., 2014; Guillén & Baeza, 2014). Both the concentration and type of contamination as well as the fungi themselves influence the microbial community in soil. It is assumed that with increasing pollution, the richness and diversity of microbial communities decreases (Agnello, Potysz, Fourdrin, Huguenot, & Chauhan, 2018; Ying et al., 2008), whereas fungal communities tend to be more stable than bacterial ones (Singh, Vaidya, Goodey, & Krumins, 2019). Mycoremediation also has an ecological impact on larger organisms such as plants. Without the utilization of mycoremediation, plants are susceptible to the accumulation of toxins and their subsequent transfer into the food chain. However, an association of plants and fungi can detoxify toxic metals, translocate them, and accumulate them in the above-ground biomass, which must be then harvested for metal recovery (Singh et al., 2018). Furthermore, fungi help to reestablish vegetation, e.g., in post-mining areas. One reason for this assistance is the improvement in water availability for plants through the hyphal network. As a result, the erosion of contaminated soils by wind and water is also reduced. Mycorrhizal fungi have been described as having a significant influence on the microbial community around the roots in contaminated soils (Bogdanova, 2022). The microbial community in contaminated soil in the rhizosphere is not significantly influenced by the tree species but by the soil parameters; therefore, the fungal community is much more stable than the bacterial (Bogdanova, 2022).

For the white-rot fungi *T. versicolor* and *P. ostreatus,* it was reported that the laccase activity increased after contact with typical soil-living microorganisms or unsterile soil (Baldrian, 2004). For *Crucibulum laeve, Irpex lacteus,* and *P. ostreatus,* in vitro experiments found that these white-rot fungi significantly influence the microbial community of contaminated soils and even contribute to its remediation by the degrading toxic substances such as polycyclic aromatic hydrocarbons (Ma et al., 2021; Stella et al., 2017).

5. Conclusion and outlook

White-rot fungi have great potential for mycoremediation (Fig. 7). Their non-specific ligninolytic enzymes and great metabolic flexibility

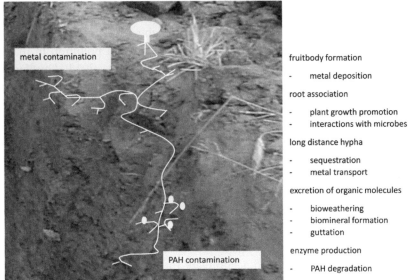

Fig. 7 Schematic drawing summarizing multiple functions of basidiomycetes in soil for mycoremediation.

make them more suitable than many other fungi. Thus, they can also be used in environments that combine different stress factors and types of contamination. However, more research is needed to apply them for mycoremediation in vivo in contaminated soils, because soil is not the natural habitat of white-rot fungi. So far, only the growth of *S. commune* without additional nutrients in natural soil has been reported, but with added nutrients, many other white-rot fungi have been shown to survive in contaminated soils and participate in their remediation. A major advantage of mycoremediation is that the hyphal network supports both plants and bacteria in stressful and especially in heterogeneously contaminated sites to build a community and actively participate in the remediation and thus stabilize the site.

Acknowledgments

We would like to thank former group members Caroline Schweichler, Alexander Mauz, Steffi Formann, Elke-Matina Jung and Soumya Madhavan for photographic illustrations and Matthias Gube for fruiting body metal contents. This work was supported by the BMBF (RENA) and DFG (EXC Balance of the microverse). We thank Alene Alder-Rangel from Alder's English Services for the English review of the manuscript.

References

Abdulsalam, O., Wagner, K., Wirth, S., Kunert, M., David, A., Kallenbach, M., ... Krause, K. (2020). Phytohormones and volatile organic compounds, like geosmin, in the ectomycorrhiza of *Tricholoma vaccinum* and Norway spruce (*Picea abies*). *Mycorrhiza.* https://doi.org/10.1007/s00572-020-01005-2.

Agerer, R. (1999). Never change the evolution of *Boletales sl* (Hymenomycetes, Basidiomycota) as seen from below-ground features. *Mitteilungen der Botanischen Staatssammlung und des Instituts für Systematische Botanik der Universitat München, 6, 5–91.*

Agnello, A. C., Potysz, A., Fourdrin, C., Huguenot, D., & Chauhan, P. S. (2018). Impact of pyrometallurgical slags on sunflower growth, metal accumulation and rhizosphere microbial communities. *Chemosphere, 208, 626–639.*

Alisi, C., Musella, R., Tasso, F., Ubaldi, C., Manzo, S., Cremisini, C., & Sprocati, A. R. (2009). Bioremediation of diesel oil in a co-contaminated soil by bioaugmentation with a microbial formula tailored with native strains selected for heavy metals resistance. *Science of the Total Environment, 407*(8), 3024–3032.

Allen, M. F. (2007). Mycorrhizal fungi: Highways for water and nutrients in arid soils. *Vadose Zone Journal, 6*(2), 291–297.

Almási, É., Sahu, N., Krizsán, K., Bálint, B., Kovács, G. M., Kiss, B., ... Nagy, I. (2019). Comparative genomics reveals unique wood-decay strategies and fruiting body development in the Schizophyllaceae. *New Phytologist, 224*(2), 902–915. https://doi.org/10.1111/nph.16032.

Anthony, M. A., Bender, S. F., & Van Der Heijden, M. G. (2023). Enumerating soil biodiversity. *Proceedings of the National Academy of Sciences, 120*(33), e2304663120.

Árvay, J., Tomáš, J., Hauptvogl, M., Kopernická, M., Kováčik, A., Bajčan, D., & Massányi, P. (2014). Contamination of wild-grown edible mushrooms by heavy metals in a former mercury-mining area. *Journal of Environmental Science and Health, Part B, 49*(11), 815–827.

Aziz, N. H., El-Fouly, M., Abu-Shady, M., & Moussa, L. (1997). Effect of gamma radiation on the survival of fungal and actinomycetal florae contaminating medicinal plants. *Applied Radiation and Isotopes, 48*(1), 71–76.

Babich, H., & Stotzky, G. (1977). Sensitivity of various bacteria, including actinomycetes, and fungi to cadmium and the influence of pH on sensitivity. *Applied and Environmental Microbiology, 33*(3), 681–695.

Babich, H., & Stotzky, G. (1979). Abiotic factors affecting the toxicity of lead to fungi. *Applied and Environmental Microbiology, 38*(3), 506–513.

Baderna, D., Maggioni, S., Boriani, E., Gemma, S., Molteni, M., Lombardo, A., ... Lodi, M. (2011). A combined approach to investigate the toxicity of an industrial landfill's leachate: Chemical analyses, risk assessment and in vitro assays. *Environmental Research, 111*(4), 603–613.

Baldrian, P. (2004). Increase of laccase activity during interspecific interactions of white-rot fungi. *FEMS Microbiology Ecology, 50*(3), 245–253.

Baldrian, P. (2006). Fungal laccases–occurrence and properties. *FEMS Microbiology Reviews, 30*(2), 215–242.

Baldrian, P., & Gabriel, J. (2002). Copper and cadmium increase laccase activity in *Pleurotus ostreatus. FEMS Microbiology Letters, 206*(1), 69–74.

Bentley, R., & Chasteen, T. G. (2002). Microbial methylation of metalloids: Arsenic, antimony, and bismuth. *Microbiology and Molecular Biology Reviews, 66*(2), 250–271.

Bhattacharya, S., Das, A., Prashanthi, K., Palaniswamy, M., & Angayarkanni, J. (2014). Mycoremediation of Benzo [a] pyrene by *Pleurotus ostreatus* in the presence of heavy metals and mediators. *3 Biotech, 4,* 205–211.

Bizo, M. L., Nietzsche, S., Mansfeld, U., Langenhorst, F., Goettlicher, J., Majzlan, J., ... Kothe, E. (2017). Response to lead pollution: Mycorrhizal Pinus sylvestris forms the biomineral pyromorphite in roots and needles. *Environmental Science and Pollution Research, 24*(16), 14455–14462. https://doi.org/10.1007/s11356-017-9020-7.

Blasi, B., Poyntner, C., Rudavsky, T., Prenafeta-Boldú, F. X., De Hoog, S., Tafer, H., & Sterflinger, K. (2016). Pathogenic yet environmentally friendly? Black fungal candidates for bioremediation of pollutants. *Geomicrobiological Journal, 33*, 308–317. https://doi.org/10.1080/01490451.2015.1052118.

Boczonádi, I., Jakab, Á., Baranyai, E., Tóth, C. N., Daróczi, L., Csernoch, L., ... Grawunder, A. (2021). Rare earth element sequestration by *Aspergillus oryzae* biomass. *Environmental Technology, 42*(24), 3725–3735.

Bogdanova, O. (2022). *Microbial processes in mycorrhizoshere of plants growing at a former uranium mining site (Dissertation)*. Germany: Friedrich Schiller University Jena.

Bogdanova, O., Kothe, E., & Krause, K. (2023). Ectomycorrhizal community shifts at a former uranium mining site. *Journal of Fungi, 9*(4), 483.

Bollag, J.-M., & Myers, C. (1992). Detoxification of aquatic and terrestrial sites through binding of pollutants to humic substances. *Science of the Total Environment, 117*, 357–366.

Borio, R., Chiocchini, S., Cicioni, R., Degli Esposti, P., Rongoni, A., Sabatini, P., ... Salvador, P. (1991). Uptake of radiocesium by mushrooms. *Science of the Total Environment, 106*(3), 183–190.

Brunsch, M., Schubert, D., Gube, M., Ring, C., Hanisch, L., Linde, J., ... Kothe, E. (2015). Dynein heavy chain, encoded by two genes in agaricomycetes, is required for nuclear migration in *Schizophyllum commune*. *PLoS One, 10*(8), e0135616.

Camel, S., Franson, R., Brown, M., Bethlenfalvay, G., Reyes-Solis, M., & Ferrera-Cerrato, R. (1991). Growth of vesicular-arbuscular mycorrhizal mycelium through bulk soil. *Soil Science Society of America Journal, 55*(2), 389–393.

Canet, R., Birnstingl, J., Malcolm, D., Lopez-Real, J., & Beck, A. (2001). Biodegradation of polycyclic aromatic hydrocarbons (PAHs) by native microflora and combinations of white-rot fungi in a coal-tar contaminated soil. *Bioresource Technology, 76*(2), 113–117.

Chen, S. H., Ng, S. L., Cheow, Y. L., & Ting, A. S. Y. (2017). A novel study based on adaptive metal tolerance behavior in fungi and SEM-EDX analysis. *Journal of Hazardous Materials, 334*, 132–141.

Chen, D. M., Taylor, A. F. S., Burke, R. M., & Cairney, J. W. G. (2001). Identification of genes for lignin peroxidases and manganese peroxidases in ectomycorrhizal fungi. *New Phytologist, 152*(1), 151–158. https://doi.org/10.1046/j.0028-646x.2001.00232.x.

Chen, L., Zhang, X., Zhang, M., Zhu, Y., & Zhuo, R. (2022). Removal of heavy-metal pollutants by white rot fungi: Mechanisms, achievements, and perspectives. *Journal of Cleaner Production, 354*, 131681.

Choi, Y.-S., Kim, G.-H., Lim, Y. W., Kim, S. H., Imamura, Y., Yoshimura, T., & Kim, J.-J. (2009). Characterization of a strong CCA-treated wood degrader, unknown Crustoderma species. *Antonie Van Leeuwenhoek, 95*, 285–293.

Claus, H., & Filip, Z. (1988). Behaviour of phenoloxidases in the presence of clays and other soil-related adsorbents. *Applied Microbiology and Biotechnology, 28*, 506–511.

Colpaert, J. V., Wevers, J. H., Krznaric, E., & Adriaensen, K. (2011). How metal-tolerant ecotypes of ectomycorrhizal fungi protect plants from heavy metal pollution. *Annals of Forest Science, 68*(1), 17–24.

Coulibaly, L., Gourene, G., & Agathos, N. S. (2003). Utilization of fungi for biotreatment of raw wastewaters. *African Journal of Biotechnology, 2*(12), 620–630.

Covino, S., Svobodová, K., Čvančarová, M., D'Annibale, A., Petruccioli, M., Federici, F., ... Cajthaml, T. (2010). Inoculum carrier and contaminant bioavailability affect fungal degradation performances of PAH-contaminated solid matrices from a wood preservation plant. *Chemosphere, 79*(8), 855–864.

Dhankhar, R., & Hooda, A. (2011). Fungal biosorption–an alternative to meet the challenges of heavy metal pollution in aqueous solutions. *Environmental Technology, 32*(5), 467–491.

Dimkpa, C., Merten, D., Svatoš, A., Büchel, G., & Kothe, E. (2009). Siderophores mediate reduced and increased uptake of cadmium by *Streptomyces tendae* F4 and sunflower (*Helianthus annuus*), respectively. *Journal of Applied Microbiology, 107*(5), 1687–1696.

Dwivedi, S. (2012). Bioremediation of heavy metal by algae: Current and future perspective. *Journal of Advanced Laboratory Research in Biology, 3*(3), 195–199.

El-Shahawi, M., Hamza, A., Bashammakh, A. S., & Al-Saggaf, W. (2010). An overview on the accumulation, distribution, transformations, toxicity and analytical methods for the monitoring of persistent organic pollutants. *Talanta, 80*(5), 1587–1597.

Elisashvili, V. (1993). Physiological regulation of ligninolytic activity in higher basidium fungi. *Microbiology (Reading, England), 62*(5), 480–487.

Erdmann, S., Freihorst, D., Raudaskoski, M., Schmidt-Heck, W., Jung, E. M., Senftleben, D., & Kothe, E. (2012). Transcriptome and functional analysis of mating in the basidiomycete *Schizophyllum commune*. *Eukaryotic Cell, 11*(5), 571–589. https://doi.org/10.1128/EC.05214-11.

Fierer, N. (2017). Embracing the unknown: Disentangling the complexities of the soil microbiome. *Nature Reviews. Microbiology, 15*(10), 579–590.

Formann, S. (2016). *Sequestration of heavy metals and radionuclides in ectomycorrhiza* (Dissertation). Germany: Friedrich Schiller University Jena.

Freihorst, D., Brunsch, M., Wirth, S., Krause, K., Kniemeyer, O., Linde, J., ... Kothe, E. (2018). Smelling the difference: Transcriptome, proteome and volatilome changes after mating. *Fungal Genetics and Biology, 112*, 2–11.

Gabriel, J., Mokrejš, M., Bílý, J., & Rychlovský, P. (1994). Accumulation of heavy metals by some wood-rotting fungi. *Folia Microbiologica, 39*(2), 115–118.

Gadd, G. M. (1992). Metals and microorganisms: A problem of definition. *FEMS Microbiology Letters, 100*(1-3), 197–203.

Gadd, G. M. (1994). *Interactions of fungi with toxic metals. The genus Aspergillus*. Springer, 361–374.

Gadd, G. M. (2000). Bioremedial potential of microbial mechanisms of metal mobilization and immobilization. *Current Opinion in Biotechnology, 11*(3), 271–279.

Gadd, G. M. (2004). Microbial influence on metal mobility and application for bioremediation. *Geoderma, 122*(2–4), 109–119. https://doi.org/10.1016/j.geoderma.2004.01.002.

Gadd, G. M. (2007). Geomycology: Biogeochemical transformations of rocks, minerals, metals and radionuclides by fungi, bioweathering and bioremediation. *Mycological Research, 111*(1), 3–49.

Gadd, G. M. (2010). Metals, minerals and microbes: Geomicrobiology and bioremediation. *Microbiology (Reading, England), 156*(Pt 3), 609–643. https://doi.org/10.1099/mic.0.037143-0.

Gadd, G. M., Bahri-Esfahani, J., Li, Q., Rhee, Y. J., Wei, Z., Fomina, M., & Liang, X. (2014). Oxalate production by fungi: Significance in geomycology, biodeterioration and bioremediation. *Fungal Biology Reviews, 28*(2-3), 36–55.

Gadd, G. M., & Griffiths, A. J. (1978). Microorganisms and heavy metal toxicity. *Microbial Ecology, 4*(4), 303–317.

Gadd, G. M., & Griffiths, A. J. (1980). Influence of pH on toxicity and uptake of copper in *Aureobasidium pullulans*. *Transactions of the British Mycological Society, 75*(1), 91–96.

Gadd, G. M., & Raven, J. A. (2010). Geomicrobiology of eukaryotic microorganisms. *Geomicrobiology Journal, 27*(6–7), 491–519.

Gadd, G. M., Rhee, Y. J., Stephenson, K., & Wei, Z. (2012). Geomycology: Metals, actinides and biominerals. *Environmental Microbiology Reports, 4*(3), 270–296.

Galván, I., Bonisoli-Alquati, A., Jenkinson, S., Ghanem, G., Wakamatsu, K., Mousseau, T. A., & Møller, A. P. (2014). Chronic exposure to low-dose radiation at Chernobyl favours adaptation to oxidative stress in birds. *Functional Ecology, 28*(6), 1387–1403.

Golian, M., Hegedűsová, A., Mezeyová, I., Chlebová, Z., Hegedűs, O., Urminská, D., ... Chlebo, P. (2021). Accumulation of selected metal elements in fruiting bodies of oyster mushroom. *Foods, 11*(1), 76.

Gray, S. (1998). Fungi as potential bioremediation agents in soil contaminated with heavy or radioactive metals. *Biochemical Society Transactions, 26*(4), 666–670. https://doi.org/10.1042/bst0260666.

Greening, J., & Moore, D. (1996). Morphometric analysis of cell size patterning involved in gravitropic curvature of the stipe of *Coprinus cinereus*. *Advances in Space Research, 17*(6-7), 83–86.

Guillén, J., & Baeza, A. (2014). Radioactivity in mushrooms: A health hazard? *Food Chemistry, 154*, 14–25.

Gundel, L. A., Lee, V. C., Mahanama, K. R., Stevens, R. K., & Daisey, J. M. (1995). Direct determination of the phase distributions of semi-volatile polycyclic aromatic hydrocarbons using annular denuders. *Atmospheric Environment, 29*(14), 1719–1733.

Günther, A., Raff, J., Merroun, M. L., Rossberg, A., Kothe, E., & Bernhard, G. (2014). Interaction of U(VI) with *Schizophyllum commune* studied by microscopic and spectroscopic methods. *Biometals: An International Journal on the Role of Metal Ions in Biology, Biochemistry, and Medicine, 27*(4), 775–785. https://doi.org/10.1007/s10534-014-9772-1.

Harms, H., Wick, L., & Schlosser, D. (2017). The fungal community in organically polluted systems. In J. Dighton, & J. F. White (Eds.). *Mycology* (pp. 459–470). Boca Raton, FL: CRC Press-Taylor & Francis Group.

Hartley, J., Cairney, J. W. G., & Meharg, A. A. (1997). Do ectomycorrhizal fungi exhibit adaptive tolerance to potentially toxic metals in the environment? *Plant and Soil, 189*, 303–319.

He, X., Pang, Y., Song, X., Chen, B., Feng, Z., & Ma, Y. (2014). Distribution, sources and ecological risk assessment of PAHs in surface sediments from Guan River Estuary, China. *Marine Pollution Bulletin, 80*(1–2), 52–58.

Huang, C., Lai, C., Xu, P., Zeng, G., Huang, D., Zhang, J., ... Wang, R. (2017). Lead-induced oxidative stress and antioxidant response provide insight into the tolerance of *Phanerochaete chrysosporium* to lead exposure. *Chemosphere, 187*, 70–77.

Hultgren, J., Pizzul, L., Castillo, M. D. P., & Granhall, U. (2009). Degradation of PAH in a creosote-contaminated soil. A comparison between the effects of willows (*Salix viminalis*), wheat straw and a nonionic surfactant. *International Journal of Phytoremediation, 12*(1), 54–66.

Jilani, K., Asgher, M., Bhatti, H. N., & Mushtaq, Z. (2011). Shake flask decolourization of direct dye solar golden yellow R by *Pleurotus ostreatus*. *Journal of The Chemical Society of Pakistan, 33*(2), 209–213.

Jung, E.-M., Kothe, E., & Raudaskoski, M. (2018). The making of a mushroom: Mitosis, nuclear migration and the actin network. *Fungal Genetics and Biology, 111*, 85–91.

Kalač, P. (2001). A review of edible mushroom radioactivity. *Food Chemistry, 75*(1), 29–35.

Kim, Y. M., Ahn, C. K., Woo, S. H., Jung, G. Y., & Park, J. M. (2009). Synergic degradation of phenanthrene by consortia of newly isolated bacterial strains. *Journal of Biotechnology, 144*(4), 293–298.

Kirtzel, J., Scherwietes, E. L., Merten, D., Krause, K., & Kothe, E. (2019). Metal release and sequestration from black slate mediated by a laccase of *Schizophyllum commune*. *Environmental Science and Pollution Research, 26*(1), 5–13. https://doi.org/10.1007/s11356-018-2568-z.

Kirtzel, J., Siegel, D., Krause, K., & Kothe, E. (2017). Stone-eating fungi: Mechanisms in bioweathering and the potential role of laccases in black slate degradation with the basidiomycete *Schizophyllum commune*. *Advances in Applied Microbiology, 99*, 83–101. https://doi.org/10.1016/bs.aambs.2017.01.002.

Kirtzel, J., Ueberschaar, N., Deckert-Gaudig, T., Krause, K., Deckert, V., Gadd, G. M., & Kothe, E. (2020). Organic acids, siderophores, enzymes and mechanical pressure for black slate bioweathering with the basidiomycete *Schizophyllum commune*. *Environmental Microbiology, 22*, 1535–1546. https://doi.org/10.1111/1462-2920.14749.

Klubicova, K., Danchenko, M., Skultety, L., Berezhna, V. V., Uvackova, L., Rashydov, N. M., & Hajduch, M. (2012). Soybeans grown in the Chernobyl area produce fertile seeds that have increased heavy metal resistance and modified carbon metabolism. *PLoS One, 7*(10), e48169.

Knabe, N., Jung, E. M., Freihorst, D., Hennicke, F., Horton, J. S., & Kothe, E. (2013). A central role for Ras1 in morphogenesis of the basidiomycete *Schizophyllum commune*. *Eukaryotic Cell, 12*(6), 941–952. https://doi.org/10.1128/EC.00355-12.

Kokkoris, V., Banchini, C., Paré, L., Abdellatif, L., Séguin, S., Hubbard, K., ... Stefani, F. (2023). *Rhizophagus irregularis*, the model fungus in arbuscular mycorrhiza research, forms dimorphic spores. *New Phytologist*. https://doi.org/10.1111/nph.19121.

Kothe, E., & Reinicke, M. (2017). Microbial communities in metal-contaminated environments adaptation and function in soil. In S. Das, & H. R. Dash (Eds.). *Handbook of metal-microbe interactions and bioremediation*. Boca Raton, FL: CRC Press.

Kovalchuk, I., Abramov, V., Pogribny, I., & Kovalchuk, O. (2004). Molecular aspects of plant adaptation to life in the Chernobyl zone. *Plant Physiology, 135*(1), 357–363.

Kraemer, S. M., Cheah, S.-F., Zapf, R., Xu, J., Raymond, K. N., & Sposito, G. (1999). Effect of hydroxamate siderophores on Fe release and Pb (II) adsorption by goethite. *Geochimica et Cosmochimica Acta, 63*(19–20), 3003–3008.

Krause, K. (2004). *Host specificity and specific gene expression in mycorrhizal fungi genus Tricholoma* (Dissertation). Germany: Friedrich Schiller University Jena.

Krause, K., Jung, E.-M., Lindner, J., Hardiman, I., Poetschner, J., Madhavan, S., ... Popp, J. (2020). Response of the wood-decay fungus *Schizophyllum commune* to co-occurring microorganisms. *PLoS One, 15*(4), e0232145.

Lamar, R. T., & Dietrich, D. M. (1990). In situ depletion of pentachlorophenol from contaminated soil by *Phanerochaete* spp. *Applied and Environmental Microbiology, 56*(10), 3093–3100.

Lee, H., Jang, Y., Choi, Y.-S., Kim, M.-J., Lee, J., Lee, H., ... Kim, J.-J. (2014). Biotechnological procedures to select white rot fungi for the degradation of PAHs. *Journal of Microbiological Methods, 97*, 56–62.

Leontievsky, A., Myasoedova, N., Golovleva, L., Sedarati, M., & Evans, C. (2002). Adaptation of the white-rot basidiomycete *Panus tigrinus* for transformation of high concentrations of chlorophenols. *Applied Microbiology and Biotechnology, 59*, 599–604.

Lloyd, D. R., & Phillips, D. H. (1999). Oxidative DNA damage mediated by copper (II), iron (II) and nickel (II) Fenton reactions: evidence for site-specific mechanisms in the formation of double-strand breaks, 8-hydroxydeoxyguanosine and putative intrastrand cross-links. *Mutation Research/Fundamental and Molecular Mechanisms of Mutagenesis, 424*(1–2), 23–36.

Maciejczyk, P., Chen, L.-C., & Thurston, G. (2021). The role of fossil fuel combustion metals in PM2. 5 air pollution health associations. *Atmosphere, 12*(9), 1086.

Madhavan, S., Krause, K., Jung, E.-M., & Kothe, E. (2014). Differential regulation of multi-copper oxidases in *Schizophyllum commune* during sexual development. *Mycological Progress, 13*, 1199–1206.

Majcherczyk, A., Johannes, C., & Hüttermann, A. (1998). Oxidation of polycyclic aromatic hydrocarbons (PAH) by laccase of *Trametes versicolor*. *Enzyme and Microbial Technology, 22*(5), 335–341.

Markowicz, A., Cycon, M., & Piotrowska, S. Z. (2016). Microbial community structure and diversity in long-term hydrocarbon and heavy metal contaminated soils. *International Journal of Environmental Research, 10*(2), 321–332.

Mazzeo, D. E. C., Fernandes, T. C. C., & Marin-Morales, M. A. (2011). Cellular damages in the *Allium cepa* test system, caused by BTEX mixture prior and after biodegradation process. *Chemosphere, 85*(1), 13–18.

Ma, X., Li, X., Liu, J., Cheng, Y., Zou, J., Zhai, F., … Han, L. (2021). Soil microbial community succession and interactions during combined plant/white-rot fungus remediation of polycyclic aromatic hydrocarbons. *Science of the Total Environment, 752*, 142224.

Michalak, I., Chojnacka, K., & Witek-Krowiak, A. (2013). State of the art for the biosorption process—A review. *Applied Biochemistry and Biotechnology, 170*, 1389–1416.

Morgan, P., Lee, S. A., Lewis, S. T., Sheppard, A. N., & Watkinson, R. J. (1993). Growth and biodegradation by white-rot fungi inoculated into soil. *Soil Biology & Biochemistry, 25*(2), 279–287. https://doi.org/10.1016/0038-0717(93)90040-I.

Morillo, E., Romero, A., Maqueda, C., Madrid, L., Ajmone-Marsan, F., Grcman, H., … Villaverde, J. (2007). Soil pollution by PAHs in urban soils: A comparison of three European cities. *Journal of Environmental Monitoring, 9*(9), 1001–1008.

Mukhopadhyay, M., Noronha, S., & Suraishkumar, G. (2007). Kinetic modeling for the biosorption of copper by pretreated *Aspergillus niger* biomass. *Bioresource Technology, 98*(9), 1781–1787.

Muthukumar, T., Priyadharsini, P., Uma, E., Jaison, S., & Pandey, R. R. (2014). Role of arbuscular mycorrhizal fungi in alleviation of acidity stress on plant growth. In M. Miransari (Vol. Ed.), *Use of microbes for the alleviation of soil stresses: Vol. 1*, (pp. 43–71). Springer.

Neubauer, U., Furrer, G., Kayser, A., & Schulin, R. (2000). Siderophores, NTA, and citrate: Potential soil amendments to enhance heavy metal mobility in phytoremediation. *International Journal of Phytoremediation, 2*(4), 353–368.

Noormohamadi, H. R., Fat'hi, M. R., Ghaedi, M., & Ghezelbash, G. R. (2019). Potentiality of white-rot fungi in biosorption of nickel and cadmium: Modeling optimization and kinetics study. *Chemosphere, 216*, 124–130.

Nriagu, J. O., & Pacyna, J. M. (1988). Quantitative assessment of worldwide contamination of air, water and soils by trace-metals. *Nature, 333*(6169), 134–139. https://doi.org/10.1038/333134a0.

Ochiai, E.-I. (1987). *General principles of biochemistry of the elements*. New York: Plenum Press.

Ohm, R. A., de Jong, J. F., Lugones, L. G., Aerts, A., Kothe, E., Stajich, J. E., … Wosten, H. A. (2010). Genome sequence of the model mushroom *Schizophyllum commune*. *Nature Biotechnology, 28*(9), 957–963. https://doi.org/10.1038/nbt.1643.

Ott, T., Fritz, E., Polle, A., & Schützendübel, A. (2002). Characterisation of antioxidative systems in the ectomycorrhiza-building basidiomycete *Paxillus involutus* (Bartsch) Fr. and its reaction to cadmium. *FEMS Microbiology Ecology, 42*(3), 359–366.

Peuke, A. D., & Rennenberg, H. (2005). Phytoremediation: Molecular biology, requirements for application, environmental protection, public attention and feasibility. *EMBO Reports, 6*(6), 497–501.

Phukhamsakda, C., Nilsson, R. H., Bhunjun, C. S., de Farias, A. R. G., Sun, Y.-R., Wijesinghe, S. N., … Tibpromma, S. (2022). The numbers of fungi: Contributions from traditional taxonomic studies and challenges of metabarcoding. *Fungal Diversity, 114*(1), 327–386.

Pozdnyakova, N. N., Nikiforova, S. V., Makarov, O. E., Chernyshova, M. P., Pankin, K. E., & Turkovskaya, O. V. (2010). Influence of cultivation conditions on pyrene degradation by the fungus *Pleurotus ostreatus* D1. *World Journal of Microbiology and Biotechnology, 26*, 205–211.

Pozdnyakova, N., Rodakiewicz-Nowak, J., & Turkovskaya, O. (2004). Catalytic properties of yellow laccase from *Pleurotus ostreatus* D1. *Journal of Molecular Catalysis B: Enzymatic, 30*(1), 19–24.

Priyadarshini, E., Priyadarshini, S. S., Cousins, B. G., & Pradhan, N. (2021). Metal-fungus interaction: Review on cellular processes underlying heavy metal detoxification and synthesis of metal nanoparticles. *Chemosphere, 274*, 129976.

Recer, G., & Boylen, C. (1986). The response of heterotrophic microorganisms to aluminum, manganese and zinc toxicity and pH in a clearwater acid lake. *BIOSIS/86/22717, Washington(0)*, 23–28.

Rhodes, C. J. (2015). Mycoremediation (bioremediation with fungi)—Growing mushrooms to clean the earth. *Chemical Speciation & Bioavailability, 26*(3), 196–198. https://doi.org/10.3184/095422914x14047407349335.

Rineau, F., Roth, D., Shah, F., Smits, M., Johansson, T., Canbäck, B., ... Tunlid, A. (2012). The ectomycorrhizal fungus *Paxillus involutus* converts organic matter in plant litter using a trimmed brown-rot mechanism involving Fenton chemistry. *Environmental Microbiology, 14*(6), 1477–1487. https://doi.org/10.1111/j.1462-2920.2012.02736.x.

Rodríguez-Rodríguez, C. E., Marco-Urrea, E., & Caminal, G. (2010). Degradation of naproxen and carbamazepine in spiked sludge by slurry and solid-phase *Trametes versicolor* systems. *Bioresource Technology, 101*(7), 2259–2266.

Saba, M., Falandysz, J., & Loganathan, B. (2020). Accumulation pattern of inorganic elements in scaly tooth mushroom (*Sarcodon imbricatus*) from Northern Poland. *Chemistry & Biodiversity, 17*(5), e2000167.

Sammer, D., Krause, K., Gube, M., Wagner, K., & Kothe, E. (2016). Hydrophobins in the life cycle of the ectomycorrhizal basidiomycete *Tricholoma vaccinum*. *PLoS One, 11*(12), e0167773. https://doi.org/10.1371/journal.pone.0167773.

Schindler, F., Gube, M., & Kothe, E. (2012). Chapter: Bioremediation and heavy metal uptake: Microbial approaches at field scale. In E. Kothe, & A. Varma (Eds.). *Bio-geo interactions in metal-contaminated soils* (pp. 365–383). Berlin Heidelberg: Springer. https://doi.org/10.1007/978-3-642-23327-2_18.

Schlunk, I., Krause, K., Wirth, S., & Kothe, E. (2015). A transporter for abiotic stress and plant metabolite resistance in the ectomycorrhizal fungus *Tricholoma vaccinum*. *Environmental Science and Pollution Research, 22*, 19384–19393.

Singh, H. (2006). *Mycoremediation: Fungal bioremediation*. John Wiley & Sons.

Singh, P. C., Srivastava, S., Shukla, D., Bist, V., Tripathi, P., Anand, V., ... Srivastava, S. (2018). Mycoremediation mechanisms for heavy metal resistance/tolerance in plants. *Mycoremediation and Environmental Sustainability, 2*, 351–381.

Singh, J. P., Vaidya, B. P., Goodey, N. M., & Krumins, J. A. (2019). Soil microbial response to metal contamination in a vegetated and urban brownfield. *Journal of Environmental Management, 244*, 313–319.

Sipos, G., Anderson, J. B., & Nagy, L. G. (2018). Armillaria. *Current Biology, 28*(7), R297–R298. https://doi.org/10.1016/j.cub.2018.01.026.

Sitaras, I. E., Bakeas, E. B., & Siskos, P. A. (2004). Gas/particle partitioning of seven volatile polycyclic aromatic hydrocarbons in a heavy traffic urban area. *Science of the Total Environment, 327*(1-3), 249–264.

Smith, M. L., Bruhn, J. N., & Anderson, J. B. (1992). The fungus *Armillaria bulbosa* is among the largest and oldest living organisms. *Nature, 356*(6368), 428–431. https://doi.org/10.1038/356428a0.

Smith, S. E., & Read, D. (2008). *Mycorrhizal symbiosis*. London: Elsevier.

Smith, M., Taylor, H., & Sharma, H. (1993). Comparison of the post-Chernobyl ^{137}Cs contamination of mushrooms from eastern Europe, Sweden, and North America. *Applied and Environmental Microbiology, 59*(1), 134–139.

Steinhauser, G. (2018). Anthropogenic radioactive particles in the environment. *Journal of Radioanalytical and Nuclear Chemistry, 318*(3), 1629–1639.

Stella, T., Covino, S., Čvančarová, M., Filipová, A., Petruccioli, M., D'Annibale, A., & Cajthaml, T. (2017). Bioremediation of long-term PCB-contaminated soil by white-rot fungi. *Journal of Hazardous Materials, 324*, 701–710.

Stork, N. E. (2018). How many species of insects and other terrestrial arthropods are there on Earth? *Annual Review of Entomology, 63*, 31–45.

Tian, J., Dungait, J. A., Lu, X., Yang, Y., Hartley, I. P., Zhang, W., ... Kuzyakov, Y. (2019). Long-term nitrogen addition modifies microbial composition and functions for slow carbon cycling and increased sequestration in tropical forest soil. *Global Change Biology, 25*(10), 3267–3281.

Tobiszewski, M., & Namieśnik, J. (2012). PAH diagnostic ratios for the identification of pollution emission sources. *Environmental Pollution, 162*, 110–119.

Torres-Cruz, T. J., Hesse, C., Kuske, C. R., & Porras-Alfaro, A. (2018). Presence and distribution of heavy metal tolerant fungi in surface soils of a temperate pine forest. *Applied Soil Ecology, 131*, 66–74. https://doi.org/10.1016/j.apsoil.2018.08.001.

Tran, N. H., Urase, T., & Kusakabe, O. (2010). Biodegradation characteristics of pharmaceutical substances by whole fungal culture *Trametes versicolor* and its laccase. *Journal of Water and Environment Technology, 8*(2), 125–140.

Traxler, L., Shrestha, J., Richter, M., Krause, K., Schäfer, T., & Kothe, E. (2022). Metal adaptation and transport in hyphae of the wood-rot fungus *Schizophyllum commune*. *Journal of Hazardous Materials, 425*, 127978.

Traxler, L., Wollenberg, A., Steinhauser, G., Chyzhevskyi, I., Dubchak, S., Großmann, S., ... Kirieiev, S. (2021). Survival of the basidiomycete *Schizophyllum commune* in soil under hostile environmental conditions in the Chernobyl Exclusion Zone. *Journal of Hazardous Materials, 403*, 124002.

Treu, R., & Falandysz, J. (2017). Mycoremediation of hydrocarbons with basidiomycetes—A review. *Journal of Environmental Science and Health. Part. B, Pesticides, Food Contaminants, and Agricultural Wastes, 52*(3), 148–155.

Van den Berg, M., Birnbaum, L., Bosveld, A., Brunström, B., Cook, P., Feeley, M., ... Kennedy, S. W. (1998). Toxic equivalency factors (TEFs) for PCBs, PCDDs, PCDFs for humans and wildlife. *Environmental Health Perspectives, 106*(12), 775–792.

Volesky, B. (1990). Removal and recovery of heavy metals by biosorption. *Biosorption of Heavy Metals, 7*–43.

Vuković, A., Schulz, W., Čamagajevac, I, Š., Gaur, A., Walther, C., & Gupta, D. K. (2020). Mycoremediation affects antioxidative status in winter rye plants grown at Chernobyl exclusion zone site in Ukraine. *Environmental Science and Pollution Research, 27*, 25818–25827.

Wagner, K., Krause, K., Gallegos-Monterrosa, R., Sammer, D., Kovács, Á. T., & Kothe, E. (2019). The ectomycorrhizospheric habitat of Norway spruce and Tricholoma vaccinum: promotion of plant growth and fitness by a rich microorganismic community. *Frontiers in Microbiology, 10*, 307.

Wallander, H., & Söderström, B. (1999). Paxillus. In J. W. G. Cairney, & S. M. Chambers (Eds.). *Ectomycorrhizal fungi: Key genera in profile* (pp. 231–252). Berlin, Germany: Springer.

Walther, C., & Denecke, M. A. (2013). Actinide colloids and particles of environmental concern. *Chemical Reviews, 113*(2), 995–1015.

Warren, N., Allan, I., Carter, J., House, W., & Parker, A. (2003). Pesticides and other micro-organic contaminants in freshwater sedimentary environments—A review. *Applied Geochemistry, 18*(2), 159–194.

Wasser, S., Grodzinska, A., & Lyugin, V. (1991). Content of [134]Cs and ^{137}Cs in higher basidiomycetes of the Ukrainian Polessie. *Ukrainian Botanical the Journal, 48*, 14–19.

Wengel, M., Kothe, E., Schmidt, C. M., Heide, K., & Gleixner, G. (2006). Degradation of organic matter from black shales and charcoal by the wood-rotting fungus *Schizophyllum commune* and release of DOC and heavy metals in the aqueous phase. *Science of the Total Environment, 367*(1), 383–393. https://doi.org/10.1016/j.scitotenv.2005.12.012.

Williamson, A. J., Folens, K., Matthijs, S., Cortes, Y. P., Varia, J., Du Laing, G., ... Hennebel, T. (2021). Selective metal extraction by biologically produced siderophores during bioleaching from low-grade primary and secondary mineral resources. *Minerals Engineering, 163*, 106774.

Wollenberg, A., Kretzschmar, J., Drobot, B., Hübner, R., Freitag, L., Lehmann, F., ... Raff, J. (2021). Uranium (VI) bioassociation by different fungi—A comparative study into molecular processes. *Journal of Hazardous Materials, 411*, 125068.

Xu, P., Liu, L., Zeng, G., Huang, D., Lai, C., Zhao, M., ... Wu, H. (2014). Heavy metal-induced glutathione accumulation and its role in heavy metal detoxification in *Phanerochaete chrysosporium*. *Applied Microbiology and Biotechnology, 98*(14), 6409–6418.

Yemendzhiev, H., Terziyska, A., Manasiev, J., & Alexieva, Z. (2011). Degradation of mixed aromatic pollutants by *Trametes versicolor* strain 1. *Biotechnology & Biotechnological Equipment, 25*(sup1), 39–40.

Ying, T., Yong-Ming, L., Chang-Yong, H., Jian, L., Zhen-Gao, L., & Christie, P. (2008). Tolerance of grasses to heavy metals and microbial functional diversity in soils contaminated with copper mine tailings. *Pedosphere, 18*(3), 363–370.

Zafar, S., Aqil, F., & Ahmad, I. (2007). Metal tolerance and biosorption potential of filamentous fungi isolated from metal contaminated agricultural soil. *Bioresource Technology, 98*(13), 2557–2561. https://doi.org/10.1016/j.biortech.2006.09.051.

Zhang, Y., Kong, H., Fang, Y., Nishinari, K., & Phillips, G. O. (2013). Schizophyllan: A review on its structure, properties, bioactivities and recent developments. *Bioactive Carbohydrates and Dietary Fibre, 1*(1), 53–71.

Zhao, M., Zhang, C., Zeng, G., Cheng, M., & Liu, Y. (2016). A combined biological removal of Cd^{2+} from aqueous solutions using *Phanerochaete chrysosporium* and rice straw. *Ecotoxicology and Environmental Safety, 130*, 87–92.

Zhdanova, N., Redchits, T., Zheltonozhsky, V., Sadovnikov, L., Gerzabek, M., Olsson, S., ... Mück, K. (2003). Accumulation of radionuclides from radioactive substrata by some micromycetes. *Journal of Environmental Radioactivity, 67*(2), 119–130.

Zhonghua, Z., Zhang, L., & Wu, J. (2016). Polycyclic aromatic hydrocarbons (PAHs) and organochlorine pesticides (OCPs) in sediments from lakes along the middle-lower reaches of the Yangtze River and the Huaihe River of China. *Limnology and Oceanography, 61*(1), 47–60.

Zhuo, R., & Fan, F. (2021). A comprehensive insight into the application of white rot fungi and their lignocellulolytic enzymes in the removal of organic pollutants. *Science of the Total Environment, 778*, 146132.

CHAPTER FIVE

Morphological responses of filamentous fungi to stressful environmental conditions

Marina Fomina[a,*], Olena Gromozova[a], and Geoffrey Michael Gadd[b,c]

[a]Zabolotny Institute of Microbiology and Virology, National Academy of Sciences of Ukraine, Kyiv, Ukraine
[b]Geomicrobiology Group, School of Life Sciences, University of Dundee, Dundee, Scotland, United Kingdom
[c]State Key Laboratory of Heavy Oil Processing, Beijing Key Laboratory of Oil and Gas Pollution Control, College of Chemical Engineering and Environment, China University of Petroleum, Beijing, P.R. China
*Corresponding author. e-mail address: M.Fomina@ukr.net

Contents

Advances in Applied Microbiology, Volume 129
ISSN 0065-2164, https://doi.org/10.1016/bs.aambs.2024.07.001

Abstract

The filamentous growth mode of fungi, with its modular design, facilitates fungal adaptation to stresses they encounter in diverse terrestrial and anthropogenic environments. Surface growth conditions elicit diverse morphological responses in filamentous fungi, particularly demonstrating the remarkable adaptability of mycelial systems to metal- and mineral-rich environments. These responses are coupled with fungal biogeochemical activity and can ameliorate hostile conditions. A tessellated agar tile system, mimicking natural environmental heterogeneity, revealed negative chemotropism to toxic metals, distinct extreme growth strategies, such as phalanx and guerrilla movements and transitions between them, and the formation of aggregated re-allocation structures (strands, cords, synnemata). Other systems showed intrahyphal growth, intense biomineralization, and extracellular hair-like structures. Studies on submerged mycelial growth, using the thermophilic fungus *Thielavia terrestris* as an example, provided mechanistic insights into the morphogenesis of two extreme forms of fungal submerged culture—pelleted and dispersed growth. It was found that the development of fungal pellets was related to fungal adaptation to unfavorable stressful conditions. The two key elements affecting morphogenesis leading to the formation of either pelleted or dispersed growth were found to be (1) a lag phase (or conidia swelling stage) as a specific period of fungal morphogenesis when a certain growth form is programmed in response to morphogenic stressors, and (2) cAMP as a secondary messenger of cell signaling, defining the implementation of the particular growth strategy. These findings can contribute to knowledge of fungal-based biotechnologies, providing a means for controllable industrial processes at both morphological and physiological levels.

1. Introduction

Fungi represent a highly diverse group of heterotrophic eukaryotes distinguished by the absence of phagotrophy and the presence of a rigid chitinous cell wall with β-glucan as a crucial component (Walker & White, 2017; Naranjo-Ortiz & Gabaldón, 2019; Ruiz-Herrera & Ortiz-Castellanos, 2019). The evolutionary success of fungi is largely attributed to the capacity of filamentous fungi for indefinite growth as cylindrical multinucleate cells, as a repeating unit called a hypha. These structural features represent the modular design and function as a fundamental element in the evolution of these microorganisms, contributing significantly to the adaptability and ecological ubiquity of fungi (Andrews, 2017).

Generally, the growth form, enclosing the mode of construction and associated morphological characteristics, establishes fundamental opportunities and constraints that govern the biological attributes of organisms, representing either unitary or modular modes. Growth form stands out as one of the most essential characteristics of living organisms, which encompasses such crucial properties as size, shape, and plasticity, as well as nutritional and reproductive modes, lifespan, and the extent and manner of mobility (Andrews, 1995). In contrast to unitary organisms, modular organisms such as fungi grow by reiteration of a basic multicellular unit of construction (module) and are much more subject to environmental influences. Modular organisms are often branched and mostly sessile or passively mobile. They typically do not show generalized aging but manifest an internal age structure, have mixed reproductive mechanisms, and often exhibit a higher chance of reproduction as they age (Andrews, 1995, 2017). The principal evolutionary implications of modular design, either directly or indirectly, arise from this iteration and sessility (Andrews, 2017). For example, sessility is related to (1) high plasticity of phenotype: size, shape, potential for reproduction, and population growth; (2) selection pressure and potential somatic mutation through the exposure to different environments of certain individual modules developed from sexual reproduction; and (3) potential indefinite longevity of the individual.

For filamentous fungi, the major evolutionary factors controlling morphology include several types of constraints. Evolutionary possibilities for fungi are shaped by historical (phylogenetic) constraints, for example, by such features as rigid microfibrillar cell walls and protoplasmic continuity in the mycelium or thalli. The porous nature of filamentous fungi can provide an unobstructed transport of material along the length of a hypha and rapid signal transmission. However, it also can pose risks of the mycelium or thallus being overtaken by foreign or mutant nuclei, which have deficient cell housekeeping (Buss, 1987). Overcoming these constraints, fungal evolution shows adaptation through sophisticated mechanisms to regulate nuclear division and migration, including synchrony, septation, and recognition systems, which safeguard genetic integrity (Andrews, 1995). Other constraints are (1) functional (adaptive), which are related to dispersal mechanisms and coexistence of two or more genetically distinct nuclei inside a in a single fungal cell (dikaryosis and heterokaryosis); (2) structural (developmental), involving limitations by physical and chemical laws and the ontogenetic program, e.g., upper and lower limits on spore size or the surface to volume relationship; (3) ecophenotypic, which involve environmentally-dependent variance in phenotype and adaptive plasticity within some boundaries; and finally (4) chance and coincidence, regarded as influencing evolutionary and contemporary contexts, also play some role in shaping fungal morphology (Andrews, 1995).

Considering early fungal evolution, since all modern terrestrial fungi share a clear monophyletic clade with a single inferred flagellum loss, ancestral fungi were probably mostly aquatic, presumably with a parasitoid lifestyle (Liu, Hodson, & Hall, 2006; Naranjo-Ortiz & Gabaldón, 2019). In support of this, indirect evidence of possible predations in Neoproterozoic eukaryotes has been found, matching zoosporic fungal parasitoid activity (Porter, 2016). The loss of the flagellum and the development of hyphal growth have defined fungal adaptation to the terrestrial environment (terrestrialization) as the most definitive evolutionary milestone in the Fungi kingdom (Naranjo-Ortiz & Gabaldón, 2019). The modular hyphal growth enabled filamentous fungi to explore and penetrate solid surfaces and provided access to various biopolymers, including cellulose, which facilitated the switch of their lifestyle from parasitism to saprotrophy. Based on circumstantial evidence and theoretical extrapolation, three possible scenarios have been suggested for fungal terrestrialization coupled with the development of filamentous growth in fungi. These include the hypotheses that (1) fungi co-evolved with the ancestors of land plants (Lücking, Huhndorf, Pfister, Plata, & Lumbsch, 2009; Beck, Divakar, Zhang, Molina, & Struwe, 2015; Delaux et al., 2015); (2) fungal co-evolution with proto-soil, which was covered with crust-like communities of other microorganisms (Taylor & Osborn, 1996; Astafieva & Rozanov, 2012; Wu, Fang, Yu, & Zhang, 2014; Kidron & Xiao, 2024); and (3) an icy environment creating micro-niches with highly saline brine channels and cryoconites caused by exclusion of solutes from the ice crystal, which could have provided nutrient-rich, spatially limited, and temporary conditions facilitating fungal terrestrialization (Boetius, Anesio, Deming, Mikucki, & Rapp, 2015; Gokul et al., 2016; Anesio, Lutz, Chrismas, & Benning, 2017; Naranjo-Ortiz & Gabaldón, 2019).

No matter what evolutionary scenario was really involved in fungal terrestrialization, success was ensured by establishing the modular architecture of hyphal growth. As a result, fungi populate all types of terrestrial ecosystems where they play crucial roles, especially in cycling carbon and other elements and nutrients (Schimel, 1995; Peay & Bruns, 2014). Besides natural terrestrial ecological niches, filamentous fungi are adapted to various anthropogenic environments including, on the one hand, human habitats, where fungi can be biodeteriorative and biopollution agents, and, on the other hand, some commercially important filamentous fungi have applications in bioreactors for biotechnological use (Alastruey-Izquierdo, Melhem, Bonfietti, & Rodriguez-Tudela, 2015; Garcia-Reyes et al., 2017; Cairns, Zheng, Zheng, Sun, & Meyer, 2019; Gadd, 2017; Gadd, Fomina, & Pinzari, 2024). In addition, some fungi have developed a wide spectrum of relationships with animals, plants, and microorganisms, which range from pathogenicity to mutualistic symbioses (Naranjo-Ortiz & Gabaldón, 2019). In any ecosystem, fungi consistently encounter a multitude of environmental challenges and stresses (Andrews, 2017). However, our understanding of the diverse range of morphological responses exhibited by filamentous fungi under different stressors is still considerably limited. This review will attempt to summarize essential findings on the diversity of morphological responses to environmental stressors manifested by filamentous fungi under both surface and submerged mycelial growth conditions.

2. The stresses that fungi encounter

Fungi, like any other living organisms, are constantly under the influence of adverse environmental factors, which can include extreme temperature and acidity values, hypertonicity and hypotonicity, lack of nutrients, and a simultaneously high concentration of metabolic end-products, the presence of harmful substances including anthropogenic pollutants, the presence of toxic metals and active forms of oxygen, ionizing radiation, hypo- and hypergravitational effects, etc. In response to various stress factors, a complex of non-specific compensatory and adaptive stress processes is launched in living organisms, which leads to their resistance to stresses (Selye, 1938; Gorban, Tyukina, Smirnova, & Pokidysheva, 2016; Voychuk & Gromozova, 2020a, 2020b).

All organisms activate a cellular stress response whenever environmental conditions surpass the range appropriate for maintaining cellular homeostasis. This response is vital for survival in critical situations as it safeguards macromolecular integrity, securing the continued functionality of cells and organisms. The cellular stress response acts against severe stress by enabling adaptation through the process of so-called stress-induced evolution. Physiologically, this process involves certain stress-induced changes in (1) rate of mutations; (2) histone post-translational modifications; (3) DNA methylation; (4) chromoanagenesis; and (5) activity of transposable elements (Mojica & Kültz, 2022). At the population level, these mechanisms generate phenotypic variations which drive natural selection and thereby enhance the chances for populations to avoid extinction and instead adapt to new environmental conditions.

While overcoming stress effects, their adaptation to stressors can occur through a number of repair and regulatory systems, for example, DNA repair systems (Abingdon, 2015); regulation of gene expression and protein synthesis (Diller, 2006; Shim et al., 2020), including heat shock proteins (Kazemi, Chang, Haserodt, McKen, & Zachara, 2010); and changes in the activity of antioxidant defense enzymes (Bansal & Kaushal, 2014). In these cases, intracellular systems are adjusted to eliminate intracellular damage caused by stress factors, which generally leads to the adaptation of organisms to environmental conditions (Fulda, Gorman, Hori, & Samali, 2010; Sano & Reed, 2013). Filamentous fungi have also evolved mechanisms to regulate cell death in response to a variety of stress factors, suggesting that the multicellular growth habits of these microorganisms have evolved diverse mechanisms of programmed cell death. This diversity includes autophagy and apoptosis, which are controlled processes for cell death that help break down and recycle cell parts systematically (e.g., during aging and reproduction), and necrosis, which is a more abrupt type of death resulting from physical or chemical injury (Gonçalves, Heller, Daskalov, Videira, & Glass, 2017).

As stress is a non-specific process by definition (Selye, 1938), some responses are common for all kinds of stresses, such as initial changes in the ultrastructural organization of cells, including an increase in the size of mitochondria; an increase in the number and size of peroxisomes; accumulation of polyphosphate granules; specific changes in cytoplasmic membrane structure; emerging of intussusceptions (under ethanol stress) or globular structures (under oxidative and heat stress). Stress responses can also include an activation of hydrolases; a decrease of synthetic processes accompanied by activation of RNA synthesis; changes in the secondary, tertiary, and quaternary structure of proteins; an increase in cytoplasmic membrane permeability resulting in the enhanced release of various substances from the cells; membrane depolarization; a decrease in cytoplasm pH; formation of free radicals as a result disrupted oxidative phosphorylation; synthesis of stress proteins, etc. The degree that such reactions are manifest is organism-dependent and defined by the dose and type of stress factor (Guan et al., 2017; Liu et al., 2019). For example, some stress-specific transcription factors are activated simultaneously with Environmental Stress Response genes, and their activity depends on the type of stressor (Gasch et al., 2000; Crawford & Pavitt, 2018). These are aimed at eliminating specific damage caused by the action of the stress factor. Several signaling pathways are specific for distinct stressors. When the cell is subjected to stress, resulting in a reconfiguration of ongoing biochemical processes, and the synthesis of specific resistance factors occurs, these signaling pathways are involved in the general response of the cell to the stressor. For example, the TORC1-pathway ("Target Of Rapamycin Complex 1"), which is the main regulator of growth and metabolism in all eukaryotic cells (Chantranupong & Sabatini, 2016), is activated in response to all types of stress factors, and along with protein kinase Sch9, ensures pH homeostasis within the cell (Eskes, Deprez, Wilms, & Winderickx, 2017). Signaling pathways, such as Ras, the cyclic AMP (cAMP) pathway, and the protein kinase A (PKA) pathway, are closely and functionally interconnected. In addition to their role in formating general responses of the cell to stress, they are involved in regulating cell growth (Besozzi et al., 2012; Cazzanelli et al., 2018; Huang, Huang, Wei, Wang, & Du, 2018). One of the most studied mechanisms is the High Osmolarity Glycerol (HOG) pathway, based on mitogen–activated protein kinases (MAPK). This was first considered to be specific to hyperosmotic stress but later was revealed to play a role in the responses to other kinds of stresses, e.g., oxidative stress, weak acid stress, and heat stress (Alonso-Monge et al., 2001; Miermont, Uhlendorf, McClean, & Hersen, 2011; Lee et al., 2016; Dunayevich et al., 2018). Along with other MAPK pathways, the HOG pathway controls the processes in morphogenesis (Hohmann, 2015). All this indicates the existence of close connections between various metabolic and transcriptomic pathways, which are involved in the formation of the cell's response to stress and contribute to our understanding of why cell exposure to certain stressors can increase resistance to other stressors, which is called the "adaptive response" phenomenon (Voychuk & Gromozova, 2020a).

In addition to regulatory factors generating a general response to stress, cells also possess specific factors and means of responding to certain specific types of stress. One such stress is oxidative stress, which has been mostly studied in yeast cells. Reactive oxygen species (ROS) are known to form in the cell normally. However, under the influence of such factors as UV and ionizing radiation or metal ions, ROS can damage DNA, lipids, and proteins (Campisi et al., 2010; Yakymenko et al., 2015; Moloney & Cotter, 2018). Resistance to oxidative stress depends on the phase of cell development, metabolic status, cell wall and cell membrane structure, and the organization of intracellular structures. Several mechanisms of ROS generation and degradation in cells may be involved in protective cell reactions against oxidation. Implementation of an adaptive response occurs due to the participation of a number of regulatory proteins (Yap1p and Yap2p; Msn2p and Msn4p; and Gcn4p and Skn7p) (Biryukova, Medentsev, Arinbasarova, & Akimenko, 2006). These proteins affect the activity of enzymes with antioxidant properties, e.g., Yap1p regulates the activity of superoxide dismutase, glucose-6-phosphate dehydrogenase, and glutathione reductase.

Cells of living organisms also constantly adapt to hyper- and hypotonic conditions associated with, e.g., changes in water availability and variations in the concentration of external and intracellular compounds. The pathways of intracellular signaling and the formation of a response to hyper- and hypoosmotic shock in yeast cells are the HOG and the PKC (Protein Kinase C) pathways, respectively. The PKC1 pathway is also known as the Cell Integrity Pathway (CIP), as it is critically important under conditions threatening the stability of the cell wall (Mensonides, Brul, Klis, Hellingwerf, & Teixeira de Mattos, 2005). Such conditions are created not only with hypotonicity (Davenport, Sohaskey, Kamada, Levin, & Gustin, 1995) but also with heat shock (Kamada, Jung, Piotrowski, & Levin, 1995) and other types of stress, leading to PKC1 pathway activation. Under the action of hyperosmotic shock, active synthesis of glycerol occurs in cells by increasing the expression of GPD1, which encodes the enzyme glycerol-3-phosphate dehydrogenase (Posas, 1997; Rep et al., 2001; Sorger & Daum, 2003; Hohmann, Krantz, & Nordlander, 2007). At the same time, regulation of the amount of glycerol during yeast growth in a hyperosmotic environment occurs due to a cascade of kinase reactions activated by the product of two other genes, SLN1 and SHO1 (Posas, 1997; Posas, Takekawa, & Saito, 1998). The expression of genes of the GRE family (GRE1–3) also increases under conditions of hyperosmotic shock and is affected by the concentration of osmolytes. These genes positively regulate the HOG pathway, and their expression increases under the action of other types of stress (thermal and oxidative). GRE1 and GRE3 genes induce the transcription of factors Msn2p and Msn4p, which are associated with yeast adaptation to temperature.

Hyper- and hypoosmotic shock is accompanied by a temporary increase of calcium ions in the cytosol due to the activation of stretch channels in intracellular stores (Batiza, Schulz, & Masson, 1996; Loukin, Kung, & Saimi, 2007). In turn, the activation of calcium channels and stimulation of the HOG and the CWI (Cell Wall Integrity) pathways also occur under the action of acid stress (Lucena, Elsztein, Simões, & Morais, 2012). This means that the mechanisms of resistance to acid stress closely overlap with resistance to osmotic shock.

Regarding acid stress, the metabolic activity of fungi causes changes in the acidity of the surrounding environment. Therefore, they have certain adaptations for living in a wide range of pH values. The intracellular pH of the cytoplasm and organelles is controlled by H^+-ATPases, such as Pma1p (cytoplasmic P2-type ATPase) and V-ATPase (vacuolar ATPase) (Kane, 2016; Collins & Forgac, 2020). Different acids may have slightly different effects on cells, depending on the reactivity of their chemical groups. At extremely low pH values, the electrochemical potential and functions of the plasma membrane may be disturbed, resulting in a loss of cytoplasmic membrane control over molecular fluxes with potentially lethal consequences (Johnston, Nallur, Gordon, Smith, & Strobel, 2020). Acid stress also induces the formation of ROS, which causes activation of defense processes combating oxidative stress. For example, the vacuolar H^+-ATPase works to maintain intracellular pH, but data also indicates its possible involvement in the neutralization of ROS (Johnston et al., 2020).

All the described regulatory pathways are directly or indirectly involved in the synthesis of components of the cell wall (Munro et al., 2007; Lucena et al., 2012; Heinisch & Rodicio, 2017), which can be the primary barrier to the stress factor.

In contrast to acid stress, alkalinization of the environment is atypical for most fungi, especially for yeast; therefore, even the smallest changes in pH towards alkaline values cause significant changes in the expression of several genes. As a result, regulatory pathways Rim101/PacC and calcineurin (Ca^{2+}-activated phosphatase), the CWI pathway, and the kinases Snf1p and PKA, as well as response pathways to oxidative stress (Serra-Cardona, Canadell, & Arino, 2015) may be activated. This is associated with the disruption of nutrient uptake processes and the corresponding mechanisms that support phosphate, iron/copper, and glucose homeostasis.

In general, acid and alkaline stresses are accompanied by an intracellular response to oxidative stress. It is currently unknown whether this is related to direct ROS entry inside the cell from the exterior, a consequence of violation of intracellular systems integrity, or a result of inactivation or blocking of individual enzymes. The variety of the processes connected by cause-and-effect relationships confirms the polyfunctionality of most cellular systems and, accordingly, the universality of some, and possibly most, protection mechanisms inherent in living organisms (Harrison, Zyla, Bardes, & Lew, 2003; Martin et al., 2015; Taymaz-Nikerel, Cankorur-Cetinkaya, & Kirdar, 2016).

By employing the physiological mechanisms described above, filamentous fungi have successfully adapted to stressful environments, resulting in significant diversity in growth form despite their comparatively simple structural organization. For example, this adaptation can manifest in the development of extensive fungal multicellular networks covering areas exceeding 15 ha (Smith, and Read, 2008). Ultimately, the structure of filamentous fungi, as modular organisms, represents a physiologically and morphologically balanced response to various environmental factors and constraints, including energy and nutrient acquisition and tolerance of abiotic conditions (Andrews, 2017). Given the significance of fungal terrestrialization in the evolution of filamentous fungi, there is a particular interest in unraveling the extent of diversity of their morphological responses to such abiotic stressful conditions as metal- and mineral-rich environments.

3. Diversity of morphological responses of mycelial systems to stress under conditions of surface growth in metal- and mineral-rich environments

Mineral- and metal-rich environments constitute a significant part of the entire terrestrial environment, which includes man-made and natural surfaces and subsurface features that are heavily populated by fungi. Most of these fungi exhibit a filamentous growth habit, which gives them the ability to increase or decrease their surface area and to adopt different growth strategies (Fomina, Burford, & Gadd, 2005a; Gadd, Fomina, & Burford, 2005; Gadd et al., 2024). In general, a nutrient-limited environment promotes filamentous growth, and the hyphal phenotype enhances survival in nature by enabling the fungus to forage for nutrients and to colonize natural surfaces (Gimeno, Ljungdahl, Styles, & Fink, 1992, Francisco, Ma, Zwyssig, McDonald, & Palma-Guerrero, 2019). Fungi predominantly employ the surface growth mode in terrestrial environments and act as important geoactive components of the biosphere. Such interactions are the subject of geomycology (Gadd, 2007). Morphological responses of geoactive filamentous fungi to metal- and mineral-rich environments are inextricably linked to underlying physiological reactions. For a comprehensive understanding of all these processes, our observations on the morphological adaptations of filamentous fungi to metal- and mineral-rich environments should be complemented by clear insights into the biogeochemical abilities of such fungi.

3.1 Fungal biogeochemical potential in metal- and mineral-rich environments

Filamentous fungi exhibit a remarkable suitability for biogeochemical activity, especially in metal- and mineral-rich environments. Fossil filamentous fungi were first discovered in the mid-late Precambrian period (1430–1542 million years ago), and at least since then, they have thrived in diverse terrestrial environments (Burford, Fomina, & Gadd, 2003). Fungi play a crucial role among the soil microbiota, possessing the largest bulk biomass and dominating in metal-polluted soils at the community level (Fomina et al., 2005a; Fomina & Gadd, 2007; Gadd et al., 2005). They can be highly resistant to extreme environmental factors such as metal toxicity, UV and solar radiation, desiccation, hypersalinity, and extreme temperature (Sterflinger, 2000; Burford et al., 2003; Gadd, 2004; Fomina et al., 2005a). Fungi can excrete protons and chelating agents, as well as perform metal mobilization from minerals faster than bacteria (Gu, Ford, Berke, Gu, & Mitchell, 1998; Fomina et al., 2007a). Additionally, they can form mutualistic symbiotic associations with plants (mycorrhizas), algae (lichens), and cyanobacteria (lichens) (Gorbushina et al., 1993; Smith & Read, 2008; Sterflinger, 2000). For mycorrhizal mycobionts, their main functions are the acquisition of phosphorus and trace metals from soil, and contribution to plant adaptation to extreme environments, including metal- and mineral-rich environments. (Meharg & Cairney, 2000; Fomina, Alexander, Hillier, & Gadd, 2004; Fomina, Alexander, Colpaert, & Gadd, 2005b; Smith & Read, 2008). Given that over 95% of land plants rely on symbiotic mycorrhizal fungi, such biogeochemical abilities make mycorrhizal fungi integral components of bio- and phytoremediation/revegetation approaches (Fomina & Gadd, 2007). Lichens are prevalent members of microbial consortia inhabiting stones and rocks, where they play a fundamental role in the initial stages of substrate colonization and development of the mineral component in soils (Banfield, Barker, Welch, & Taunton, 1999; Chen, Blume, & Beyer, 2000). Numerous fungi are adapted to mineral-rich environments, surviving on stone and rock surfaces because of their oligotrophic abilities to extract nutrients from the air, rainwater, etc. (Gadd et al., 2024). Inhabitants of subaerial mineral surfaces include fungi adapted to sudden extreme changes in microclimatic conditions, including salinity, pH, solar radiation, and water potential, due to the formation of antioxidant protectors inside the cell and in cell walls (for example, mycosporines, as well as melanins and other pigments), and surrounding colonies with a matrix of extracellular polymeric substances (EPS), often containing clay particles (Gorbushina et al., 2003; Volkmann, Whitehead, Rutters, Rullkotter, & Gorbushina, 2003).

Fungi come into contact with metals as normal components of their natural habitat, as well as with metals that have been introduced and redistributed in this environment due to human activities (Fomina & Skorochod, 2020). Metal toxicity can be mitigated by physicochemical environmental conditions, and the fungi can be inherently tolerant and resistant to metals; therefore, microscopic fungi can be isolated from any habitat contaminated with toxic metals. Over three-quarters of the elements in the Periodic table fall under the classification of metals. Elements like K, Na, Ca, Mn, Mg, Fe, Cu, Zn, Co, and Ni play crucial roles as macro- and microelements in biological processes. On the other hand, such elements as Rb, Cs, Al, Cd, Ag, Au, Sr, U, Hg, and Pb are not physiologically essential for the normal functioning of living organisms (Gadd, 1993, 2007). Nonetheless, all metals can interact with fungi and exert toxicity at concentrations above a certain threshold, depending on the organism, the metal, and environmental factors. The anthropogenic redistribution of toxic metals in the biosphere has become an important process in the biogeochemical cycles of these metallic elements.

In fungi, metal toxicity is exhibited in various ways, i.e., blocking of functional groups in crucial biological molecules like enzymes, displacing or replacing vital metal ions, damaging cell membranes and organelles, and compromising the systems designed to protect the cell from the harmful effects of free radicals formed under normal metabolic conditions (Gadd, 1993; Howlett & Avery, 1997). Nevertheless, various mechanisms are manifested at biochemical, physiological, and morphological levels that facilitate fungal tolerance and their ability to thrive in metal-contaminated and mineral-rich environments.

Generally, biochemical strategies for reducing metal toxicity in fungi can be divided into two main groups: (1) processes that limit the entry of metals into the cell, known as avoidance of metal toxicity by the cell, and (2) intracellular processes that assure fungal tolerance to metals and cell survival at high internal concentrations of metals (Fomina & Gadd, 2007). Both free-living and mycorrhizal fungi can employ the metal toxicity avoidance strategy at the biochemical-physiological level through active transport and compartmentalization, including suppression of metal influx and enhanced metal efflux from the cytosol or cellular compartments. Extracellularly, metals can be bound by cell wall components (e.g., through biosorption) and exopolysaccharides and other metabolites, which can lead to extensive metal precipitation. In addition, fungi have the capacity to transform metalloids into volatile forms (Baldrian, 2003; Fomina & Gadd, 2007, 2014; Gadd et al., 2014). Intracellular processes involve metal chelation, e.g., by metallothioneins, low molecular weight phenolic compounds, glutathione, and phytochelatins in cytosol and vacuoles (Jacob, Courbot, Martin, Brun, & Chalot, 2004; Fomina & Gadd, 2007; Bellion,

Courbot, Jacob, Blaudez, & Chalot, 2006). Furthermore, toxic metal binding by vacuolar polyphosphates may play a role in maintaining intracellular zinc homeostasis during its excess in ectomycorrhizal fungi (Bucking & Heyser, 1999). Additionally, an essential biochemical defence mechanism against oxidative stress induced by toxic metals and radionuclides involves the fungal capacity to scavenge free radicals via the activity of antioxidant enzymes, superoxide dismutase, and catalase or through the formation of glutathione (Bellion et al., 2006).

As an integral part of the fungal manifestation of metal- and mineral-tolerance at biochemical/physiological levels, the biogeochemical aspect plays a key role in fungal interactions with metal- and mineral-rich environments and significantly contributes to such tolerance.

Two primary reactions—dissolution and precipitation—underlie biogeochemical transformations of metals and minerals that are integral to Earth's elemental cycling (Burford et al., 2003; Gadd et al., 2005; Fomina & Gadd, 2007; Gadd, 2007, 2017; Fomina & Skorochod, 2020). Through dissolution reactions, the activities of geoactive free-living filamentous fungi and their symbiotic associations can lead to mineral dissolution primarily through heterotrophic leaching, resulting in the release of mobile metal species and associated elements (e.g., phosphorus) into the environment (Fig. 1). The two main mechanisms of heterotrophic leaching are (1) proton-mediated or acidification, where fungi acidify their microenvironment through the excretion of protons and organic acids and the formation of respiratory carbonic acid; and (2) ligand-promoted or complexation, through the excretion of metal-complexing metabolites, such as carboxylic acids, amino acids, Fe (III)-binding siderophores, and phenolic compounds. Low molecular weight carboxylic acids with strong chelating properties, such as oxalic and citric acids, can shift the mechanism of mineral dissolution from proton-promoted to ligand-promoted, leading to an aggressive attack on the mineral surface and a considerable increase in metal mobilization (Gadd, 1993; Fomina et al., 2004, 2005c; Fomina, Bowen, Charnock, Podgorsky, & Gadd, 2017; Gadd et al., 2014). The biochemical processes involved in the decay of rocks and minerals by fungi are also facilitated by biomechanical actions, which are especially characteristic of filamentous hyphal growth and lichens. These actions can involve (1) a direct interaction with the mineral surface and thigmotropic penetration into the mineral matrix by fungal cells or (2) an indirect interaction occurring through the fungal production of the EPS matrix and its biomechanical consequences (Burford et al., 2003; Gadd, 2017; Fomina & Skorochod, 2020). The consequences of such fungal mineral dissolution leads

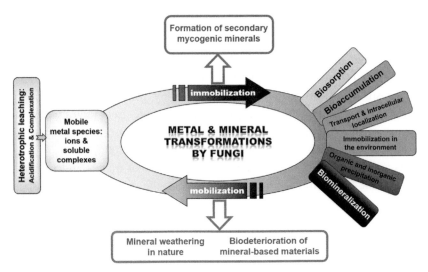

Fig. 1 The main biogeochemical processes in fungal transformations of metals and minerals. *Adapted from Fomina, M., & Skorochod, I. (2020). Microbial interaction with clay minerals and its environmental and biotechnological implications. Minerals, 10(10), 861. https://doi.org/10.3390/min10100861.*

to the delivery of bioavailable nutrients for plants and other terrestrial microbiota; to soil genesis, function, and fertility; and to the formation of sediments resulting from the weathering and decay of rocks and minerals. The negative outcomes of mineral dissolution by fungi include the biodeterioration of mineral-based building and barrier materials, as well as historic and cultural heritage (Fomina et al., 2007a, 2022; Gadd, 2017; Gadd et al., 2024).

In contrast, geoactive fungi can immobilize metals through biogeo-chemical processes, such as biosorption; transport; intracellular localization and bioaccumulation by living cells; redox immobilization; organic and inorganic precipitation of metals; and, finally, biomineralization, resulting in the formation of secondary mycogenic minerals (Fig. 1). The capacity of free-living and mycorrhizal fungi for metal immobilization and biomi-neralization is of great environmental importance as a means of deconta-mination by stabilizing/immobilizing metal toxicants, as well as bior-emediating metal-polluted soils (Fomina & Gadd, 2007; Gadd, 2007; Fomina et al., 2008).

In the following subsection, we will focus on the morphological flex-ibility of filamentous fungi in response to metal- and mineral-rich envir-onments, primarily based on our own observations.

3.2 Morphological flexibility of filamentous fungi in response to metals and minerals

With the establishment of a modular architecture for filamentous fungi as sessile organisms, habitable space became a critical factor in defining their survival and reproductive potential (Andrews, 2017). This implies adaptations for the inevitable creation of resource depletion zones and the necessity to employ exploitation strategies to search for new resources from relatively fixed positions, along with the dispersal of progeny through growth or transport (e.g., through spore dispersal). For metal- and mineral-rich environments, it also implies specific problems relating to escaping adverse stressful environments necessitating effective adaptation *in situ* (e.g., through dormancy). The strategies employed by filamentous fungi to arrange their individual units, i.e., the hyphae, define their morphology and determine how they colonize heterogeneous terrestrial environments. The extreme cases of this arrangement, in terms of modular growth, are represented by two different tendencies in the hyphal branching pattern: the first, with prevalence of explorative dissociative growth characterized by scarce branching, is called "guerrilla", and the second, with prevalence of exploitative associative growth characterized by dense branching and hyphal aggregation, is called phalanx (Boddy, 1993; Carlile, 1995; Rayner, Griffith, & Ainsworth, 1995; Andrews, 2017).

3.2.1 Explorative strategies

In the terrestrial environment, both metal-polluted soils and mineral surfaces are usually characterized by a spatially diverse distribution of metal concentrations and nutritional resources. Most studies on filamentous fungi traditionally use simple microcosms with a homogeneous metal distribution, such as Petri dishes containing a certain concentration of toxic metal in an agar medium. However, an experimental microcosm based on tessellated agar tiles can simulate the natural heterogeneity of the terrestrial environment by using agar tiles with different concentrations of metal toxicants and nutrients. Such a microcosm allows the capture of the starting point of fungal adaptation *in situ* and reveals the spatial development of morphological reactions of the mycelial system in response to metal and mineral stressors (Fomina, Ritz, & Gadd, 2000; Fomina, Ritz, & Gadd, 2003; Fomina et al., 2005a).

As mentioned above, one of the two extremes of modular organism growth is the guerrilla or explorative pattern of mycelial system development (Fig. 2). In our experiments using a tessellated agar tile system,

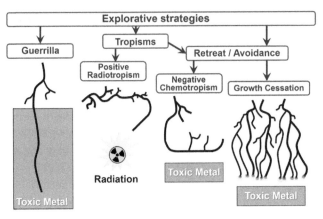

Fig. 2 Explorative strategies of filamentous fungi in mineral-rich and toxic metal- and radionuclide-contaminated environments.

filamentous fungi employed this strategy while attempting to colonize spatially discrete metal-containing domains represented by an agar tile (Fomina et al., 2000, 2003, 2005a). Under poor nutritional conditions, fungi approaching the toxic metal-containing areas often produced long, sparsely branched, or branchless hyphae, which are characteristic of the guerrilla growth strategy (Fomina et al., 2003) (Figs. 2 and 3A). In addition, we observed such guerrilla hyphae in filamentous fungi colonizing the surface of the uranyl-oxide hydroxy-hydrate mineral schoepite in soil previously contaminated with metallic depleted uranium, as well as on calcite-containing buon frescoes in a medieval cathedral (Fomina et al., 2008, 2022) (Fig. 3B and C).

This strategy enables the exploration of hostile environments, either by finding adjacent domains with conditions suitable for existence or by escaping/retreating from areas that are too stressful for the fungus to overcome. In addition to the guerrilla growth pattern, explorative strategies include growth cessation and tropism reactions related to directional hyphal growth. Both growth cessation and negative chemotropism represent a retreating strategy, avoiding the areas severely contaminated with toxic metals. This was well observed in the gaps between control and metal-laden agar tiles using the tessellated-agar-tile approach (Fomina et al., 2000, 2003) (Figs. 2 and 3D–H). The manifestation of fungal negative chemotropism away from metal-laden domains depended on metal and nutrient concentrations in this domain and was enhanced with increasing toxic metal concentration and reduced with increasing sugar concentration (Fomina et al., 2000).

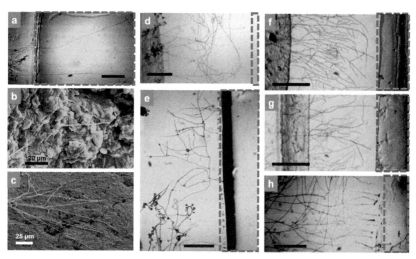

Fig. 3 Examples of explorative strategies of filamentous fungi in response to metal- and mineral-rich environments. Morphological responses occurring (A, D–H) in the experimental system with tessellated agar tiles either (A) within a metal-laden tile or (D–H) in the gap between metal-free and metal-containing tiles (light microscopy images where metal-polluted tiles are indicated by dashed-line boxes on the right-hand side); and (B,C) on the mineral surfaces (scanning electron microscopy images). Long sparsely branched or branchless guerrilla hyphae (A) of *Clonostachys rosea* on a tile containing 2 mM Cu with 15 gL^{-1} sucrose, and of unidentified filamentous fungi (C) on the U-containing mineral schoepite in soil, and (D) on a calcite-rich wall of a medieval cathedral. Negative chemotropic responses to toxic metals by *Trichoderma virens* growing toward the tile with (D) 2 mM Cu and 1 gL^{-1} sucrose and (E) 1 mM Cu and 1 gL^{-1} sucrose; and (F, G) by hyphae of *Geotrichum candidum* growing towards the tiles containing 2 mM Cu and 30 gL^{-1} sucrose. (H) Cessation of growth, swelling, and lysis of hyphal tips of *T. virens* growing toward the tile containing 2 mM Cu and 1 gL^{-1} sucrose. Bar marker for the light microscopy images is 0.5 mm. *Images (A, D–H) were adapted from Fomina et al. (2000, 2003). Image (B): Fomina and Gadd, unpublished, related to Fomina, Hong, and Gadd (2020). Image (C): Fomina, unpublished, related to Fomina et al. (2022).*

Other tropism reactions contribute to fungal exploration of mineral surfaces, as well as environments contaminated with toxic metals and radionuclides. These reactions include positive radiotropism and thigmotropism. Positive radiotropism is defined as the directed growth of fungal tips towards a source of radiation (e.g., a low gamma–irradiation source), which was observed in many fungal isolates from the region of the Chernobyl nuclear power plant after the accident in 1986 (Zhdanova, Fomina, Redchitz, & Olsson, 2001; Zhdanova, Tugay, Dighton, Zheltonozhsky, &

Mcdermott, 2004) (Fig. 2). Particularly relevant for the colonization of solid surfaces, such as rocks and minerals, thigmotropism is a common ability in filamentous fungi for contact guidance, directing hyphal growth along grooves, ridges, pores, etc. (Gadd et al., 2024). Both positive radiotropism and thigmotropism enable fungi to proceed with explorative growth in hostile environments.

Those filamentous fungi that employ explorative strategies and the guerrilla growth pattern with often very sparse mycelium can ameliorate the hostile microenvironment through their biogeochemical activity. The application of an experimental system with tessellated agar tiles allowed visualization of the amorphous precipitation of toxic metals in the microenvironment surrounding guerrilla hyphae colonizing the toxic metal domain. For example, both saprotrophic fungi (such as *Aspergillus flavipes*, *Cladosporium cladosporioides*, and *Trichoderma virens*), and ecto- and ericoid mycorrhizal fungi (*Rhizopogon rubescens* and *Hymenoscyphus ericae*) were able to produce droplet-like precipitates of Cd, Cu, and Sr in tiles laden with soluble metal species (Fomina & Gadd, 2007). Elemental mapping conducted by energy dispersive X-ray (EDX) microanalysis coupled with scanning electron microscopy (SEM) revealed the concentration of toxic metals and their chemical immobilization within these droplets (Fig. 4). For instance, precipitated copper was found to be associated with phosphorus within such droplets, suggesting coordination of copper by phosphate ligands.

3.2.2 Exploitative and aggregative strategies

Another extreme case of modular unit arrangements in filamentous fungi is phalanx growth which represents exploitative and aggregative strategies (Rayner et al., 1995; Fomina et al., 2005a; Andrews, 2017) (Fig. 5).

Various manifestations of mycelial aggregation often occur in response to metal- and mineral-related stress in filamentous fungi. In a broad sense, any hyphal aggregation could be considered an example of phalanx growth. On solid surfaces, the growth of microorganisms, including fungi, primarily occurs in the form of biofilms of varying complexity, representing an aggregated mode of microbial cell existence (Pinna, 2021; Gadd et al., 2024). This strategy universally provides survival advantages for aggregated microbial cells, including protection from stressful environmental factors. These advantages are particularly pertinent to phalanx growth of filamentous fungi in mineral- and metal-rich environments.

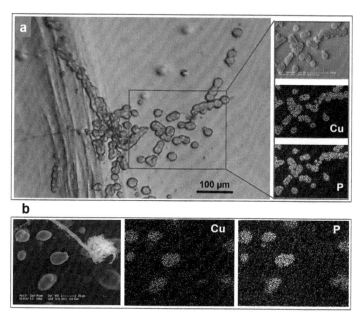

Fig. 4 Example of biogeochemical activity of explorative guerrilla hyphae colonizing metal-laden agar tiles and mediating the extracellular precipitation of amorphous copper-containing precipitates, Droplet-like Cu precipitates formed by thin sparsely-branched or branchless hyphae of (A) the ectomycorrhizal fungus *Rhizopogon rubescens*, and (B) the saprotrophic fungus *Aspergillus flavipes*. (A) Light microscopy image and (A,B) SEM images with EDX elemental mapping for copper and phosphorus are presented. *Adapted from Fomina, M., & Gadd, G. M. (2007). Metal and mineral transformations: A mycoremediation perspective. In G.D. Robson, P. Van West & G.M. Gadd (Eds.), Exploitation of fungi (pp. 236–254). Cambridge: Cambridge University Press. https://doi.org/10.1017/CBO9780511902451.014.*

The phalanx growth mode, characterized by the formation of highly dense mycelium, was reported for *Trichoderma viride* and *Rhizopus arrhizus* exposed to Cu and Cd (Gadd, Ramsay, Crawford, & Ritz, 2001). In these experiments, which used a large-scale mycelial mapping technique, the distribution of dense phalanx mycelium in *Trichoderma viride* colonies depended on the toxic metal cadmium, which caused a shift to phalanx growth in the central area of colony, whereas copper exposure resulted in denser mycelium at the colony periphery. Cadmium was also observed to cause an increase in the frequency of hyphal branching in *Schizophyllum commune, Daedalea quercina,* and *Paxillus involutus* (Ramsay, Sayer, & Gadd, 1999). Other studies on the interactions of a range of soil fungi with precious metals revealed that the development of the phalanx mode of

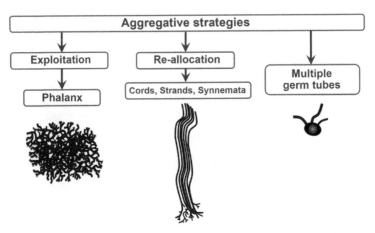

Fig. 5 Aggregative growth strategies of filamentous fungi in metal- and mineral-rich environments: exploitative/consolidative dense growth ("phalanx"), linearly aggregated hyphal structures (cords, strands, and synnemata) representing "re-allocation" within the mycelial system, and the formation of multiple germ tubes on spore germination.

growth increased with rising gold concentrations up to 150 mgL^{-1} for colonies of *Aspergillus niger* and *Fusarium solani*, and silver concentrations up to 300 mgL^{-1} for *Hypocrea lixii*, *Fusarium oxysporum*, and *F. solani* colonies (Argumedo-Delira, Gómez-Martínez, & Uribe-Kaffure, 2020).

Experiments using a tessellated agar tile system enabled observation of the genesis of the phalanx growth mode, i.e., the formation of very dense mycelium or mycelial "bushes" (also termed "point growth") during colonization of metal-contaminated domains (Fomina et al., 2003) (Fig. 6A–D). Tessellated agar tile microcosms also revealed the phenomenon of a mycelial system shift to aggregative growth of longitudinally aligned hyphae (the "re-allocation" strategy), facilitating fungal colonization of toxic metal-contaminated domains, in fungal species which typically do not produce cords, strands, rhizomorphs, or synnemata under normal conditions (Figs. 5 and 6A–D). Synnemata formation in response to Mn or tributyltin compounds has been observed in *Penicillium* spp. (Tinnell, Jefferson, & Benoit, 1977; Newby & Gadd, 1987). The presence of calcium or strontium induced rhizomorph and synnemata development in *Sphaerostilbe repens* (Botton, 1978). The formation of synnemata in response to metal- and mineral-related stress results in a broader spatial distribution between the conidia and the substrate compared to a non-synnematal mycelium. This re-allocation strategy facilitates more extensive

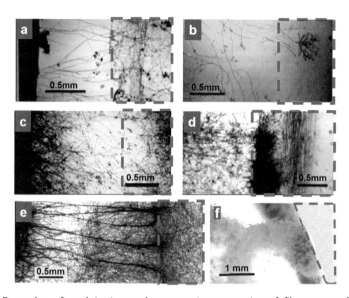

Fig. 6 Examples of exploitative and aggregative strategies of filamentous fungi in response to metal- and mineral-rich environments observed in a tessellated agar tiles system (light microscopy images where metal-containing tiles are indicated by dashed-line boxes on the right-hand side): (A, B) *Trichoderma virens*, (C, D) *Clonostachys rosea*; (A, C) control growth on the metal-free inoculum agar tiles, (B, D) formation of dense mycelial bushes representing phalanx growth on metal-containing agar tiles containing (B) 2 mM Cd and 1 gL^{-1} sucrose and (D) 1 mM Cd and 5 gL^{-1} sucrose. (E, F) Formation of aggregated and longitudinally aligned hyphae in response to toxic metal stress in the gap between the metal-free inoculum tile and metal-containing tile: (E) *Cladosporium cladosporioides* melanized strands colonizing a tile containing 1 mM Sr and 5 gL^{-1} sucrose; (F) Large synnema (= coremium) of *Aspergillus flavipes*, covered with conidiophores, colonizing a tile with 1 mM Cu and 30 gL^{-1} sucrose. *Images (A–D) adapted from Fomina, M., Ritz, K., & Gadd, G. M. (2003). Nutritional influence on the ability of fungal mycelia to penetrate toxic metal-containing domains. Mycological Research, 107, 861–871. https://doi.org/10.1017/s095375620300786x. Images (E, F) adapted from Fomina, M., Burford, E. P., & Gadd, G. M. (2005a). Toxic metals and fungal communities. In J. Dighton, J. White & Oudemans P. (Eds.), The fungal community: Its organization and role in the ecosystem (pp. 733–758). Boca Raton: CRC Press, Taylor & Francis Group.*

and secure fungal dispersal including conidial development away from hostile, metal-polluted environments, while also benefiting from the translocation of necessary resources from the parental colony and from the protective role resulting from hyphal aggregation (Fomina et al., 2005a).

The protective role of hyphal aggregation also includes the production of high local concentrations of extracellular enzymes, antibiotics, and other metabolites necessary for colonization of new domains. In mineral- and metal-rich environments, aggregated filamentous fungi can generate high local concentrations of many biogeochemically active extracellular products, including: (1) chelating and sequestering agents such as low molecular weight organic acids, siderophores, polyphenolic compounds; (2) metal-precipitating agents such as oxalate; and (3) polysaccharides and pigments with metal-binding abilities (Gadd, 1993, 2017; Baldrian, 2003; Gadd et al., 2014). For aggregated fungal growth, the production of extracellular polymeric substances (EPS) facilitates and enhances the protective effect of aggregation, serving as a means of hyphal aggregation and as a matrix for biogeochemical reactions involving the mobilization and immobilization of metals, and the dissolution and re-precipitation of minerals (Fomina et al., 2005c; Mittelmann, 2018).

Studies with the geoactive fungus *Beauveria caledonica*, which overproduces oxalic acid, observed that toxic metal minerals induced phalanx growth in the marginal zones of the colonies (Fomina et al., 2005c) (Fig. 7A–C). When exposed to copper phosphate, uranium oxides, or metallic depleted uranium, *B. caledonica* formed atypical mycelial cords encrusted with numerous secondary mycogenic toxic metal minerals, embedded in a thick hydrated mucilaginous EPS sheath (Fomina et al., 2005c; Fomina, Charnock, Hillier, Alvarez, & Gadd, 2007b; Fomina et al., 2008) (Fig. 7D–G).

Multiple germ tube formation during the germination of fungal spores can also be considered a fungal aggregative strategy (Fig. 5). Generally, conidial germination in filamentous fungi has not been sufficiently explored, particularly regarding the effects of various stressors on the process (Danion et al., 2021). The number of germ tubes that a conidium forms can depend on the fungal species and can vary even for the same strain. For example, a study of the germination process of the clinically important soil fungus *Aspergillus fumigatus* demonstrated that many conidia produced two or three germ tubes per conidium (51% and 44%, respectively). Single germ tubes were produced by 4% of conidia, while four germ tubes were produced by only 2% of conidia (Danion et al., 2021). Our study on the growth of *Cladosporium cladosporioides* strains, isolated from unpolluted soil in the late 1950s and from radioactive soil from an industrial site at the Chernobyl nuclear power plant after the accidents in 1986 and 1992, found that exposure to a fixed collimated source of gamma-irradiation ([109]Cd) resulted in a significant increase in the occurrence of double-germ-tubed conidia (~2.5 times) compared to the non-irradiated control where over 80% of conidia produced only single germ tube (Zhdanova et al., 2001). This radiostimulation occurred regardless of

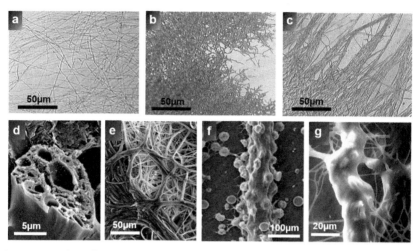

Fig. 7 Examples of exploitative and aggregative strategies of the geoactive fungus *Beauveria caledonica* in response to metal- and mineral-rich environments: (A–C) light microscopy images of (A) loose control mycelium on mineral-free agar, and (B, C) phalanx growth in marginal zones of the colonies induced by the presence of (B) lead phosphate ($Pb_3(PO_4)_2$) and (C) cuprite (Cu_2O); (D–G) SEM images obtained in (D–F) high-vacuum-mode and (G) ESEM wet-mode, showing cord-like hyphal aggregates formed by *B. caledonica* in response to (D) metallic depleted uranium (DU) and (E) triuranium octoxide (U_3O_8), and (F, G) copper phosphate. Images (D, F, G) show the biogeochemical transformations by *B. caledonica*: (D) cross-fractured cord revealing numerous crystals of uranyl phosphate minerals within the EPS surrounding the cord hyphae, (F) a hyphal cord encrusted by copper oxalate hydrate (moolooite) crystals and (G) a similar sample observed in the hydrated state revealing a thick mucilaginous sheath covering the mycelium and crystals. *Images (A–C, F, G) adapted from Fomina et al. (2005c). Images (D, E): Fomina & Gadd, unpublished, related to (D) Fomina et al. (2008) and (E) Fomina et al. (2007b).*

the source of isolation or the ability to exhibit positive radiotropism inherent in the Chernobyl strains. The increasing number of germ tubes per conidium was clearly associated with the phalanx strategy at a very early stage of fungal development, enhancing the chances of survival and successful colonization and assimilation of resources.

3.2.3 Exploitation-exploration transition

Mycelial systems can adjust the balance between assimilation and exploration, resource conservation, and redistribution, enabling effective adaptation to changes in local conditions (Rayner et al., 1995). To cope with the challenges posed by toxic metals and limited nutrients, filamentous fungi exhibit remarkable flexibility, undergoing multiple phase shifts between phalanx and guerrilla growth states. This adaptability was clearly demonstrated in the tessellated agar tile system, where fungi, including *Trichoderma virens* and *Cladosporium cladosporioides,* exhibited repeated phalanx/guerrilla/phalanx phase shifts while colonizing a domain of poor nutritional status and high metal content (Fomina et al., 2003, 2005a) (Fig. 8). When entering the toxic metal-rich domains, the fungi initially developed border mycelial "bushes" characterized by slow but dense growth, then extending long, branchless exploring hyphae, representing fast/effuse growth, and, finally, reversion to forming "bushes" at the tips of these explorative hyphae. A given mycelial system can employ guerrilla or phalanx strategies as needed, although inherent traits of the species may impose constraints on such adaptations. However, the concentration of the carbon source decreased, and the concentration of toxic metal increased, the growth responses of the different fungal species (e.g., *Trichoderma virens* and *Clonostachys rosea*) became more alike and the observed morphological features became more uniform (Fomina et al., 2003).

The capability of filamentous fungi to employ the phalanx and guerrilla extremes of modular growth, shifting between assimilative and non-assimilative physiological states, allows the mycelial system to explore physico-chemically hostile regions and, in certain situations, to ameliorate conditions for subsequent exploitative phases (Rayner et al., 1995).

3.2.4 Recycling, specialization, and other strategies

While colonizing hostile metal- and mineral-rich environments, filamentous fungi can also employ morphological adaptations related to recycling and specialization, as well as some other growth patterns.

The strategy of recycling in fungi involves the reuse of cellular resources, in particular the cell wall, leading to intrahyphal growth. Intrahyphal hyphae can be defined as hyphae growing inside a hypha. It has been suggested that the formation of the cell wall of intrahyphal hyphae is initiated from the cell wall of the enclosing hyphae (Kim, Hyun, & Park, 2004). In a study of this phenomenon in *Aspergillus nidulans,* the formation of intrahyphal hyphae resulted from disruption of the CsmA gene encoding a class V chitin synthase with a myosin motor-like domain and playing important roles in polarized cell wall synthesis and maintenance of cell wall integrity (Horiuchi, Fujiwara, Yamashita, Ohta, & Takagi, 1999). The phenomenon of intrahyphal growth is considered to be widespread in fungi. It has been observed in plant pathogenic, arbuscular mycorrhizal, and microcolonial rock-dwelling fungi and is thought to provide shelter

Phalanx-Guerrilla Transitions

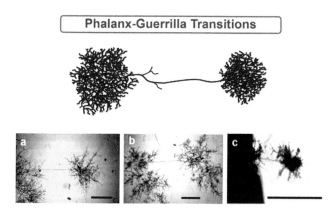

Fig. 8 Examples of exploitation-exploration transitions observed in a tessellated agar tile system (light microscopy images): repeated phalanx-guerrilla "phase shifts" during colonization of domains containing 1 mM Cu and 1 gl^{-1} sucrose by (A, B) *Trichoderma virens* and (C) *Cladosporium cladosporioides*. Scale bar = 0.5 mm. *Images adapted from Fomina, M., Ritz, K., & Gadd, G. M. (2003). Nutritional influence on the ability of fungal mycelia to penetrate toxic metal-containing domains. Mycological Research, 107, 861–871. https://doi.org/10.1017/s095375620300786x and Fomina, M., Burford, E. P., & Gadd, G. M. (2005a). Toxic metals and fungal communities. In J. Dighton, J. White & Oudemans P. (Eds.), The fungal community: Its organization and role in the ecosystem (pp. 733–758). Boca Raton: CRC Press, Taylor & Francis Group.*

under limiting growth conditions and protection from extreme environments (Lim, Fineran, & Cole, 1983; Horiuchi et al., 1999; Gorbushina et al., 2003; Szentivanyi & Kiss, 2003; Kim et al., 2004; Suzuki, et al., 2017). In our research, we observed intrahyphal hyphal growth of the mycobiont *Rhizopogon rubescens* in ectomycorrhizal association with pine grown in zinc-containing mesocosms (Fig. 9).

Intrahyphal hyphae can occupy most of the lumen of the enclosing hyphae and grow out of the degenerated hyphae that are almost devoid of cellular content (simple intrahyphal hyphae). In addition, enclosing hyphae can contain multiple intrahyphal hyphae (Kim et al., 2004). We also observed the formation of both single and multiple intrahyphal hyphae in the ectomycorrhizal fungus *Rhizopogon rubescens* in response to zinc phosphate (Fig. 9).

The biological role of intrahyphal hyphae has been suggested to enable fungal survival under such hostile conditions that can result in hyphal degeneration (Lim et al., 1983; Kim et al., 2004). This involves the accumulation of hyphal cell wall components, particularly chitin, to create multiple-walled entities within an enclosing hypha, thus providing a physical barrier and

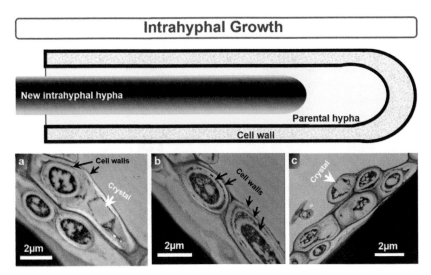

Fig. 9 Formation of intrahyphal hyphae as a recycling growth strategy of filamentous fungal growth. (A–C) TEM images of cross-sections of intrahyphal hyphae in the extraradical mycelium of the ectomycorrhizal mycobiont in a *Rhizopogon rubescens/* pine system grown on Ingestad agar containing 5 mM $Zn_3(PO_4)_2$. Black arrows indicate double and triple cell walls, and white arrows show tetragonal zinc-containing crystals precipitated within the extramatrical mycelium (Fomina, Charnock, Bowen, & Gadd, 2007c). *Images by Fomina and Gadd, unpublished, related to Fomina, M., Charnock, J., Bowen, A. D., & Gadd, G. M. (2007c). X-ray absorption spectroscopy (XAS) of toxic metal mineral transformations by fungi. Environmental Microbiology, 9(2), 308–321. https://doi.org/10.1111/j.1462-2920.2006.01139.x.*

protective layer. In our study, the lumen of the enclosing hyphae contained precipitated large tetragonal zinc-containing crystals.

A similar protective role, which can lead to ameliorating a hostile environment, is a characteristic of the specialization growth strategy of filamentous fungi in response to toxic metal stress. Specialization can be defined as the partitioning of tasks within a mycelial system, facilitating the survival and colonization of a metal-polluted domain. We observed that certain segments of the fungal mycelium showed enhanced biomineralization of toxic metals, thereby ameliorating the hostile environment for other parts of the colony to thrive and colonize detoxified regions (Fomina et al., 2007b, 2008). This phenomenon was most evident for fungal growth in the presence of uranium oxides or metallic depleted uranium (Fig. 10).

Another adaptation of filamentous fungi to a mineral-rich environment combined with very low humidity is the formation of extracellular hair-like structures covering the hyphae. This is a characteristic of the remarkable

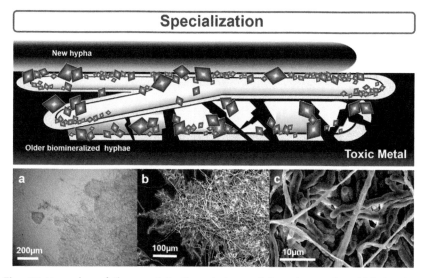

Fig. 10 Examples of the specialization strategy of fungal growth, which involves spatial separation of specific parts of the mycelial system, exhibiting different extents of metal immobilization and biomineralization. (A–C) Introduction of metallic depleted uranium in an agar plate microcosm, which resulted in heavy biomineralization of uranium on hyphae located at the bottom part of the fungal colony of *B. caledonica*: (A) light microscopy image of the marginal zone of the colony; (B, C) Cryo-SEM images of biomineralized mycelium at the colony bottom and thin non-biomineralized new hyphae on the top. *Images by Fomina and Gadd, unpublished, related to Fomina, M., Charnock, J. M., Hillier, S., Alvarez, R., Livens, F., & Gadd, G. M. (2008). Role of fungi in the biogeochemical fate of depleted uranium. Current Biology, 18(9), R375–R377.*

extremophilic fungal species *Aspergillus halophilicus* (teleomorph *Eurotium halophilicum*) (Micheluz, Pinzari, Rivera-Valentín, Manente, & Hallsworth, 2022; Gadd et al., 2024). This fungus is one of the most extreme xerophiles in the biosphere, and is capable of growth at water activity down to 0.651, and at very high NaCl concentrations. This species is a common agent of biodeterioration of many heritage materials, including mineral calcite-rich frescoes (Fomina et al., 2022) (Fig. 11). It is suggested that the hair-like microfilaments covering *A. halophilicus* hyphae play an important role as a biogeochemically active epihyphal matrix in providing a suitable water activity for fungal growth (Micheluz et al., 2022).

In addition to common filamentous fungi dwelling on the surfaces of minerals and rocks, a specific group of microcolonial fungi grow as spherical clusters of tightly packed thick-dark pigmented-walled cells and as moniliform thick-walled hyphae (Egidi et al., 2014). This growth pattern can be

Hair-like Extracellular Structures

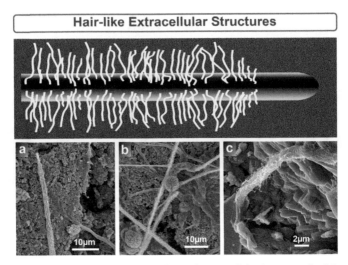

Fig. 11 Hair-like structures covering hyphae of a xerophilic fungus colonizing calcite-based mineral surfaces: (A–C) SEM images of *Aspergillus halophilicus* (= *Eurotium halophilicum*) on painted walls in the medieval St. Sophia cathedral, Kyiv, Ukraine. *Images by M. Fomina, unpublished, related to Fomina, M., Cuadros, J., Pinzari, F., Hryshchenko, N., Najorka, J., Gavrilenko, M., ... Gadd, G. M. (2022). Fungal transformation of mineral substrata of biodeteriorated medieval murals in Saint Sophia's cathedral, Kyiv, Ukraine. International Biodeterioration and Biodegradation, 175, 105486. https://doi.org/10.1016/j.ibiod.2022.105486 and Gadd, G. M., Fomina, M., & Pinzari, F. (2024). Fungal biodeterioration and preservation of cultural heritage, artwork, and historical artifacts: Extremophily and adaptation. Microbiology and Molecular Biology Reviews, e0020022. https://doi.org/10.1128/mmbr.00200-22.*

regarded as another manifestation of fungal morphological flexibility in colonizing mineral- and metal-rich domains in the terrestrial environment.

In conclusion, all observations of the diverse morphological responses of filamentous fungi under surface growth conditions emphasize the remarkable flexibility of mycelial systems that ensures successful adaptation to stressful metal- and mineral-rich environments.

4. Morphological responses to stress in filamentous fungi grown under submerged conditions

4.1 The importance of mycelial morphology in submerged cultivation of filamentous fungi

Due to the outstanding biochemical potential of fungi for the industrial production of various commercially important metabolites and biomass, filamentous fungi have long been cultivated in bioreactors under submerged fermentation conditions (Cairns et al., 2019). Filamentous fungi demonstrate remarkable adaptability to artificial environments, adjusting to different fermentation conditions at both physiological and morphological levels. Despite the significant importance of fungal biotechnology, our fundamental understanding of the role of mycelial growth morphology in bioreactors and its impact on fermentation processes and productivity remains incomplete. The growing attention to the mechanisms of mycelium formation in submerged culture is linked with cultivation conditions and their consequences at the biochemical level for biosynthetic processes (Futagami et al., 2011; Veiter, Rajamanickam, & Herwig, 2018; Cairns et al., 2019; Yoshimi, Miyazawa, Kawauchi, & Abe, 2022). Hence, the majority of research focuses on biotechnological aspects of producing biologically-active products, where mycelial macromorphology is considered to be a determining parameter of the process. For mycelial morphology, the two modern approaches for enhancing fungal-based synthesis of biologically-active compounds are the optimization of cultivation conditions (Veiter, Kager, & Herwig, 2020) and the application of synthetic biology (Maruyama, 2021). Several genetic studies have been conducted to manipulate the morphology of submerged mycelium to optimize process productivity (Cai et al., 2014). Proteomic analysis indicated the role of low molecular weight proteins in the formation of pelleted growth of filamentous fungi (Kumar, Rajput, & Dubey, 2023). Genomic-wide metabolic models (GWMM) have been developed for various fungi to predict phenotypes with maximum product yield by associating them with certain genes (Cairns et al., 2019). The creation of a genetic library containing information on thousands of mutant producer strains was also considered promising (Brandl et al., 2018; Meyer et al., 2021).

Through modern research methods, it has become possible to study the structural features of fungal macromorphology. For example, to describe pelleted growth, an analysis of their fractal image has been used (Panankar, Liu, & Oolman, 1993; Papagianni & Mattey, 2006). X-ray micro-computed tomography (μCT) has also been employed for the non-destructive 3-dimensional reconstruction of fungal pellets (Schmideder et al., 2019). Computer analysis of stereomicroscopic images enabled the rapid acquisition of information on the spore inoculum as it developed into mycelial structures (Müller et al., 2022, 2023). A modified flow cytometry method has been used to study mycelial growth (Ehgartner, Herwig, & Fricke, 2017; Veiter & Herwig, 2019; Schrinner et al., 2020; Veiter, Kager, &

Herwig, 2020) as well as the application of neural networks and cluster analysis to characterize mycelial morphology (Papagianni & Mattey, 2006). Mathematical modeling has also been applied to study processes taking place inside the hyphae (King, 2015). An integrated approach to comprehensively study fungal producers requires the development of adequate tools for controlling fungal biotechnological processes in a field known as metabolic engineering (Meyer, 2021).

However, a unified understanding of the connection between mycelial morphology, metabolism, and productivity for different species of fungi and a general technology for controlling mycelial morphology has not yet been developed (Grimm, Kelly, Krull, & Hempel, 2005; Dinius, Kozanecka, Hoffmann, Kevin, & Rainer, 2023). The choice of morphological form for a given biotechnological process is usually made empirically. Although the important roles of pH and inoculum concentration have been demonstrated in pelleted form development and determining methods to control the reproducibility of such macromorphology, the molecular mechanisms underlying this process still need to be elucidated (Kheirkhah, Neubauer, & Junne, 2023; Zhang, Liu, Zhang, Chen, & Wang, 2023). One new direction in the use of pelleted mycelium is the development of bioengineered living materials (ELMs) (Li et al., 2023). The application of mycelial pellets as a carrier in wastewater treatment has been described (Zhang, Wang, Wang, & Wang, 2024). Clearly, the regulation of fungal macromorphology in submerged culture is relevant to both fundamental and applied science. The multifactorial and, therefore, difficult-to-study nature of submerged mycelium makes such knowledge essential for the effective use of fungi in biotechnological processes (Žnidaršič-Plazl, & Pavko, 2001).

Signal transmission represents a complex integrated system of numerous cellular processes, which have been insufficiently explored in filamentous microscopic fungi at the physiological, cellular, and molecular levels (Gadd, 1995; Cairns et al., 2019). Despite the hundred-year history of studying hyphal growth, our knowledge about this issue remains superficial (Gow, 1994; Meyer et al., 2016). In particular, the nature of the formation of vegetative mycelium, as a self-organizing process, in connection with the artificial method of fungal cultivation under submerged conditions must still be elucidated. It is also necessary to consider the lability of the system and the multifaceted nature of ontogenesis with significant dependence of development on environmental conditions. Thus, the morphology of mycelium under submerged growth conditions can be considered as an adaptation to a complex of cultivation factors.

4.2 Macromorphology as a form of adaptation to stressful environmental conditions

Mycelial macromorphology in submerged culture is diverse and strain dependent. However, filamentous fungi have two forms of submerged growth: pelleted and dispersed growth (Garcia-Reyes et al., 2017) (Fig. 12). Most works record the morphological features of the mycelium from a practical point of view when obtaining the target product (Sun et al., 2018; Maumela, Rose, van Rensburg, Chimphango, & Görgens, 2021). At the same time, the macromorphology of the submerged mycelium could reflect adaptation to the cultivation conditions. Considering mycelial development in submerged culture in terms of energy balance as a basis for the adaptation process, a spherical pellet shape can be assumed to be the most optimal under stressful conditions. Filamentous fungi grow in the form of spheres because these have a minimal surface area limiting the volume that minimizes possible adverse effects on cells. Experimental evidence of this was found in studies on the formation of pellets by the thermophilic fungus *Thielavia terrestris* (optimal T = 37–42 °C, pH 4.0) and the mesophile *Chaetomium globosum* (optimal T = 25–28 °C, pH 7.0) (Shemshur, Gromozova, & Podgorsky, 1989). The provision of optimal cultivation conditions for each fungus promoted the dispersed growth form, while stresses, e.g., temperature, pH, and inoculum concentration, resulted in growth in the form of pellets. Nair, Gmoser, Lennartsson, and Taherzadeh (2018) also suggested that pellet growth may be a response to stressful conditions.

In the morphogenesis of submerged mycelium under the simultaneous action of several morphogenic factors, the defining role in the process was played by unfavorable factor(s) operating on the "bottleneck" principle (Shemshur et al., 1989). To explain this phenomenon, it was necessary to determine the stage(s) of fungal development sensitive to the morphogenic signals.

4.3 The role of the lag phase in the morphogenesis of pelleted and dispersed growth forms

It was demonstrated that the main stage determining the path of fungal morphogenesis was the lag phase, corresponding to the stage of conidia swelling. Only this stage was clearly sensitive to morphogenic signals, resulting in the regulation of either pelleted or dispersed growth in *T. terrestris* (Gromozova et al., 1989). Changing cultivation conditions during later stages of development, such as germination (germ tube emergence) and initial branching, did not lead to morphological restructuring, and instead, the growth program selected earlier in the lag phase was implemented

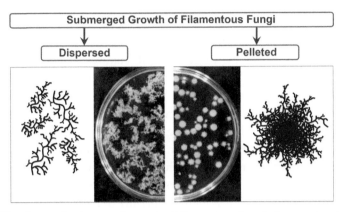

Fig. 12 Two forms of submerged growth of filamentous fungi: dispersed and pelleted growth. Image inserts show examples of dispersed and pelleted forms of *Thielavia terrestris* harvested after submerged cultivation and displayed in the standard 90-mm Petri dishes. *Images by Gromozova and Fomina unpublished, related to Gromozova, E. N., Shemshur, T. V., & Fomina, M. A. (1989). Formation of mycelial structures under the influence of cultivation conditions at different stages of micromycete development. Mykologiya i Fitopatologiya, 23(3), 202–205. (in Russian).*

(Gromozova et al., 1989) (Fig. 13). During the lag phase or conidial swelling, the organism is known to utilize endogenous substrates and be less dependent on nutrients from the environment (Papagianni, Mattey, & Kristiansen, 1999). This was experimentally confirmed in tests on the germination of conidia of *T. terrestris* in tap water (Gromozova & Blazchuk, 1989). The same conclusion was also reached by other authors studying the asynchrony of germination in *Aspergillus fumigatus* conidia (Danion et al., 2021). The asynchrony of conidial germination can explain the formation of mixed dispersed-pelleted growth in response to exposure to stressful conditions at the stage of germ tube emergence in *T. terrestris* (Fig. 13).

A change in the concentration of carbon-, nitrogen- and phosphorus-containing substrates within physiological values, as well as dissolved oxygen, did not affect the choice of morphological status in *T. terrestris* (Gromozova & Blazchuk, 1989). However, an increase in glucose concentration to over 30 g/L in the lag phase resulted in a switch from pH-programmed pelleted growth to dispersed growth. Program switching was observed only when high glucose concentrations caused the pelleted form to turn into a dispersed form, not vice versa. High concentrations of glucose cause a rapid increase in cAMP and activation of protein phosphorylation in yeast (Mbonyi, van Aelst, Argüelles, Jans, & Thevelein,

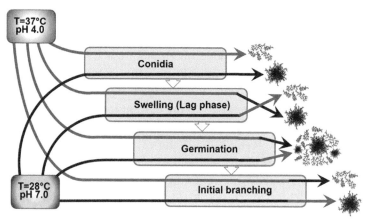

Fig. 13 The key role of the lag phase (or conidial swelling stage) in programming morphogenesis for pelleted or dispersed growth of the thermophilic fungus *Thielavia terrestris* in submerged culture. The environmental conditions programming the development of certain growth form, if applied initially to conidia, were a combination of temperature and pH values: (i) optimal for *T. terrestris* growth (37 °C and pH 4.0) leading to dispersed growth (shown in the figure as red or black [for grayscale] arrows), and (ii) more stressful for this fungus (28 °C and pH 7.0) resulting in formation of pellets (shown in figure as blue or gray [for grayscale] arrows). If these conditions were switched at the stage of conidial swelling (lag phase), the switch from 28 °C/pH 7.0 to 37 °C/pH 4.0 led to dispersed growth, and the change from 37 °C/pH 4.0 to 28 °C/pH 7.0 resulted in the formation of pellets. If exposure to opposing conditions occurred at the germination stage, after germ tube emergence, it led to the formation of mixed dispersed-pelleted growth for both combinations of conditions. Switching conditions at the initial branching stage did not lead to any change in morphogenesis, and the dispersed form developed under initial optimal conditions, whereas pellets were formed under more stressful conditions. *Image adapted from Gromozova, E. N. (2016). The regulation of the progresses of form creation of micromycetes in submerged culture. In N. Zaimenko,G. Grodzynska, T. Bugatnko & S. Syrchin (Eds.), Botany and mycology: Modern horizons (pp. 408–425). Kyiv: Nash Format.*

1990; Rolland, Winderickx, & Thevelein, 2002). The role of cAMP in the morphogenesis of submerged-growing mycelium of *T. terrestris* will be discussed later.

The lack of predetermined development of incubated conidia into dispersed or pelleted forms, i.e., conidia are initially totipotent, was experimentally demonstrated (Gromozova, 1998). Additionally, the prior generation of the mycelial inoculum is important in determining the mycelial form (Gromozova & Sadovskii, 2006). The use of fragments of pelleted or dispersed mycelia as inoculating material affects the macromorphology of the next generation. The transition point from one form to another shifts (for example, when the pH changes) along with the boundaries of the form's distribution. The phenomenon of hysteresis, or the dependence of a system's state on its past, was confirmed by experimental data and mathematical modeling of the transition between these adaptive fungal growth forms (Gromozova & Sadovskii,1992). This information is of special interest given the current trend of replacing the use of spore inoculum with the more beneficial approach of pellet dispersion (Liu, Hu, Zhao, Lv, & Ren, 2017; Wang et al., 2017).

The participation of polysaccharide bridges in the aggregation of conidia has been suggested (Gerin, Dufrene, Bellon-Fontaine, Asther, & Rouxhet, 1993; Zhang & Zhang, 2016). This was confirmed in studies on pelleted and dispersed forms of *T. terrestris*, based on our mathematical models, as well as atomic force microscopy (AFM) and transmission electron microscopy (TEM) coupled with EDX using gold-labeled concanavalin A, where the swelling conidia were found to be surrounded by polysaccharides containing D-glucose, D-mannose, and N-acetyl-D glucosamine (Gromozova, 2003, 2016).

An AFM study of the ultrastructure and forces of molecular interactions on the surface of *T. terrestris* conidia demonstrated that changes in these indicators occurred not only due to swelling of the conidia in the lag phase but also depended on the direction of the dimorphic morphogenic processes. During this period, the adhesive and viscoelastic properties of conidia decreased, with the effect being more pronounced in the dispersed form of growth (Gromozova, Litvin, & Podgorsky, 2002). This may be because the surface of conidia becomes more plastic during the swelling process. However, despite the general decrease in interaction forces responsible for surface plasticity, there could still be areas of increased adhesion on the conidial surface (Dufrêne, Boonaert, Gerin, Asther, & Rouxhet, 1999).

From a developmental biology perspective, the significance of the lag phase in the morphogenesis of fungal mycelium under submerged conditions may have an analogy with the stage of competence known from embryology. Both represent a period of genesis during which a future developmental strategy is programmed in response to specific environmental conditions (Gromozova, 1998).

4.4 Biochemical and physiological features of pelleted and dispersed growth forms of *Thielavia terrestris*

The direction of morphogenesis defines specific biochemical and physiological features of the resulting mycelial growth form. According to physiological and biochemical indicators, pellets corresponded to the behavior of the organism under conditions of non-specific stress as described for unicellular microorganisms (Rabotnova, 1980). However, in comparative studies of the two growth forms, the analysis of growth and metabolic characteristics of fungal producers often focused on wild-type versus recombinant strains (Chen, Zhou, Ding, Luo, & Liu, 2019; Chai et al., 2020). Therefore, studies on the kinetics and stoichiometry of pelleted and dispersed growth forms of a non-genetically modified fungus, *T. terrestris*, were of special interest.

The energy costs for maintaining submerged *T. terrestris* growth were almost two-fold higher for the pelleted form compared to the dispersed form 0.047 $\pm$ 0.007 g/g·h and 0.024 $\pm$ 0.007 g/g·h, respectively. The pelleted form had a lower economic coefficient of growth than dispersed mycelium (30.5 $\pm$ 1.9% and 43.6 $\pm$ 2.5%, respectively), and a higher metabolic coefficient (0.41 $\pm$ 0.08 h^{-1} and 0.16 $\pm$ 0.01 h^{-1}, respectively). The dispersed form exhibited a higher level of constructive anabolic processes but a lower resistance to adverse environmental factors compared to pellets. The affinity of the dispersed form for glucose (K_s) was considerably lower (0.09 $\pm$ 0.01 g/L) than for pellets (0.19 $\pm$ 0.01 g/L). The substrate (glucose) inhibition constant (K_i) for dispersed mycelium was lower (30.0 $\pm$ 5.0 g/L) than for pellets (112.7 $\pm$ 6.4 g/L). An exception to these differences is that both pelleted and dispersed forms exhibited similar specific growth rates (μ): 0.106 $\pm$ 0.017 h^{-1} and 0.116 $\pm$ 0.012 h^{-1}, respectively. However, inhibition of growth processes does occur in pellets, and the apical growth rate for the pelleted form was lower than that for the dispersed form (Gromozova, Fomina, Blazchuk, & Podgorsky, 1989; Gromozova, Blazchuk, & Podgorsky, 1995).

The ATP pool in pelleted mycelium was found to be almost twice that of the dispersed form: 2.5 pmol/mg DW biomass and 1.3 pmol/mg DW biomass, respectively (Gromozova, Blazchuk, & Galunina, 1990). The pelleted mycelium of *T. terrestris* had a decrease in protein and nucleic acid content and an increase in reserve substances such as lipids and carbohydrates (Grube, Zagreba, Gromozova, & Fomina, 1999). The protective function of the pelleted growth form was further illustrated by optimizing the conditions for obtaining protoplasts from dispersed and pelleted mycelia of *T. terrestris* using a complex of lytic enzymes. This study clearly demonstrated the greater strength of the pellets in releasing protoplasts (Blazchuk, Gromozova, & Stepanyuk, 1991).

A comparison of the energy efficiency of *T. terrestris* growth on different substrates found that, in addition to the general patterns in the assimilation of glucose, ethanol, and lactose, the two growth forms exhibited characteristic form-specific features during the utilization of substrates such as acetate and xylose. In other words, growth in the dispersed or pelleted form resulted in not only quantitative but also qualitative differences in metabolic pathways (Gromozova et al., 1991). This was confirmed by a comparative study on the energy efficiency of *T. terrestris* growth in dispersed and pelleted forms, which included assessment of biological oxygen demand using respiration intensity curves and revealed a large number of incomplete oxidation products, such as acetic acid in the pellets (Zeltina, Shvinka, Blaschuk, & Gromozova, 1990; Gromozova et al., 1991). Studies of the physiological, biochemical, and energy characteristics of pelleted and dispersed mycelia, using *T. terrestris* as an example, have therefore shown different directions of biosynthetic processes for the two forms. If the pelleted form of growth is considered as an adaptation of the fungus to stressful environmental conditions, this could be primarily attributed to diffusion restrictions. However, examination of the early stage of fungal development (lag phase/swelling of conidia) suggested that the process of mycelium formation under submerged conditions represents the development of a self-organizing system where the pelleted form expresses a reaction to stressful cultivation conditions.

4.5 Molecular model of perception of a morphogenic signal by conidia of *Thielavia terrestris*

In studies of the basic mechanisms involved in the regulation of structural transformations of filamentous fungal morphology, attention has focused on the main intracellular mediators, cAMP, cGMP, and Ca^{2+}. It has been established that known morphogenic signals (pH, T, inoculum concentration) affect the level of cAMP in *T. terrestris* cells during the lag phase. An increase in the content of intracellular cAMP in the lag phase was associated with the development of dispersed mycelium; however, the cAMP content was significantly lower for pelleted growth. In addition, the introduction of exogenous cAMP in the lag phase switched development from pelleted to dispersed growth (Gromozova et al., 1995). Data on the exogenous and intracellular cAMP content during *T. terrestris* morphogenesis correlated with data regarding cAMP content in morphological mutants of *Neurospora crassa* (Terenzi, Flawia, & Torres, 1974; Terenzi, Flawia, Tellez-Inon, & Torres, 1979). Unfortunately, information on the influence of cAMP levels on the macromorphology of mycelium is scarce. However, it is generally accepted that cAMP plays an active role in determining the path of fungal morphogenesis (Nair et al., 2018).

The specific participation of calcium ions in morpho-physiological development in *T. terrestris* has also shown high concentrations of exogenous calcium (15–30 mM) changed the pelleted growth form to the dispersed form. The participation of calcium ions in the morphogenic process was confirmed in control studies with other cations and blocking calcium entry into the cell with veropamil. Blocking calmodulin with chlorpromazine did not lead to visible changes in morphogenesis, as did verapamil addition, which excluded the possible effects of trace amounts of calcium in the medium. However, at high concentrations of chlorpromazine, exogenous calcium exhibited a morphogenic effect, apparently due to involvement in other metabolic pathways (Gromozova & Blazchuk, 1996). It is known that calcium signal transmission through calmodulin and calcineurin plays an important role in the stress reactions of fungi (Kraus & Heitman, 2003). However, given the great diversity in the molecular mechanisms underlying this process, it is necessary to study this for each individual organism (Valero et al., 2021). Examination of the cGMP pool during the development of pelleted and dispersed mycelia, along with changes associated with conidial germination, did not reveal any differences in the mycelial growth forms, confirming its non-involvement in the selection of developmental programs. However, studies on the participation of exogenous calcium in this process suggested the possible role of cGMP in calcium signal transduction (Gromozova, unpublished data).

Experimental data indicated that the *T. terrestris* signaling system at the first stage of response to a morphogenic factor includes depolarization of the cytoplasmic membrane followed by activation of adenylate cyclase which, in turn, leads to the formation of cAMP (Fig. 14). It is possible to predict a whole cascade of enzymatic reactions including activation of protein kinase C, phosphorylation of nuclear proteins, expression of corresponding genes, etc., which all result in the development of a certain form of submerged growth. The main assumption of this model is that different modes of functioning of this signaling system and resulting cAMP levels determine the two forms of mycelial growth: pelleted and dispersed (Fig. 15) (Gromozova, 2016).

Fig. 14 shows similarity to representations of the signaling system proposed in early work on *Mucorales dimorphism* and to subsequent modern conceptions of external signal reception by fungi based on genetic and

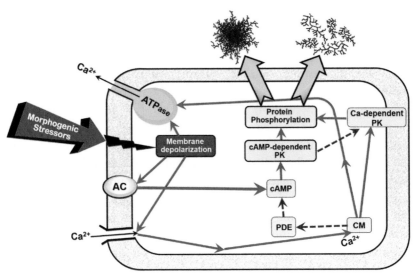

Fig. 14 The effect of environmental factors on the choice of morphogenic program for *Thielavia terrestris* growth (red/solid arrows indicate activation processes, purple/dashed arrows show inhibition). Morphogenic signals (pH, T, inoculum concentration) cause depolarization of the cytoplasmic membrane which activates adenylate cyclase (AC). AC catalyzes the conversion of ATP to cyclic adenosine monophosphate (cAMP), and cAMP binds to cAMP-dependent protein kinase (cAMP-dependent PK), which phosphorylates a number of proteins (including nuclear proteins), changing their activity. This leads to the expression of a number of genes and the development of a certain morphology of the fungus. Calcium ions (Ca^{2+}), entering the cell through ion channels, bind to calmodulin (CM). CM activates calcium-dependent protein kinase (Ca-dependent PK), which also affects protein phosphorylation. Along with AC, the level of cAMP in the cell is also determined by cAMP phosphodiesterase (PDE), which destroys cAMP. CM can inhibit PDE, thereby increasing the content of cAMP in the cell. *Image adapted from Gromozova, E. N. (2016). The regulation of the progresses of form creation of micromycetes in submerged culture. In N. Zaimenko,G. Grodzynska, T. Bugatnko & S. Syrchin (Eds.), Botany and mycology: Modern horizons (pp. 408–425). Kyiv: Nash Format.*

molecular approaches (Steward & Rogers, 1983; Lengeler et al., 2000; Biswas, Van Dijck, & Datta, 2007; Valiante, Macheleidt, Föge, & Brakhage, 2015; Cairns et al., 2019; Chow, Pang, & Wang, 2021).

It is worth mentioning the connections between the phenomenon of pelleted and dispersed forms of filamentous fungal growth under submerged cultivation and classical yeast-hyphal dimorphism (Boyce & Andrianopoulos, 2015). In both cases, (1) the original cells do not have a predetermined programmed choice of form, i.e., from the same conidium

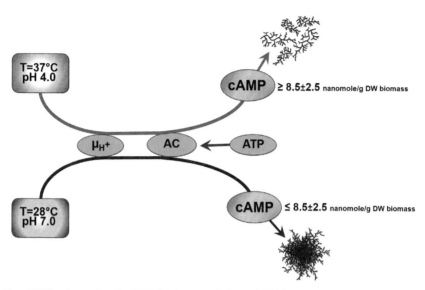

Fig. 15 The key role of cAMP in the regulation of *Thielavia terrestris* mycelial macromorphology. The environmental conditions leading the development of a certain growth form are a combination of values of temperature and pH: 37 °C/pH 4.0 results in dispersed growth (shown as red or black [for grayscale] arrows); whereas 28 °C/pH 7.0 leads to the formation of pellets (shown as blue or gray [for grayscale] arrows). Culture conditions (T, pH) act as primary agents, depolarizing the cytoplasmic membrane (μ_{H+}). This leads to activation of adenylate cyclase (AC) located on the inner side of the cytoplasmic membrane. AC catalyzes the synthesis of 3',5'-cyclic adenosine monophosphate (cAMP). Being a universal second messenger of the signaling system, cAMP contributes to implementation of a dispersed or pelleted growth strategy.

or cell, depending on the conditions, one or another morphological form can develop; (2) the choice of form depends on the cultivation conditions; (3) morphogenic factors are significant only in a certain period of cell development—the G1 phase for yeast-hyphal dimorphism and the lag phase for pelleted-dispersed growth; (4) intracellular mediators, like cAMP and calcium ions, participate in the morphogenesis; (5) cGMP does not seem to play a key role in morphogenesis; and (6) morphologically different vegetative structures are accompanied by pronounced physiological and biochemical features. Such a comparison allows expansion of the concepts from classical dimorphism to include mycelial pelleted-dispersed dimorphism.

Our proposed scheme for the cellular entry of a morphogenic signal shows that cAMP is subject to small fluctuations in physicochemical environmental factors while remaining monofunctional regarding morphogenesis.

Filamentous fungi can exist in several alternative, stable modes, such as dispersed or pelleted mycelium, regardless of the replacement of some signals by others. The cAMP/PKA signaling pathway controls not only fungal morphology but also the synthesis of target products (Cairns et al., 2019; Liu, Du, Ma, Pei, & Li, 2020; Zheng et al., 2022). Another important regulator of the cellular response to stress is the calcium–calcineurin pathway (Liu et al., 2015). Considering the complexity of the regulatory pathways and the coexpression of genes involved in morphogenesis and synthesis of secondary metabolites in filamentous fungi, it can be assumed that these particular signaling systems play a role in genome minimization.

4.6 Concluding remarks on dispersed-pelleted mycelial dimorphism in submerged culture

Traditionally, the study of mycelial forms in submerged cultivation has focused on the production of biologically active substances and assessment of the main cultivation parameters: pH, temperature, concentration of inoculum, oxygenation, and medium composition (Papagianni, 2004, Meyer et al., 2021). In applied studies on the formation of dispersed and pelleted mycelia, particular attention has been paid to hydrodynamic conditions, which not only provide good mass transfer of nutrients and oxygen but also mechanically affect the morphogenic process (Wucherpfennig, Kiep, Driouch, Wittmann, & Krull, 2010; Buffo, Esperança, Farinas, & Badino, 2020). The issues of so-called coagulation and non-coagulation pathways of pellet formation with the participation of electrostatic and hydrophobic forces, as well as the possible role of polysaccharides in these processes, have been discussed (Zhang & Zhang, 2016) although most studies are descriptive (Gibbs, Seviour, & Schmid, 2000). In recent years, evidence has increased about the functional heterogeneity of submerged mycelium which enables survival under stressful conditions (Nair et al., 2018; Lyu et al., 2023). In the evolutionary process, living organisms acquire several adaptive reactions that allow survival under stressful conditions, and environmental conditions are a decisive factor in the genesis of their form and function. During exposure of fungi to cultivation conditions, their morphogenesis cannot be considered only from the standpoint of the physical properties of matter. While the size of spores corresponds to colloidal particles, their interactions are governed by biological processes, and physical processes are only one of the components. Despite advances in genetic research, the process of mycelium formation is still only partially understood (Meyer et al., 2021). The physiological responses of filamentous fungi to stressful conditions in bioreactors have not been well studied. One of the approaches to solving this complex problem may be to study taxonomically different species of filamentous fungi during submerged cultivation. This would allow identification of general patterns and establish differences in morphogenetic processes. Considering the metabolic diversity and the wide range of ecological niches of fungi, it can sometimes be difficult to unambiguously determine which conditions in an artificial system are indeed stressful. It is, therefore, necessary to apply radical new approaches to the dynamic theory of form in such studies, constructing adequate models for the implementation of signals into corresponding structures with various properties and developing general approaches to the controlled cultivation of a given morphological form of mycelial growth. The form of mycelium that develops under conditions of submerged cultivation is not only of interest to fungal-based biotechnology but also provides a good model for fundamental studies on the phenomenon of morphogenesis.

5. General conclusions

The modular design of fungal filamentous growth has enabled the great adaptability of fungi in natural terrestrial ecological niches and different anthropogenic environments as well as secured the evolutionary success and importance of this group of microorganisms. Mainly based on our own studies, we have generalized findings on the diversity of morphological responses to environmental stressors manifested by filamentous fungi under conditions of surface growth in metal- and mineral-rich environments, as well as under conditions of submerged cultivation. It should be highlighted that our observations are not solely about the morphological responses of filamentous fungi to stressful environments because morphological changes are only one of several kinds of adaptation responses which also occur at biochemical, genetic, and physiological levels.

The diverse morphological responses of filamentous fungi under surface growth conditions demonstrate the remarkable flexibility of mycelial systems under metal- and mineral-rich environments. Such morphological responses are inextricably complemented by fungal biogeochemical activities, resulting in amelioration of a hostile environment and ensuring successful adaptation to stress. In the terrestrial environment, both nutritional resources and the presence of metals and minerals are unevenly spatially distributed. The use of a tessellated agar tile system simulated such natural heterogeneity and its application enabled visualization of negative fungal chemotropism to toxic metals, and clearly distinguished phalanx (aggregative exploitative strategy) and guerrilla (explorative strategy) growth forms, and multiple phalanx-guerrilla transitions, as well as the formation of linearly aggregated cords, strands, and synnemata (re-allocation strategy). In other experimental or natural surface growth systems, morphological features such as recycling (or intrahyphal growth), specialization (related to intense biomineralization providing amelioration of toxic conditions), and extracellular hair-like structures are also encountered.

Because submerged cultivation of filamentous fungi is fundamental in fungal biotechnology, the two main forms of mycelial growth occurring in bioreactors—pelleted and dispersed growth—are very important for fungal fermentations. This is because the morphological form can significantly impact the productivity of industrial-significant metabolites. Some new insights into mechanisms of morphogenesis and functioning of pelleted and dispersed fungal growth forms under submerged conditions have been revealed using the thermophilic ascomycetous fungus *Thielavia terrestris*. These mechanisms included identification of the importance of the lag phase (or conidia swelling stage) for fungal development, representing a period of morphogenesis when the future developmental strategy is programmed in response to specific environmental conditions or morphogenic stressors. The key role of cAMP in the regulation of pelleted-dispersed dimorphism for *Thielavia terrestris* mycelial growth under submerged conditions was also identified in determining the future implementation of a particular growth strategy. Comparative data on the functioning of pelleted and dispersed growth forms under submerged conditions show that the formation of pellets is a specific adaptive reaction to stressful environmental conditions.

How to apply the basic morphological strategies of a mycelial system during surface growth to dispersed-pelleted growth dimorphism under submerged conditions? The phenomenon of dispersed and pelleted growth is probably more complicated than the guerrilla and phalanx growth extremes because the morphology of pellets can be diverse in terms of branching indicating some transitions between explorative and exploitative strategies within the pelleted form. Undoubtedly, the formation of pellets is a higher developmental level in organizational terms of a protective aggregative strategy of the mycelial system compared to the dispersed growth form, and a reflection of complex biochemical and physiological processes. The lower apical growth rates of pelleted mycelium of *T. terrestris* are more related to the slower extension growth characteristics for the phalanx exploitative strategy. However, in pellets, physiological compensatory mechanisms ensure the same specific growth rate for total fungal biomass as for dispersed mycelium. Further interdisciplinary research using a wider taxonomic range of fungi is needed for mechanistic insights into how different filamentous fungi can adapt to various environmental stresses. Such knowledge will contribute to both fundamental mycology and applied microbiology enabling better control of fungal-based biotechnological processes.

Acknowledgments

OG and MF gratefully acknowledge Dr. Iryna Blazchuk, an expert in biochemistry, who worked at Zabolotny Institute of Microbiology and Virology of National Academy of Sciences of Ukraine in Kyiv; Dr Peter Lytvyn, an expert in atomic force microscopy from Lashkaryov Institute of Semiconductor Physics of National Academy of Sciences of Ukraine in Kyiv; and Dr Mara Grube, an expert in Infrared Spectroscopy at Institute of Microbiology and Biotechnology of University of Latvia in Riga, for their helpful discussions and invaluable technical support. We are also grateful to Alene Alder-Rangel from Alder's English Services for helpful language amendments.

References

Abingdon, J. E. H. (2015). Genome stability: DNA repair and recombination. *Yale Journal of Biology and Medicine, 88*, 199–202.

Alastruey-Izquierdo, A., Melhem, M. S. C., Bonfietti, L. X., & Rodriguez-Tudela, J. L. (2015). Susceptibility test for fungi: Clinical and laboratorial correlations in medical mycology. *Revista Do Instituto de Medicina Tropical de São Paulo, 57*(Suppl. 19), 57–64. https://doi.org/10.1590/s0036-46652015000700011.

Alonso-Monge, R., Real, E., Wojda, I., Bebelman, J.-P., Mager, W. H., & Siderius, M. (2001). Hyperosmotic stress response and regulation of cell wall integrity in *Saccharomyces cerevisiae* share common functional aspects. *Molecular Microbiology, 41*(3), 717–730. https://doi.org/10.1046/j.1365-2958.2001.02549.x.

Andrews, J. H. (1995). Fungi and the evolution of growth form. *Canadian Journal of Botany, 73*(S1), 1206–1212. https://doi.org/10.1139/b95-380.

Andrews, J. H. (2017). *Comparative ecology of microorganisms and macroorganisms* (2nd ed.). Springer Science + Business Media LLC. https://doi.org/10.1007/978-1-4939-6897-8.

Anesio, A. M., Lutz, S., Chrismas, N. A. M., & Benning, L. G. (2017). The microbiome of glaciers and ice sheets. *NPJ Biofilms and Microbiomes, 3*, 10. https://doi.org/10.1038/s41522-017-0019-0.

Argumedo-Delira, R., Gómez-Martínez, M. J., & Uribe-Kaffure, R. (2020). Fungal tolerance: An alternative for the selection of fungi with potential for the biological recovery of precious metals. *Applied Sciences, 10*(22), 8096. https://doi.org/10.3390/app10228096.

Astafieva, M. M., & Rozanov, A. Y. (2012). Bacterial paleontological study of early Precambrian weathering crusts. *Earth Science Research, 1*, 163–170.

Baldrian, P. (2003). Interaction of heavy metals with white-rot fungi. *Enzyme and Microbial Technology, 32*, 78–91. https://doi.org/10.1016/S0141-0229(02)00245-4.

Banfield, J. P., Barker, W. W., Welch, S. A., & Taunton, A. (1999). Biological impact on mineral dissolution: Application of the lichen model to understanding mineral weathering in the rhizosphere. *Proceedings of the National Academy of Sciences (PNAS) USA, 96*, 3404–3411.

Bansal, M., & Kaushal, N. (2014). *Oxidative stress mechanisms and their modulation*. New Delhi: Springer India. https://doi.org/10.1007/978-81-322-2032-9.

Batiza, A. F., Schulz, T., & Masson, P. H. (1996). Yeast respond to hypotonic shock with a calcium pulse. *Journal of Biological Chemistry, 271*(38), 23357–23362. https://doi.org/10.1074/jbc.271.38.23357.

Beck, A., Divakar, P. K., Zhang, N., Molina, M. C., & Struwe, L. (2015). Evidence of ancient horizontal gene transfer between fungi and the terrestrial alga Trebouxia. *Organisms, Diversity and Evolution, 15*, 235–248.

Bellion, M., Courbot, M., Jacob, C., Blaudez, D., & Chalot, M. (2006). Extracellular and cellular mechanisms sustaining metal tolerance in ectomycorrhizal fungi. *FEMS Microbiology Letters, 254*, 173–181.

Besozzi, D., Cazzaniga, P., Pescini, D., Mauri, G., Colombo, S., & Martegani, E. (2012). The role of feedback control mechanisms on the establishment of oscillatory regimes in the Ras/cAMP/PKA pathway in *S. cerevisiae. EURASIP Journal on Bioinformatics and Systems Biology, 1.* https://doi.org/10.1186/1687-4153-2012-10.

Biryukova, E. N., Medentsev, A. G., Arinbasarova, A. Y., & Akimenko, V. K. (2006). Tolerance of the yeast *Yarrowia lipolytica* to oxidative stress. *Mikrobiologiia, 75*(3), 293–298.

Biswas, S., Van Dijck, P., & Datta, A. (2007). Environmental sensing and signal transduction pathways regulating morphopathogenic determinants of *Candida albicans. Microbiology and Molecular Biology Reviews, 71*(2), 348–376. https://doi.org/10.1128/MMBR.00009-06.

Blazchuk, I. S., Gromozova, E. N., & Stepanyuk, V. V. (1991). Features of obtaining and characterizing protoplasts from various parts of mycelial structures *Thielavia terrestris. Mikrobiologichnyi Zhurnal, 53*(6), 21–27 (in Russian).

Boddy, L. (1993). Saprotrophic cord-forming fungi: Warfare strategies and other ecological aspects. *Mycological Research, 97*, 641–655.

Boetius, A., Anesio, A. M., Deming, J. W., Mikucki, J. A., & Rapp, J. Z. (2015). Microbial ecology of the cryosphere: Sea ice and glacial habitats. *Nature Reviews. Microbiology, 13*, 677–690.

Botton, B. (1978). Influence of calcium on the differentiation and growth of aggregated organs in *Sphaerostilbe repens. Canadian Journal of Microbiology, 24*, 1039–1047.

Boyce, K. J., & Andrianopoulos, A. (2015). Fungal dimorphism: The switch from hyphae to yeast is a specialized morphogenetic adaptation allowing colonization of a host. *FEMS Microbiology Reviews, 39*(6), 797–811. https://doi.org/10.1093/femsre/fuv035.

Brandl, J., Aguilar-Pontes, M. V., Schäpe, P., Noerregaard, A., Arvas, M., Ram, A. F. J., et al. (2018). A community-driven reconstruction of the *Aspergillus niger* metabolic network. *Fungal Biology and Biotechnology, 5*(1), 1–12. https://doi.org/10.1186/s40694-018-0060-7.

Bucking, H., & Heyser, W. (1999). Elemental composition and function of polyphosphates in ectomycorrhizal fungi and X-ray microanalytical study. *Mycological Research, 103*, 31–39.

Buffo, M. M., Esperança, M. N., Farinas, C. S., & Badino, A. C. (2020). Relation between pellet fragmentation kinetics and cellulolytic enzymes production by *Aspergillus niger* in conventional reactor with different impellers. *Enzyme and Microbial Technology, 139*, 109587. https://doi.org/10.1016/j.enzmictec.2020.109587.

Burford, E. P., Fomina, M., & Gadd, G. M. (2003). Fungal involvement in bioweathering and biotransformation of rocks and minerals. *Mineralogical Magazine, 67*, 1127–1155.

Buss, L. W. (1987). *The evolution of individuality.* Princeton NJ: Princeton University Press.

Cai, M., Zhang, Y., Hu, W., Shen, W., Yu, Z., Zhou, W., et al. (2014). Genetically shaping morphology of the filamentous fungus *Aspergillus glaucus* for production of antitumor polyketide aspergiolide A. *Microbial Cell Factories, 13*(1), 73. https://doi.org/10.1186/1475-2859-13-73.

Cairns, T. C., Zheng, X., Zheng, P., Sun, J., & Meyer, V. (2019). Moulding the mould: Understanding and reprogramming filamentous fungal growth and morphogenesis for next generation cell factories. *Biotechnology for Biofuels, 12*(1), 77. https://doi.org/10.1186/s13068-019-1400-4.

Campisi, A., Gulino, M., Acquaviva, R., Bellia, P., Raciti, G., ... Triglia, A. (2010). Reactive oxygen species levels and DNA fragmentation on astrocytes in primary culture after acute exposure to low intensity microwave electromagnetic field. *Neuroscience Letters, 473*(1), 52–55. https://doi.org/10.1016/j.neulet.2010.02.018.

Carlile, M. J. (1995). The success of the hypha and mycelium. In N. A. R. Gow, & G. M. Gadd (Eds.). *The growing fungus* (pp. 3–19). London: Chapman & Hall.

Cazzanelli, G., Pereira, F., Alves, S., Francisco, R., Azevedo, L., ... Preto, A. (2018). The yeast *Saccharomyces cerevisiae* as a model for understanding RAS proteins and their role in human tumorigenesis. *Cells, 7*(2), 14. https://doi.org/10.3390/cells7020014.

Chai, X., Ai, Z., Liu, J., Guo, T., Wu, J., ... Lin, Q. (2020). Effects of pigment and citrinin biosynthesis on the metabolism and morphology of *Monascus purpureus* in submerged fermentation. *Food Science and Biotechnology, 29*(7), 927–937. https://doi.org/10.1007/s10068-020-00745-3.

Chantranupong, L., & Sabatini, D. M. (2016). The TORC1 pathway to protein destruction. *Nature, 536*(7615), 155–156. https://doi.org/10.1038/nature18919.

Chen, X., Zhou, J., Ding, Q., Luo, Q., & Liu, L. (2019). Morphology engineering of *Aspergillus oryzae* for l-malate production. *Biotechnology & Bioengineering, 116*, 2662–2673.

Chen, J., Blume, H.-P., & Beyer, L. (2000). Weathering of rocks induced by lichen colonization—A review. *CATENA, 39*, 121–146.

Chow, E. W. L., Pang, L. M., & Wang, Y. (2021). From Jekyll to Hyde: The yeast-hyphal transition of *Candida albicans*. *Pathogens, 10*(7), 859. https://doi.org/10.3390/pathogens10070859.

Collins, M. P., & Forgac, M. (2020). Regulation and function of V-ATPases in physiology and disease. *Biochimica et Biophysica Acta—Biomembranes, 183341*. https://doi.org/10.1016/j.bbamem.2020.183341.

Crawford, R. A., & Pavitt, G. D. (2018). Translational regulation in response to stress in *Saccharomyces cerevisiae*. *Yeast (Chichester, England), 36*(1), 5–21. https://doi.org/10.1002/yea.3349.

Danion, F., van Rhijn, N., Dufour, A. C., Legendre, R., Sismeiro, O., ... Latgé, J.-P. (2021). *Aspergillus fumigatus*, one uninucleate species with disparate offspring. *Journal of Fungi, 7*(1), 30. https://doi.org/10.3390/jof7010030.

Davenport, K. R., Sohaskey, M., Kamada, Y., Levin, D. E., & Gustin, M. C. (1995). A second osmosensing signal transduction pathway in yeast. Hypotonic shock activates the PKC1 protein kinase-regulated cell integrity pathway. *Journal of Biological Chemistry, 270*(50), 30157–30161. https://doi.org/10.1074/jbc.270.50.30157.

Delaux, P.-M., Radhakrishnan, G. V., Jayaraman, D., Cheema, J., Malbreil, M., Volkening, J. D., et al. (2015). Algal ancestor of land plants was preadapted for symbiosis. *Proceedings of the National Academy of Sciences of the United States of America, 112*, 13390–13395.

Diller, K. R. (2006). Stress protein expression kinetics. *Annual Review of Biomedical Engineering, 8*(1), 403–424. https://doi.org/10.1146/annurev.bioeng.7.060804.100449.

Dinius, A., Kozanecka, Z. J., Hoffmann, J., Kevin, P. H., & Rainer, K. (2023). Intensification of bioprocesses with filamentous microorganisms. *Physical Science Reviews*. https://doi.org/10.1515/psr-2022-0112.

Dunayevich, P., Baltanás, R., Clemente, J. A., Couto, A., Sapochnik, D., ... Colman-Lerner, A. (2018). Heat-stress triggers MAPK crosstalk to turn on the hyperosmotic response pathway. *Scientific Reports, 8*(1), https://doi.org/10.1038/s41598-018-33203-6.

Dufrêne, Y. F., Boonaert, C. J., Gerin, P. A., Asther, M., & Rouxhet, P. G. (1999). Direct probing of the surface ultrastructure and molecular interactions of dormant and germinating spores of *Phanerochaete chrysosporium*. *Journal of Bacteriology, 181*(17), 5350–5354. https://doi.org/10.1128/JB.181.17.5350-5354.1999.

Egidi, E., De Hoog, G. S., Isola, D., Onofri, S., Quaedvlieg, W., De Vries, M., et al. (2014). Phylogeny and taxonomy of meristematic rock-inhabiting black fungi in the Dothideomycetes based on multi-locus phylogenies. *Fungal Diversity*, *65*(1), 127–165. https://doi.org/10.1007/s13225-013-0277-y.

Ehgartner, D., Herwig, C., & Fricke, J. (2017). Morphological analysis of the filamentous fungus *Penicillium chrysogenum* using flow cytometry-the fast alternative to microscopic image analysis. *Applied Microbiology and Biotechnology*, *101*(20), 7675–7688. https://doi. org/10.1007/s00253-017-8475-2.

Eskes, E., Deprez, M.-A., Wilms, T., & Winderickx, J. (2017). pH homeostasis in yeast; the phosphate perspective. *Current Genetics*, *64*(1), 155–161. https://doi.org/10.1007/ s00294-017-0743-2.

Fomina, M., & Gadd, G. M. (2007). Metal and mineral transformations: A mycor-emediation perspective. In G. D. Robson, P. Van West, & G. M. Gadd (Eds.). *Exploitation of fungi* (pp. 236–254). Cambridge: Cambridge University Press. https://doi. org/10.1017/CBO9780511902451.014.

Fomina, M., & Gadd, G. M. (2014). Biosorption: Current perspectives on concept, definition and application. *Bioresource Technology*, *160*, 3–14. https://doi.org/10.1016/j. biortech.2013.12.102.

Fomina, M., & Skorochod, I. (2020). Microbial interaction with clay minerals and its environmental and biotechnological implications. *Minerals*, *10*(*10*), 861. https://doi. org/10.3390/min10100861.

Fomina, M., Ritz, K., & Gadd, G. M. (2000). Negative fungal chemotropism to toxic metals. *FEMS Microbiology Letters*, *193*, 207–211. https://doi.org/10.1111/j.1574-6968. 2000.tb09425.x.

Fomina, M., Ritz, K., & Gadd, G. M. (2003). Nutritional influence on the ability of fungal mycelia to penetrate toxic metal-containing domains. *Mycological Research*, *107*, 861–871. https://doi.org/10.1017/s095375620300786x.

Fomina, M., Alexander, I. J., Hillier, S., & Gadd, G. M. (2004). Zinc phosphate and pyromorphite solubilization by soil plant-symbiotic fungi. *Geomicrobiology Journal*, *21*(5), 351–366. https://doi.org/10.1080/01490450490462066.

Fomina, M., Burford, E. P., & Gadd, G. M. (2005a). Toxic metals and fungal communities. In J. Dighton, J. White, & P. Oudemans (Eds.). *The fungal community: Its organization and role in the ecosystem* (pp. 733–758). Boca Raton: CRC Press, Taylor & Francis Group.

Fomina, M. A., Alexander, I. J., Colpaert, J. V., & Gadd, G. M. (2005b). Solubilization of toxic metal minerals and metal tolerance of mycorrhizal fungi. *Soil Biology and Biochemistry*, *37*(5), 851–866. https://doi.org/10.1016/j.soilbio.2004.10.013.

Fomina, M., Hillier, S., Charnock, J. M., Melville, K., Alexander, I. J., & Gadd, G. M. (2005c). Role of oxalic acid overexcretion in transformations of toxic metal minerals by *Beauveria caledonica*. *Applied and Environmental Microbiology*, *71*(1), 371–381. https://doi. org/10.1128/AEM.71.1.371-381.2005.

Fomina, M., Podgorsky, V. S., Olishevska, S. V., Kadoshnikov, V. M., Pisanska, I. R., ... Gadd, G. M. (2007a). Fungal deterioration of barrier concrete used in nuclear waste disposal. *Geomicrobiology Journal*, *24*, 643–653. https://doi.org/10.1080/01490450701672240.

Fomina, M., Charnock, J. M., Hillier, S., Alvarez, R., & Gadd, G. M. (2007b). Fungal transformations of uranium oxides. *Environmental Microbiology*, *9*(7), 1696–1710. https:// doi.org/10.1111/j.1462-2920.2007.01288.x.

Fomina, M., Charnock, J., Bowen, A. D., & Gadd, G. M. (2007c). X-ray absorption spectroscopy (XAS) of toxic metal mineral transformations by fungi. *Environmental Microbiology*, *9*(2), 308–321. https://doi.org/10.1111/j.1462-2920.2006.01139.x.

Fomina, M., Charnock, J. M., Hillier, S., Alvarez, R., Livens, F., & Gadd, G. M. (2008). Role of fungi in the biogeochemical fate of depleted uranium. *Current Biology*, *18*(9), R375–R377.

Fomina, M., Bowen, A. D., Charnock, J. M., Podgorsky, V. S., & Gadd, G. M. (2017). Biogeochemical spatio-temporal transformation of copper in *Aspergillus niger* colonies grown on malachite with different inorganic nitrogen sources. *Environmental Microbiology, 19*(3), 1310–1321. https://doi.org/10.1111/1462-2920.13664.

Fomina, M., Hong, J. W., & Gadd, G. M. (2020). Effect of depleted uranium on a soil microcosm fungal community and influence of a plant-ectomycorrhizal association. *Fungal Biology, 124*(5), 289–296.

Fomina, M., Cuadros, J., Pinzari, F., Hryshchenko, N., Najorka, J., Gavrilenko, M., ... Gadd, G. M. (2022). Fungal transformation of mineral substrata of biodeteriorated medieval murals in Saint Sophia's cathedral, Kyiv, Ukraine. *International Biodeterioration and Biodegradation, 175*, 105486. https://doi.org/10.1016/j.ibiod.2022.105486.

Francisco, C. S., Ma, X., Zwyssig, M. M., McDonald, B. A., & Palma-Guerrero, J. (2019). Morphological changes in response to environmental stresses in the fungal plant pathogen *Zymoseptoria tritici*. *Scientific Reports, 9*(1), 9642. https://doi.org/10.1038/s41598-019-45994-3.

Fulda, S., Gorman, A. M., Hori, O., & Samali, A. (2010). Cellular stress responses: Cell survival and cell death. *International Journal of Cell Biology, 214074*, 1–23. https://doi.org/10.1155/2010/214074.

Futagami, T., Nakao, S., Kido, Y., Oka, T., Kajiwara, Y., Takashita, H., et al. (2011). Putative stress sensors WscA and WscB are involved in hypo-osmotic and acidic pH stress tolerance in *Aspergillus nidulans*. *Eukaryotic Cell, 10*(11), 1504–1515. https://doi.org/10.1128/EC.05080-11.

Gadd, G. M. (1993). Interactions of fungi with toxic metals. *New Phytologist, 124*, 25–60.

Gadd, G. M. (1995). Signal transduction in fungi. In N. A. R. Gow, & G. M. Gadd (Eds.). *The growing fungus* (pp. 183–210). Dordrecht: Springer. https://doi.org/10.1007/978-0-585-27576-5_9.

Gadd, G. M. (2004). Microbial influence on metal mobility and application for bioremediation. *Geoderma, 122*(2–4), 109–119.

Gadd, G. M. (2007). Geomycology: Biogeochemical transformations of rocks, minerals, metals and radionuclides by fungi, bioweathering and bioremediation. *Mycological Research, 111*, 3–49.

Gadd, G. M. (2017). Geomicrobiology of the built environment. *Nature Microbiology, 2*, 16275. https://doi.org/10.1038/nmicrobiol.2016.275.

Gadd, G. M., Ramsay, L., Crawford, J. W., & Ritz, K. (2001). Nutritional influence on fungal colony growth and biomass distribution in response to toxic metals. *FEMS Microbiology Letters, 204*(2), 311–316. https://doi.org/10.1111/j.1574-6968.2001.tb10903.x.

Gadd, G. M., Fomina, M., & Burford, E. P. (2005). Fungal roles and functions in rock, mineral and soil transformations. In G. M. Gadd, K. T. Semple, & H. M. Lappin-Scott (Eds.). *Microorganisms and Earth systems—Advances in geomicrobiology* (pp. 201–232). Cambridge: Cambridge University Press. https://doi.org/10.1017/CBO9780511754852.011.

Gadd, G. M., Bahri-Esfahani, J., Li, Q., Rhee, Y. J., Wei, Z., Fomina, M., et al. (2014). Oxalate production by fungi: Significance in geomycology, biodeterioration and bioremediation. *Fungal Biology Reviews, 28*(2–3), 36–55.

Gadd, G. M., Fomina, M., & Pinzari, F. (2024). Fungal biodeterioration and preservation of cultural heritage, artwork, and historical artifacts: Extremophily and adaptation. *Microbiology and Molecular Biology Reviews*, e0020022. https://doi.org/10.1128/mmbr.00200-22.

Gasch, A. P., Spellman, P. T., Kao, C. M., Carmel-Harel, O., Eisen, M., ... Brown, P. O. (2000). Genomic expression programs in the response of yeast cells to environmental changes. *Molecular Biology of the Cell, 11*(12), 4241–4257. https://doi.org/10.1091/mbc.11.12.4241.

García-Reyes, M., Beltran-Hernandez, R. I., Vazquez-Rodriguez, G. A., Coronel-Olivares, C., Medina-Moreno, S. A., Juarez-Santillan, L. F., et al. (2017). Formation, morphology and biotechnological applications of filamentous fungal pellets: A review. *Revista Mexicana de Ingeniería Química, 16*(3), 703–720.

Gerin, P. A., Dufrene, Y., Bellon-Fontaine, M. N., Asther, M., & Rouxhet, P. G. (1993). Surface properties of conidiospores of *Phanerochaete chrysosporium* and their relevance to pellet formation. *Journal of Bacteriology, 175*(16), 5135–5144. https://doi.org/10.1128/jb.175.16.5135-5144.1993.

Gibbs, P. A., Seviour, R. J., & Schmid, F. (2000). Growth of filamentous fungi in submerged culture: Problems and possible solutions. *Critical Reviews in Biotechnology, 20*(1), 17–48. https://doi.org/10.1080/07388550091144177.

Gimeno, C. J., Ljungdahl, P. O., Styles, C. A., & Fink, G. R. (1992). Unipolar cell divisions in the yeast *S. cerevisiae* lead to filamentous growth: Regulation by starvation and RAS. *Cell, 68*, 1077–1090. https://doi.org/10.1016/0092-8674(92)90079-r.

Gokul, J. K., Hodson, A. J., Saetnan, E. R., Irvine-Fynn, T. D. L., Westall, P. J., Detheridge, A. P., et al. (2016). Taxon interactions control the distributions of cryoconite bacteria colonizing a High Arctic ice cap. *Molecular Ecology, 25*, 3752–3767.

Gonçalves, A. P., Heller, J., Daskalov, A., Videira, A., & Glass, N. L. (2017). Regulated forms of cell death in fungi. *Frontiers in Microbiology, 8*, 1837. https://doi.org/10.3389/fmicb.2017.01837.

Gorban, A. N., Tyukina, T. A., Smirnova, E. V., & Pokidysheva, L. I. (2016). Evolution of adaptation mechanisms: Adaptation energy, stress, and oscillating death. *Journal of Theoretical Biology, 405*, 127–139. https://doi.org/10.1016/j.jtbi.2015.12.017.

Gorbushina, A. A., Krumbein, W. E., Hamann, R., Panina, L., Soucharjevsky, S., & Wollenzien, U. (1993). On the role of black fungi in colour change and biodeterioration of antique marbles. *Geomicrobiology Journal, 11*, 205–221.

Gorbushina, A. A., Whitehead, K., Dornieden, T., Niesse, A., Schulte, A., & Hedges, J. I. (2003). Black fungal colonies as units of survival: Hyphal mycosporines synthesized by rock-dwelling microcolonial fungi. *Canadian Journal of Botany, 81*, 131–138.

Gow, N. A. (1994). Growth and guidance of the fungal hypha. *Microbiology (Reading, England), 140*(Pt 12), 3193–3205. https://doi.org/10.1099/13500872-140-12-3193.

Grimm, L. H., Kelly, S., Krull, R., & Hempel, D. C. (2005). Morphology and productivity of filamentous fungi. *Applied Microbiology and Biotechnology, 69*(4), 375–384. https://doi.org/10.1007/s00253-005-0213-5.

Gromozova, E. N. (1998). The role of the lag phase in the development of various forms of mycelium of some fungi in submerged culture. *Ontogenez, 29*(5), 362–365 (in Russian).

Gromozova, E. N. (2003). Adsorption-structural mechanisms of the fungal mycelium formation in submerged culture. *Dopovidi NANU (Reports of National Academy of Sciences of Ukraine), 12*, 141–145 (in Russian).

Gromozova, E. N. (2016). The regulation of the progresses of form creation of micromycetes in submerged culture. In N. Zaimenko, G. Grodzynska, T. Bugatnko, & S. Syrchin (Eds.). *Botany and mycology: Modern horizons* (pp. 408–425). Kyiv: Nash Format.

Gromozova, E. N., & Blazchuk, I. S. (1989). The influence of some factors on the growth pattern of *Thielavia* sp. in submerged culture. *Mikrobiologichnyi Zhurnal, 51*(4), 30–31. (in Russian).

Gromozova, E. N., & Blazhchuk, I. S. (1996). Participation of calcium ions in the process of mycelial dimorphism of *Theielavia terrestris* (in Russian) *Biopolymers and Cell, 12*(3), 91–96. https://doi.org/10.7124/bc.000432.

Gromozova, E. N., & Sadovskii, G. M. (1992). *Evolutionary mechanisms of adaptive changes in the morphology of some fungi. Evolutionary modeling and kinetics.* Novosibirsk: Nauka, 161–176.

Gromozova, E. N., & Sadovskii, M. G. (2006). Relationship of the morphogenesis of *Thielavia terrestris* (Apinis) Malloch et Cain with previous culture conditions. *Doklady Biological Sciences, 406*(1–6), 41–43. https://doi.org/10.1134/s001249660601011x.

Gromozova, E. N., Shemshur, T. V., & Fomina, M. A. (1989). Formation of mycelial structures under the influence of cultivation conditions at different stages of micromycete development. *Mykologiya i Fitopatologiya, 23*(3), 202–205. (in Russian).

Gromozova, E. N., Fomina, M. A., Blazchuk, I. S., & Podgorsky, V. S. (1989). Physiological features of the growth of various mycelial structures of *Thielavia terrestris* on a medium with glucose. *Mikrobiologichnyi Zhurnal, 51*(1), 43–46. (in Russian).

Gromozova, E. N., Blazchuk, I. S., & Galunina, N. I. (1990). ATP level during the development of various mycelial structures of *Thielavia terrestris*. *Mikrobiologichnyi Zhurnal, 52*(5), 48–51. (in Russian).

Gromozova, E. N., Fomina, M. A., Podgorsky, V. S., Leite, V. S., Zeltina, M., & Shvinka, Y. E. (1991). Growth efficiency of *Thielavia terrestris* mycelial structures. *Acta Biotechnologica, 11*(4), 325–329.

Gromozova, O. M., Blazchuk, I. S., & Podgorsky, V. S. (1995). Pool of adenosine-3-5 monophosphate in the process of *Thielavia terrestris* mycelial structures formation in submerged culture. *Ukrainian Botanical Journal, 52*(5), 615–620. (in Ukrainian).

Gromozova, E. N., Litvin, P. M., & Podgorsky, V. S. (2002). Peculiarities of *Th. terrestris* spores surface ultrastructure investigated by AFM. In E. Buzanova, & P. Scharff (Eds.), *Frontiers of multifunctional nanosystems* (pp. 341–346). Netherlands: Kluwer Academic Publishers.

Grube, M., Zagreba, E., Gromozova, E., & Fomina, M. (1999). Comparative investigation of the macromolecular composition of mycelia forms *Thielavia terrestris* by infrared spectroscopy. *Vibrational Spectroscopy, 19*(2), 301–306. https://doi.org/10.1016/s0924-2031(98)00074-5.

Gu, J. D., Ford, T. E., Berke, N. S., Gu, J. D., & Mitchell, R. (1998). Biodeterioration of concrete by the fungus *Fusarium*. *International Biodeterioration and Biodegradation, 41*, 101–109.

Guan, N., Li, J., Shin, H., Du, G., Chen, J., & Liu, L. (2017). Microbial response to environmental stresses: From fundamental mechanisms to practical applications. *Applied Microbiology and Biotechnology, 101*(10), 3991–4008. https://doi.org/10.1007/s00253-017-8264-y.

Harrison, J. C., Zyla, T. R., Bardes, E. S. G., & Lew, D. J. (2003). Stress-specific activation mechanisms for the "Cell Integrity" MAPK pathway. *Journal of Biological Chemistry, 279*(4), 2616–2622. https://doi.org/10.1074/jbc.m306110200.

Heinisch, J. J., & Rodicio, R. (2017). Protein kinase C in fungi—More than just cell wall integrity. *FEMS Microbiology Reviews, 42*(1), https://doi.org/10.1093/femsre/fux051.

Hohmann, S. (2015). An integrated view on a eukaryotic osmoregulation system. *Current Genetics, 61*(3), 373–382. https://doi.org/10.1007/s00294-015-0475-0.

Hohmann, S., Krantz, M., & Nordlander, B. (2007). Yeast osmoregulation. Osmosensing and osmosignaling. In D. Häussinger, & H. Sies (Vol. Eds.), *Methods in enzymology. Vol. 428. Methods in enzymology* (pp. 29–45). Academic Press. https://doi.org/10.1016/s0076-6879(07)28002-4.

Horiuchi, H., Fujiwara, M., Yamashita, S., Ohta, A., & Takagi, M. (1999). Proliferation of intrahyphal hyphae caused by disruption of csmA, which encodes a class V chitin synthase with a myosin motor-like domain in *Aspergillus nidulans*. *Journal of Bacteriology, 181*(12), 3721–3729. https://doi.org/10.1128/JB.181.12.3721-3729.1999.

Howlett, N. G., & Avery, S. V. (1997). Relationship between cadmium sensitivity and degree of plasma membrane fatty acid unsaturation in *Saccharomyces cerevisiae*. *Applied Microbiology and Biotechnology, 48*, 539–545.

Huang, G., Huang, Q., Wei, Y., Wang, Y., & Du, H. (2018). Multiple roles and diverse regulation of the Ras/cAMP/protein kinase A pathway in *Candida albicans*. *Molecular Microbiology, 111*(1), 6–16. https://doi.org/10.1111/mmi.14148.

Jacob, C., Courbot, M. L., Martin, F., Brun, A., & Chalot, M. (2004). Transcriptomic responses to cadmium in the ectomycorrhizal fungus *Paxillus involutus*. *FEBS Letters, 576*, 423–427.

Johnston, N., Nallur, S., Gordon, P. B., Smith, K. D., & Strobel, S. (2020). Genome-wide identification of genes involved in general acid stress and fluoride toxicity in *Saccharomyces cerevisiae*. *Frontiers in Microbiology*, *11*, 1410. https://doi.org/10.3389/fmicb.2020.01410.

Kamada, Y., Jung, U. S., Piotrowski, J., & Levin, D. E. (1995). The protein kinase C-activated MAP kinase pathway of *Saccharomyces cerevisiae* mediates a novel aspect of the heat shock response. *Genes & Development*, *9*(13), 1559–1571. https://doi.org/10.1101/gad.9.13.1559.

Kane, P. M. (2016). Proton transport and pH control in fungi. In J. Ramos, (Ed.). *Yeast membrane transport, advances in experimental medicine and biology* (pp. 33–68). Switzerland: Springer International Publishing. https://doi.org/10.1007/978-3-319-25304-6_3.

Kazemi, Z., Chang, H., Haserodt, S., McKen, C., & Zachara, N. E. (2010). O-linked β-N-acetylglucosamine (O-GlcNAc) regulates stress-induced heat shock protein expression in a GSK-3β-dependent manner. *Journal of Biological Chemistry*, *285*(50), 39096–39107. https://doi.org/10.1074/jbc.m110.131102.

Kheirkhah, T., Neubauer, P., & Junne, S. (2023). Controlling *Aspergillus niger* morphology in a low shear-force environment in a rocking-motion bioreactor. *Biochemical Engineering Journal*, *195*, 108905. https://doi.org/10.1016/j.bej.2023.108905.

Kidron, G. J., & Xiao, B. (2024). A false paradigm? Do biocrust types necessarily reflect "successional stages"? *Ecohydrology*, *17*(1), e2610. https://doi.org/10.1002/eco.2610.

Kim, K. W., Hyun, J. W., & Park, E. W. (2004). Cytology of cork layer formation of citrus and limited growth of *Elsinoe fawcettii* in scab lesions. *European Journal of Plant Pathology*, *110*, 129–138. https://doi.org/10.1023/B:EJPP.0000015330.21280.4c.

King, R. (2015). A framework for an organelle-based mathematical modeling of hyphae. *Fungal Biology and Biotechnology*, *2*(1), 5. https://doi.org/10.1186/s40694-015-0014-2.

Kraus, P. R., & Heitman, J. (2003). Coping with stress: Calmodulin and calcineurin in model and pathogenic fungi. *Biochemical and Biophysical Research Communications*, *311*(4), 1151–1157. https://doi.org/10.1016/s0006-291x(03)01528-6.

Kumar, P. K., Rajput, D., & Dubey, K. K. (2023). Insights into the mechanism of mycelium transformation of *Streptomyces toxytricini* into pellet. *FEMS Microbes*, *4*, xtad017. https://doi.org/10.1093/femsmc/xtad017.

Lee, Y. M., Kim, E., An, J., Lee, Y., Choi, E., ... Kim, W. (2016). Dissection of the HOG pathway activated by hydrogen peroxide in *Saccharomyces cerevisiae*. *Environmental Microbiology*, *19*(2), 584–597. https://doi.org/10.1111/1462-2920.13499.

Lengeler, K. B., Davidson, R. C., Dsouza, C., Harashima, T., Shen, W.-C., Wang, P., et al. (2000). Signal transduction cascades regulating fungal development and virulence. *Microbiology and Molecular Biology Reviews*, *64*, 746–785. https://doi.org/10.1128/mmbr.64.4.746-785.2000.

Li, K., Wei, Z., Jia, J., Xu, Q., Liu, H., Zhong, C., et al. (2023). Engineered living materials grown from programmable *Aspergillus niger* mycelial pellets. *Materials Today Bio*, *19*, 100545. https://doi.org/10.1016/j.mtbio.2023.100545.

Lim, L. L., Fineran, B. A., & Cole, A. L. J. (1983). Ultrastructure of intrahyphal hyphae of *Glomus fasciculatum* (Thaxter) Gerdemann and Trappe in roots of white clover (*Trifolium repens* L.). *New Phytologist*, *95*, 231–239. https://doi.org/10.1111/j.1469-8137.1983.tb03489.x.

Liu, Y. J., Hodson, M. C., & Hall, B. D. (2006). Loss of the flagellum happened only once in the fungal lineage: Phylogenetic structure of kingdom Fungi inferred from RNA polymerase II subunit genes. *BMC Evolutionary Biology*, *6*, 74. https://doi.org/10.1186/1471-2148-6-74.

Liu, S., Hou, Y., Liu, W., Lu, C., Wang, W., & Sun, S. (2015). Components of the calcium-calcineurin signaling pathway in fungal cells and their potential as antifungal targets. *Eukaryotic Cell*, *14*, 324–334.

Liu, Y., Hu, T., Zhao, J., Lv, Y., & Ren, R. (2017). Simultaneous removal of carbon and nitrogen by mycelial pellets of a heterotrophic nitrifying fungus—*Penicillium* sp. L1. *Journal of Bioscience and Bioengineering, 123*(2), 223–229. https://doi.org/10.1016/j.jbiosc.2016.08.009.

Liu, D., Keiblinger, K. M., Leitner, S., Wegner, U., Zimmermann, M., & Zechmeister-Boltenstern, S. (2019). Response of microbial communities and their metabolic functions to drying–rewetting stress in a temperate forest soil. *Microorganisms, 7*(5), 129. https://doi.org/10.3390/microorganisms7050129.

Liu, J., Du, Y., Ma, H., Pei, X., & Li, M. (2020). Enhancement of *Monascus* yellow pigments production by activating the cAMP signalling pathway in *Monascus purpureus* HJ11. *Microbial Cell Factories, 19*, 224. https://doi.org/10.1186/s12934-020-01486-y.

Loukin, S. H., Kung, C., & Saimi, Y. (2007). Lipid perturbations sensitize osmotic downshock activated Ca^{2+} influx, a yeast "deletome" analysis. *FASEB Journal, 21*(8), 1813–1820. https://doi.org/10.1096/fj.06-7898com.

Lucena, R. M., Elsztein, C., Simões, D. A., & Morais, M. A. (2012). Participation of CWI, HOG and calcineurin pathways in the tolerance of *Saccharomyces cerevisiae* to low pH by inorganic acid. *Journal of Applied Microbiology, 113*(3), 629–640. https://doi.org/10.1111/j.1365-2672.2012.05362.x.

Lücking, R., Huhndorf, S., Pfister, D. H., Plata, E. R., & Lumbsch, H. T. (2009). Fungi evolved right on track. *Mycologia, 101*(6), 810–822. https://doi.org/10.3852/09-016.

Lyu, J., Tegelaar, M., Post, H., Moran Torres, J., Torchia, C., Altelaar, A. F., et al. (2023). Heterogeneity in spore aggregation and germination results in different sized, cooperative microcolonies in an *Aspergillus niger* culture. *mBio, 14*(1), e0087022. https://doi.org/10.1128/mbio.00870-22.

Martin, H., Shales, M., Fernandez-Pinar, P., Wei, P., Molina, M., Fiedler, D., et al. (2015). Differential genetic interactions of yeast stress response MAPK pathways. *Molecular Systems Biology, 11*(4), 800. https://doi.org/10.15252/msb.20145606.

Maruyama, J.-I. (2021). Genome editing technology and its application potentials in the industrial filamentous fungus *Aspergillus oryzae*. *Journal of Fungi, 7*(8), 638. https://doi.org/10.3390/jof7080638.

Maumela, P., Rose, S., van Rensburg, E., Chimphango, A. F. A., & Görgens, J. F. (2021). Bioprocess optimisation for high cell density endoinulinase production from recombinant *Aspergillus niger*. *Applied Biochemistry and Biotechnology, 193*(10), 3271–3286. https://doi.org/10.1007/s12010-021-03592-y.

Mbonyi, K., van Aelst, L., Argüelles, J. C., Jans, A. W., & Thevelein, J. M. (1990). Glucose-induced hyperaccumulation of cyclic AMP and defective glucose repression in yeast strains with reduced activity of cyclic AMP-dependent protein kinase. *Molecular and Cellular Biology, 10*(9), 4518–4523. https://doi.org/10.1128/mcb.10.9.4518-4523.1990.

Meharg, A. A., & Cairney, J. W. G. (2000). Co-evolution of mycorrhizal symbionts and their hosts to metal-contaminated environments. *Advances in Ecological Research, 30*, 69–112. https://doi.org/10.1016/S0065-2504(08)60017-3.

Mensonides, F. I. C., Brul, S., Klis, F. M., Hellingwerf, K. J., & Teixeira de Mattos, M. J. (2005). Activation of the protein kinase C1 pathway upon continuous heat stress in *Saccharomyces cerevisiae* is triggered by an intracellular increase in osmolarity due to trehalose accumulation. *Applied and Environmental Microbiology, 71*(8), 4531–4538. https://doi.org/10.1128/aem.71.8.4531-4538.2005.

Meyer, V., Andersen, M. R., Brakhage, A. A., Braus, G. H., Caddick, M. X., Cairns, T. C., et al. (2016). Current challenges of research on filamentous fungi in relation to human welfare and a sustainable bio-economy: A white paper. *Fungal Biology and Biotechnology, 3*, 6. https://doi.org/10.1186/s40694-016-0024-8.

Meyer, V. (2021). Metabolic engineering of flamentous fungi. In J. Nielsen, G. Stephanopoulos, & S. Y. Lee (Eds.). *Metabolic engineering—Concepts and applications* (pp. 765–801)(1st ed.). Weinheim: Wiley-VCH. https://doi.org/10.1002/9783527823468.ch20.

Meyer, V., Cairns, T., Barthel, L., King, R., Kunz, P., Schmideder, S., et al. (2021). Understanding and controlling flamentous growth of fungal cell factories: Novel tools and opportunities for targeted morphology engineering. *Fungal Biology and Biotechnology, 8*, 8. https://doi.org/10.1186/s40694-021-00115-6.

Micheluz, A., Pinzari, F., Rivera-Valentín, E. G., Manente, S., & Hallsworth, J. E. (2022). Biophysical manipulation of the extracellular environment by *Eurotium halophilucum*. *Pathogens, 11*(12), 1462. https://doi.org/10.3390/pathogens11121462.

Miermont, A., Uhlendorf, J., McClean, M., & Hersen, P. (2011). The dynamical systems properties of the HOG signaling cascade. *Journal of Signal Transduction, 930940*, 1–12. https://doi.org/10.1155/2011/930940.

Mittelmann, M. W. (2018). The importance of microbial biofilms in the deterioration of heritage materials. In R. Mitchell, & J. Clifford (Eds.). *Biodeterioration and preservation in art, archaeology and architecture* (pp. 3–15). London: Archetype Publications.

Mojica, E. A., & Kültz, D. (2022). Physiological mechanisms of stress-induced evolution. *The Journal of Experimental Biology, 225*(Suppl. 1), jeb243264. https://doi.org/10.1242/jeb.243264.

Moloney, J. N., & Cotter, T. G. (2018). ROS signalling in the biology of cancer. *Seminars in Cell & Developmental Biology, 80*, 50–64. https://doi.org/10.1016/j.semcdb.2017.05.023.

Müller, H., Barthel, L., Schmideder, S., Schütze, T., Meyer, V., & Briesen, H. (2022). From spores to fungal pellets: A new high-throughput image analysis highlights the structural development of *Aspergillus niger*. *Biotechnology and Bioengineering, 119*(8), 2182–2195. https://doi.org/10.1002/bit.28124.

Müller, H., Deffur, C., Schmideder, S., Barthel, L., Friedrich, T., Mirlach, L., et al. (2023). Synchrotron radiation-based microcomputed tomography for three-dimensional growth analysis of *Aspergillus niger* pellets. *Biotechnology and Bioengineering, 120*, 3244–3260. https://doi.org/10.1002/bit.28506.

Munro, C. A., Selvaggini, S., de Bruijn, I., Walker, L., Lenardon, M. D., ... Gow, N. A. R. (2007). The PKC, HOG and Ca^{2+} signalling pathways co-ordinately regulate chitin synthesis in *Candida albicans*. *Molecular Microbiology, 63*(5), 1399–1413. https://doi.org/10.1111/j.1365-2958.2007.05588.x.

Nair, R. B., Gmoser, R., Lennartsson, P. R., & Taherzadeh, M. J. (2018). Does the second messenger cAMP have a more complex role in controlling filamentous fungal morphology and metabolite production? *MicrobiologyOpen, 7*(4), e00627. https://doi.org/10.1002/mbo3.627.

Naranjo-Ortiz, M. A., & Gabaldón, T. (2019). Fungal evolution: Major ecological adaptations and evolutionary transitions. *Biological Reviews of the Cambridge Philosophical Society, 94*(4), 1443–1476. https://doi.org/10.1111/brv.12510.

Newby, P. J., & Gadd, G. M. (1987). Synnema induction in *Penicillium funiculosum* by tributyltin compounds. *Transactions of the British Mycological Society, 89*, 381–384.

Panankar, D., Liu, N., & Oolman, T. (1993). A fractal mjdel for the characterixation of mycelial morphology. *Biotechnology and Bioengineering, 42*, 571–578.

Papagianni, M. (2004). Fungal morphology and metabolite production in submerged mycelial processes. *Biotechnology Advances, 22*(3), 189–259. https://doi.org/10.1016/j.biotechadv.2003.09.005.

Papagianni, M., & Mattey, M. (2006). Morphological development of *Aspergillus niger* in submerged citric acid fermentation as a function of the spore inoculum level. Application of neural network and cluster analysis for characterization of mycelial morphology. *Microbial Cell Factories, 5*(1), 3. https://doi.org/10.1186/1475-2859-5-3.

Papagianni, M., Mattey, M., & Kristiansen, B. (1999). The influence of glucose concentration on citric acid production and morphology of *Aspergillus niger* in batch and continous culture. *Enzyme and Microbial Technology, 25*, 710–717.

Peay, K. G., & Bruns, T. D. (2014). Spore dispersal of basidiomycete fungi at the landscape scale is driven by stochastic and deterministic processes and generates variability in plant-fungal interactions. *New Phytologist, 204,* 180–191.

Pinna, D. (2021). Microbial growth and its effects on inorganic heritage materials. In E. Joseph (Ed.). *Microorganisms in the deterioration and preservation of cultural heritage* (pp. 3–35). Springer International Publishing. https://doi.org/10.1007/978-3-030-69411-1_1.

Porter, S. M. (2016). Tiny vampires in ancient seas: Evidence for predation via perforation in fossils from the 780–740 million-year-old Chuar Group, Grand Canyon, USA. *Proceedings. Biological Sciences, 283*(1831), 20160221. https://doi.org/10.1098/rspb.2016.0221.

Posas, F. (1997). Osmotic activation of the HOG MAPK pathway via Ste11p MAPKKK: Scaffold role of Pbs2p MAPKK. *Science (New York, N. Y.), 276*(5319), 1702–1705. https://doi.org/10.1126/science.276.5319.1702.

Posas, F., Takekawa, M., & Saito, H. (1998). Signal transduction by MAP kinase cascades in budding yeast. *Current Opinion in Microbiology, 1*(2), 175–182. https://doi.org/10.1016/s1369-5274(98)80008-8.

Rabotnova, I. (1980). *Growth inhibitors and metabolism of microorganisms. Limitation and inhibition of microbiological processes.* Pushchino: IBPHM3–22. (in Russian).

Ramsay, L. M., Sayer, J. A., & Gadd, G. M. (1999). Stress responses of fungal colonies towards metals. In N. A. R. Gow, G. D. Robson, & G. M. Gadd (Eds.). *The fungal colony* (pp. 178–200). Cambridge: Cambridge University Press.

Rayner, A. D. M., Griffith, G. S., & Ainsworth, A. M. (1995). Mycelial interconnectedness. In N. A. R. Gow, & G. M. Gadd (Eds.). *The growing fungus* (pp. 21–40). London: Chapman & Hall.

Rep, M., Proft, M., Remize, F., Tamas, M., Serrano, R., ... Hohmann, S. (2001). The *Saccharomyces cerevisiae* Sko1p transcription factor mediates HOG pathway-dependent osmotic regulation of a set of genes encoding enzymes implicated in protection from oxidative damage. *Molecular Microbiology, 40*(5), 1067–1083. https://doi.org/10.1046/j.1365-2958.2001.02384.x.

Rolland, F., Winderickx, J., & Thevelein, J. M. (2002). Glucose-sensing and -signalling mechanisms in yeast. *FEMS Yeast Research, 2*(2), 183–201. https://doi.org/10.1111/j.1567-1364.2002.tb00084.x.

Ruiz-Herrera, J., & Ortiz-Castellanos, L. (2019). Cell wall glucans of fungi. A review. *Cell Surface (Amsterdam, Netherlands), 5,* 100022. https://doi.org/10.1016/j.tcsw.2019.100022.

Sano, R., & Reed, J. C. (2013). ER stress-induced cell death mechanisms. *Biochimica et Biophysica Acta, 1833*(12), 3460–3470. https://doi.org/10.1016/j.bbamcr.2013.06.028.

Schimel, D. S. (1995). Terrestrial ecosystems and the carbon cycle. *Global Change Biology, 1,* 77–91. https://doi.org/10.1111/j.1365-2486.1995.tb00008.x.

Schmideder, S., Barthel, L., Friedrich, T., Thalhammer, M., Kovačević, T., Niessen, L., et al. (2019). An X-ray microtomography-based method for detailed analysis of the three-dimensional morphology of fungal pellets. *Biotechnology and Bioengineering, 116*(6), 1355–1365. https://doi.org/10.1002/bit.26956.

Schrinner, K., Veiter, L., Schmideder, S., Doppler, P., Schrader, M., Münch, N., et al. (2020). Morphological and physiological characterization of filamentous *Lentzea aerocolonigenes*: Comparison of biopellets by microscopy and flow cytometry. *PLoS One, 15*(6), e0234125. https://doi.org/10.1371/journal.pone.0234125.

Selye, H. (1938). Experimental evidence supporting the conception of "adaptation energy". *American Journal of Physiology-Legacy Content, 123*(3), 758–765. https://doi.org/10.1016/10.1152/ajplegacy.1938.123.3.758.

Serra-Cardona, A., Canadell, D., & Arino, J. (2015). Coordinate responses to alkaline pH stress in budding yeast. *Microbial Cell, 2*(6), 182–196. https://doi.org/10.15698/mic2015.06.205.

Shemshur, T. V., Gromozova, E. N., & Podgorsky, V. S. (1989). Influence of cultivation conditions on the growth pattern of micromycetes in submerged culture. *Mikrobiologichnyi Zhurnal, 51*(3), 30–33. (in Russian).

Shim, A. R., Nap, R. J., Huang, K., Almassalha, L. M., Matusda, H., ... Szleifer, I. (2020). Dynamic crowding regulates transcription. *Biophysical Journal, 118*(9), https://doi.org/10.1016/j.bpj.2019.11.007.

Smith, S. E., & Read, D. J. (2008). *Mycorrhizal symbiosis* (3rd ed.). London, UK: Academic Press. https://doi.org/10.1016/B978-0-12-370526-6.X5001-6.

Sorger, D., & Daum, G. (2003). Triacylglycerol biosynthesis in yeast. *Applied Microbiology and Biotechnology, 61*(4), 289–299. https://doi.org/10.1007/s00253-002-1212-4.

Sterflinger, K. (2000). Fungi as geologic agents. *Geomicrobiology Journal, 17*, 97–124.

Steward, P. R., & Rogers, P. J. (1983). *Fungal dimorphism. Fungal differentiation.* New York: Dekker, 267–313.

Sun, X., Wu, H., Zhao, G., Li, Z., Wu, X., Liu, H., et al. (2018). Morphological regulation of *Aspergillus niger* to improve citric acid production by chsC gene silencing. *Bioprocess and Biosystems Engineering, 41*(7), 1029–1038. https://doi.org/10.1007/s00449-018-1932-1.

Suzuki, T., Maeda, A., Hirose, M., Ichinose, Y., Shiraishi, T., & Toyoda, K. (2017). Ultrastructural and cytological studies on *Mycosphaerella pinodes* infection of the model legume *Medicago truncatula. Frontiers in Plant Science, 8*, 1132. https://doi.org/10.3389/fpls.2017.01132.

Szentivanyi, O., & Kiss, L. (2003). Overwintering of *Ampelomyces* mycoparasites on apple trees and other plants infected with powdery mildews. *Plant Pathology, 52*, 737–746. https://doi.org/10.1111/j.1365-3059.2003.00937.x.

Taylor, T. N., & Osborn, J. M. (1996). The importance of fungi in shaping the paleoecosystem. *Review of Palaeobotany and Palynology, 90*, 249–262.

Taymaz-Nikerel, H., Cankorur-Cetinkaya, A., & Kirdar, B. (2016). Genome-wide transcriptional response of *Saccharomyces cerevisiae* to stress-induced perturbations. *Frontiers in Bioengineering and Biotechnology, 4*. https://doi.org/10.3389/fbioe.2016.00017.

Terenzi, H. F., Flawia, M. M., & Torres, H. N. (1974). A *Neurospora crassa* morphological mutant showing reduce adenylate cyclase activity. *Biochemical and Biophysical Research Communications, 58*(4), 990–996.

Terenzi, H. F., Flawia, M. M., Tellez-Inon, M. T., & Torres, H. N. (1979). Control of *Neurospora crassa* morphology of cyclic adenosine-3',5'-monophosphate and dibutyryl cyclic adenosine-3',5'-monophosphate. *Journal of Bacteriology, 126*(1), 91–99.

Tinnell, W. H., Jefferson, B. L., & Benoit, R. E. (1977). Manganese mediated morphogenesis in *Penicillium claviforme* and *Penicillium clavigerum. Canadian Journal of Microbiology, 23*, 209–212.

Valero, C., Colabardini, A. C., de Castro, P. A., Silva, L. P., Ries, L. N. A., Pardeshi, L., et al. (2021). *Aspergillus fumigatus* ZnfA, a novel zinc finger transcription factor involved in calcium metabolism and caspofungin tolerance. *Frontiers in Fungal Biology, 2*, 689900. https://doi.org/10.3389/ffunb.2021.689900.

Valiante, V., Macheleidt, J., Föge, M., & Brakhage, A. A. (2015). The *Aspergillus fumigatus* cell wall integrity signaling pathway: drug target, compensatory pathways, and virulence. *Frontiers in Microbiology, 6*, 325. https://doi.org/10.3389/fmicb.2015.00325.

Veiter, L., Rajamanickam, V., & Herwig, C. (2018). The filamentous fungal pellet—relationship between morphology and productivity. *Applied Microbiology and Biotechnology, 102*, 2997–3006. https://doi.org/10.1007/s00253-018-8818-7.

Veiter, L., & Herwig, C. (2019). The filamentous fungus *Penicillium chrysogenum* analysed via flow cytometry—A fast and statistically sound insight into morphology and viability. *Applied Microbiology and Biotechnology, 103*, 6725–6735. https://doi.org/10.1007/s00253-019-09943-4.

Veiter, L., Kager, J., & Herwig, C. (2020). Optimal process design space to ensure maximum viability and productivity in *Penicillium chrysogenum* pellets during fed-batch cultivations through morphological and physiological control. *Microbial Cell Factories, 19*, 33. https://doi.org/10.1186/s12934-020-1288-5.

Volkmann, M., Whitehead, K., Rutters, H., Rullkotter, J., & Gorbushina, A. A. (2003). Mycosporine-glutamicol-glucoside: a natural UV-absorbing secondary metabolite of rock-inhabiting microcolonial fungi. *Rapid Communications in Mass Spectrometry, 17*, 897–902.

Voychuk, S., & Gromozova, E. N. (2020a). The functional role of PPN1 and PPX1 polyphosphatases under stresses action and for adaptive response development. *Mikrobiolohichnyi Zhurnal, 82*(1), 3–12. https://doi.org/10.15407/microbiolj82.01.003.

Voychuk, S. I., & Gromozova, O. M. (2020b). The role of PPN1 and PPX1 polyphosphatases in the stress-induced changes of the polysaccharide composition of cell wall and extracellular matrix of *Saccharomyces cerevisiae* cells. *Mikrobiolohichnyi Zhurnal, 82*(2), 3–13. https://doi.org/10.15407/microbiolj82.02.003.

Walker, G. M., & White, N. A. (2017). Introduction to fungal physiology. In K. Kavanagh (Ed.). *Fungi: Biology and applications* (pp. 1–35). Hoboken: John Wiley & Sons, Inc.

Wang, B., Chen, J., Li, H., Sun, F., Li, Y., & Shi, G. (2017). Pellet-dispersion strategy to simplify the seed cultivation of *Aspergillus niger* and optimize citric acid production. *Bioprocess and Biosystems Engineering, 40*(1), 45–53. https://doi.org/10.1007/s00449-016-1673-y.

Wu, H., Fang, Y., Yu, J., & Zhang, Z. (2014). The quest for a unified view of bacterial land colonization. *ISME Journal, 8*, 1358–1369.

Wucherpfennig, T., Kiep, K., Driouch, H., Wittmann, C., & Krull, R. (2010). Morphology and rheology in filamentous cultivations. *Advances in Applied Microbiology, 72*, 89–136. https://doi.org/10.1016/S0065-2164(10)72004-9.

Yakymenko, I., Tsybulin, O., Sidorik, E., Henshel, D., Kyrylenko, O., & Kyrylenko, S. (2015). Oxidative mechanisms of biological activity of low-intensity radiofrequency radiation. *Electromagnetic Biology and Medicine, 35*(2), 186–202. https://doi.org/10.3109/15368378.2015.1043557.

Yoshimi, A., Miyazawa, K., Kawauchi, M., & Abe, K. (2022). Cell wall integrity and its industrial applications in filamentous fungi. *Journal of Fungi, 8*(5), 435. https://doi.org/10.3390/jof8050435.

Zeltina, M. O., Shvinka, Y. E., Blaschuk, I. S., & Gromozova, E. N. (1990). Energy utilization of a number of carbon-containing substrates by various mycelial structures of *Thielavia terrestris*. *Mikrobiologichnyi Zhurnal, 52*(6), 65–69.

Zhang, J., & Zhang, J. (2016). The filamentous fungal pellet and forces driving its formation. *Critical Reviews in Biotechnology, 36*(6), 1066–1077. https://doi.org/10.3109/07388551.2015.1084262.

Zhang, X., Liu, H., Zhang, M., Chen, W., & Wang, C. (2023). Enhancing *Monascus* pellet formation for improved secondary metabolite production. *Journal of Fungi, 9*(11), 1120. https://doi.org/10.3390/jof9111120.

Zhang, S., Wang, Y., Wang, B., & Wang, S. (2024). A review on superiority of mycelial pellets as bio-carriers: Structure, surface properties, and bioavailability. *Journal of Water Process Engineering, 58*, 104745. https://doi.org/10.1016/j.jwpe.2023.104745.

Zheng, X., Cairns, T. C., Ni, X., Zhang, L., Zhai, H., Meyer, V., et al. (2022). Comprehensively dissecting the hub regulation of PkaC on high-productivity and pellet macromorphology in citric acid producing *Aspergillus niger*. *Microbial Biotechnology, 15*(6), 1867–1882. https://doi.org/10.1111/1751-7915.14020.

Zhdanova, N., Fomina, M., Redchitz, T., & Olsson, S. (2001). Chernobyl effect: growth characteristics of soil fungi *Cladosporium cladosporioides* [Fresen] de Vries with and without positive radiotropism. *Polish Journal of Ecology, 49*(4), 309–318.

Zhdanova, N. N., Tugay, T., Dighton, J., Zheltonozhsky, V., & Mcdermott, P. (2004). Ionizing radiation attracts soil fungi. *Mycological Research, 108*(9), 1089–1096.

Žnidaršič-Plazl, P., & Pavko, A. (2001). The morphology of filamentous fungi in submerged cultivations as a bioprocess parameter. *Food Technology and Biotechnology, 39*(3), 237–252.

CHAPTER SIX

Remediation of toxic metal and metalloid pollution with plant symbiotic fungi

Qing Zhen, Xinru Wang, Xianxian Cheng, and Weiguo Fang[*]
MOE Key Laboratory of Biosystems Homeostasis & Protection, Institute of Microbiology, College of Life Science, Zhejiang University, Hangzhou, P.R. China
*Corresponding author. e-mail address: wfang1@zju.edu.cn

Contents

Abstract

Anthropogenic activities have dramatically accelerated the release of toxic metal(loid)s into soil and water, which can be subsequently accumulated in plants and animals, threatening biodiversity, human health, and food security. Compared to physical and chemical remediation, bioremediation of metal(loid)-polluted soil using plants and/or plant symbiotic fungi is usually low-cost and environmentally friendly. Mycorrhizal fungi and endophytic fungi are two major plant fungal symbionts. Mycorrhizal fungi can immobilize metal(loid)s via constitutive mechanisms, including intracellular sequestration with vacuoles and vesicles and extracellular immobilization by cell wall components and extracellular polymeric substances such as glomalin. Mycorrhizal fungi can improve the efficacy of phytoremediation by promoting plant symplast and apoplast pathways. Endophytic fungi also use constitutive cellular components to immobilize metal(loid)s and to reduce the accumulation of metal(loid)s in plants by modifying plant physiological status. However, a specific mechanism for the removal of methylmercury pollution was recently discovered in the endophytic fungi *Metarhizium*, which could be acquired from bacteria via horizontal gene transfer. In contrast to mycorrhizal fungi that are obligate biotrophs, some endophytic fungi, such as *Metarhizium* and *Trichoderma*, can be massively and cost-effectively produced, so they seem to be well-placed for remediation of metal(loid)-polluted soil on a large scale.

Advances in Applied Microbiology, Volume 129
ISSN 0065-2164, https://doi.org/10.1016/bs.aambs.2024.04.001

1. Introduction

Some metals and metalloids, referred to collectively as metal(loid)s in this paper, are essential for life, such as zinc (Zn), copper (Cu), nickel (Ni), and cobalt (Co); however, they are toxic at high concentrations. In contrast, arsenic (As), cadmium (Cd), lead (Pb), and mercury (Hg) are non-essential and highly toxic even at relatively low concentrations. These toxic metal(loid)s occur naturally in the earth's crust and are found in soils, rocks, sediments, and waters with background concentrations. Since industrialization, the release of naturally occurring toxic metal(loid)s has dramatically accelerated due to anthropogenic activities of mining, metallurgical, and energy production as well as agricultural and medical practices, resulting in metal(loid) pollution in soil and water (Al-Sulaiti, Soubra, & Al-Ghouti, 2022; Dhaliwal, Singh, Taneja, & Mandal, 2020; Zhao, Tang, Song, Huang, & Wang, 2022). Globally, an estimated five million sites have been contaminated by toxic metal(loid)s (He et al., 2015; Zhang, Xu, Kanyerere, Wang, & Sun, 2022). These metal(loid)s are non-degradable in the environment and can accumulate in plants and animals, threatening biodiversity, human health, and food security (Chen & Dong, 2022; Gong & Tian, 2019; Huang et al., 2018).

Currently, metal(loid)-polluted soil can be physically, chemically, and biologically remediated. Physical remediation methods include soil replacement, soil washing, and vitrification. These methods are usually costly and time-consuming; thus, they cannot be scaled up to remediate metal (loid)-polluted agricultural soil (Dhaliwal et al., 2020; Liu, Li, Song, & Guo, 2018). Chemical remediation involves immobilization or extraction of toxic metal(loid)s from polluted soil using chemical agents (Dhaliwal et al., 2020). Chemical remediation is usually expensive, can produce large amounts of sludge and secondary pollution, and results in loss of soil environmental functions (Liu et al., 2018; Zhang et al., 2022).

Biological remediation involves the use of plants and/or microorganisms to remediate polluted soil, which is usually low-cost and environmentally friendly (Liu et al., 2018; Pande, Pandey, Sati, Bhatt, & Samant, 2022, Rangel, Finlay, Hallsworth, Dadachova, & Gadd, 2018). Around 0.2% of all known plant species, called hyperaccumulators (Alfred stonecrop *Sedum alfredii* for Cd, centipede grass *Eremochloa* for Pb), are currently known to be able to actively uptake metal(loid)s in large amounts from the soil without showing any traits of toxicity; thus, they are used to remediate metal(loid)-polluted soil, i.e., phytoremediation (Rascio & Navari-Izzo, 2011). The efficacy of phytoremediation can be improved by plant symbionts including fungi (Li, Xu, et al., 2022; Riaz et al., 2021). Plant symbiotic fungi are also able to remediate polluted soil, which is called mycoremediation (Wu et al., 2022). In this article, we review the current knowledge of how plant symbiotic fungi remediate metal(loid)-polluted soil and the underlying cellular, molecular, biochemical, and evolutionary mechanisms.

2. Plant symbiotic fungi

Over 90% of terrestrial plants have symbiotic associations with fungi (Malloch, Pirozynski, & Raven, 1980). These relationships include parasitism and mutualism. In a parasitic association, fungi live in or on plants, obtain their nutrients from them, and cause illness to their plant hosts. In contrast, both plants and fungi benefit from the mutualistic associations, in which fungi promote plant growth and health by helping absorb water and nutrients and antagonizing plant pathogens; in exchange, they obtain various products (carbon) of photosynthesis.

The mutualistic associations between fungi and plants are either obligate or facultative. Mycorrhizae are the most commonly found and widely known obligate symbiotic associations. In mycorrhizae, carbon flows to the fungi and inorganic nutrients move to the plants, providing a critical linkage between plant roots and soil. Mycorrhizal plants are often more competitive and more able to tolerate environmental stresses than non-mycorrhizal plants (Drijber & McPherson, 2021). Although multiple types of mycorrhizae have been documented (Riaz et al., 2021; Siddiqui & Pichtel, 2008), they are generally classified into ectomycorrhizae and endomycorrhizae. In ectomycorrhizae, fungal hyphae grow between root cortical and epidermal cells, forming a Hartig net and providing a large surface area for the interchange of nutrients between hosts and fungi (Teste, Jones, & Dickie, 2020). Only around 2% of terrestrial plant species are associated with ectomycorrhizal fungi (Tedersoo et al., 2012), but the boreal and temperate forest biome that these plant species dominate occupies a disproportionately large global area; therefore, ectomycorrhizae have a significant influence on global biogeochemical cycling (Averill, Turner, & Finzi, 2014).

In endomycorrhizae, fungal hyphae penetrate and establish in plant cells. The most commonly recorded endomycorrhizae is arbuscular mycorrhizae, in which characteristic vesicles and arbuscules are formed in the root cortical cells. The plasmalemma of the host cell invaginates and encloses arbuscules. The arbuscular mycorrhizal fungi develop mutualistic associations with more than 80% of terrestrial plant species and are regarded as "the mother of plant root endosymbiosis" as they probably facilitated plant territorialization during the Mid-Ordovician (475 Ma) (Parniske, 2008; Smith & Read, 1997).

In addition to the obligate biotrophic mycorrhizal fungi, many free-living saprotrophic fungi develop facultative mutualistic and unsymptomatic associations with their plant hosts. Most of these fungi are endosymbiotic microorganisms and are called fungal endophytes. Some endophytes are capable of internal migration within plant tissue to colonize foliage, stems, and bark, as well as roots (Behie & Bidochka, 2014). In addition to colonizing of plant tissues, some endophytes can also proliferate in rhizospheric soil nourished by root exudates. Therefore, the definition of "endophyte" is being challenged as more information about their important ecological roles is being discovered (Behie & Bidochka, 2014). Many fungal endophytes have multiple habits; for example, *Metarhizium* are known to be decomposers of organic detritus (saprophytism) and pathogens of insects (parasitism) (St Leger & Wang, 2020). In contrast to mycorrhizal fungi, with their strong saprophytic ability some beneficial endophytic fungi, such as *Metarhizium* and *Trichoderma*, can be massively produced to develop biological agents for controlling plant pathogens and insect pests (Harman, Howell, Viterbo, Chet, & Lorito, 2004; Zhao, Lovett, & Fang, 2016).

3. Remediation of toxic metal(loid)-polluted soil by mycorrhizal fungi

In the metal(loid)-contaminated soil, mycorrhizal fungi use various strategies to immobilize the toxic elements in the rhizosphere soil, providing physical barriers that limit the access of the metal(loid)s to plant hosts. Depending on the fungal and plant species and the properties of metal(loid)s, mycorrhizal fungi either promote or reduce the accumulation of toxic metal(loid)s in plant hosts.

Mycorrhizal fungi can extracellularly and intracellularly immobilize metal(loid)s using constitutively expressed cellular components (Fig. 1). In the extracellular strategy, metal(loid)s are adsorbed by the cell wall components or extracellular polymeric substances (EPS). The protein, glucan, chitin, and melanin in fungal cell walls contain amino, hydroxyl, carboxyl, and other groups, which can bind and sequestrate several toxic metal(loid)s (Kapoor & Viraraghavan, 1995). The extracellular glycoprotein glomalin is

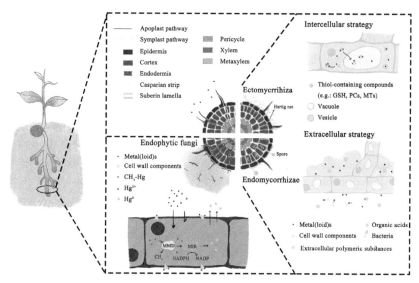

Fig. 1 A schematic diagram to show remediation of metal(loid)-polluted soil by mycorrhizal and endophytic fungi associated with their plant hosts. Note: for endophytic fungi, removal of methylmercury (CH$_3$-Hg) and Hg^{2+} with MMD and MIR was only found in *Metarhizium*. *GSH*, Glutathione; *PCs*, phytochelatins; *MTs*, metallothioneins; *MMD*, methylmercury demethylase; *MIR*, mercury ion reductase.

water-insoluble, thermostable, and highly recalcitrant. It is produced by arbuscular mycorrhizal fungi and deposited in soil in large amounts (Ghasemi Siani, Fallah, Pokhrel, & Rostamnejadi, 2017; Wang et al., 2019). Glomalin significantly contributes to maintaining the stability of soil aggregates and the store of soil C and N pools, as well as alleviating various plant stresses (Rillig, 2004; Wang et al., 2018). This protein is also able to sequestrate or chelate toxic metal(loid)s (Sayin, Khalvati, Erdinçler, & Soils, 2019). Some mycorrhizal fungi produce other EPSs, and their carboxyl, amine, phosphoric, and hydroxyl groups are also able to chelate metal(loid)s (Li & Yu, 2014; More, Yadav, Yan, Tyagi, & Surampalli, 2014). In addition to EPSs, mycorrhizal fungi secret organic acids, such as citric, malic, oxalic, and amino acids, into the rhizospheric soil (Jones, 1998). This acidifies the rhizosphere to increase the mobility of some metal(loid)s or to immobilize others through precipitation and complexation (Fomina, Alexander, Colpaert, & Gadd, 2005). The exudants from the mycorrhizal fungi even modify the microenvironments around the hyphae, promoting sequestration of metal(loid)s by the bacteria in the hyposphere (Ma, Rajkumar, Oliveira, Zhang, & Freitas, 2019).

In the intracellular strategy, mycorrhizal fungi use several mechanisms to sequestrate toxic metal(loid)s transported into the cells, which results in detoxification and provides fungal tolerance (Fig. 1). The vacuolar compartmentalization is the major mechanism to detoxify metal(loid)s within fungal cells (González-Guerrero et al., 2008). In the cytosol, thiol-containing compounds, including GSH (glutathione), PCs (phytochelatins), and MTs (metallothioneins), chelate the metal(loid)s to maintain homeostasis, and the resulting complexes are then compartmented into the vacuoles (González-Guerrero et al., 2010; Liao, O'Brien, Fang, & St Leger, 2014; Mishra et al., 2009). Polyphosphates in the form of insoluble granules in the vacuoles can also complex with a variety of metal(loid)s (Ashford, Peterson, Dwarte, & Chilvers, 1986; Martin, Rubini, Côté, & Kottke, 1994), thus sequestrating them in the organelle. Similar to vacuoles, vesicles may be also involved in storing toxic compounds, thereby, providing an additional dimension of detoxification (Göhre & Paszkowski, 2006). Toxic metal(loid)s usually induce oxidative stress by producing reactive oxygen species (ROSs), which are usually scavenged by superoxide dismutases (SODs) (Gonzalez-Guerrero, Benabdellah, Valderas, Azcon-Aguilar, & Ferrol, 2009).

4. Impact of mycorrhizal fungal colonization on phytoremediation

In addition to impacting the availability of metal(loid)s to plant hosts, colonization by mycorrhizal fungi substantially influences the physiological and biochemical statuses of their hosts, which could either increase or decrease uptake of toxic metal(loid)s by plants. Mycorrhizal fungi-mediated promotion of toxic metal(loid)s uptake can facilitate phytoremediation, which aims to remove the toxins from soil via phytoextraction or phytostabilization; however, this could increase bioaccumulation in crops and vegetables to threaten human health (Li, Gao, et al., 2022).

Uptake and transportation of toxic metal(loid)s by plants are achieved via symplast or apoplast pathways (Fig. 1). For symplast pathway, toxic metal (loid) uptake and partitioning (to distribute in tissues) are largely determined by transporters and cellular components including PCs and MTs that can chelate and sequestrate the toxins in cytosol (Geldner, 2013; Lei, Yamaji, & Ma, 2021; Ueno et al., 2010). In rice, the silicon transporters (Lsi1 and Lsi2) can uptake arsenite [As(III)], dimethylarsinic acid (DMA), and mono-methylarsonic acid (MMA) (Li et al., 2009; Ma et al., 2008). In the presence of As(III) in soil, the mycorrhizal fungus *Glomus intraradices* reduced the expression of Lsi1 and Lsi2 in rice roots, reducing As(III) accumulation in the plants and increasing in the plant biomass (Chen et al., 2012). The higher As (III)/As(V) concentration ratio in mycorrhizal roots indicated that mycorrhizal fungi could prevent transformation of As(III) oxidation to As(V) (Chen et al., 2012; González-Chávez et al., 2011).

Plants usually have a variety of phosphate transporters (PTs) that can uptake and transport arsenate [As(V)] (Zhao, Ma, Meharg, & McGrath, 2009). Symbiotic association with mycorrhizal fungi substantially affects the expression of PTs in plant hosts and, in turn, presumably impacts the uptake and distribution of As(V) in the plants. The opposite impacts of mycorrhizal fungal colonization on phosphate transporter expression were found in *Medicago* species. The fungus *Rhizophagus irregularis* colonization increased expression of phosphate transporter MtPT4 in roots, resulting in increased As accumulation in the roots (Li, Sun, Jiang, Chen, & Zhang, 2018). However, the fungus *G. versiforme* decreased the expression of the phosphate transporters MtPT1 and MtPT2 in the roots of *M. truncatula* (Liu, Trieu, Blaylock, & Harrison, 1998), but its impact on As accumulation has not been documented. In rice, 13 OsPts have been characterized, some of which are expressed in tissue-specific patterns. OsPT1 is expressed in epidermal cells, which thus transports As(V) from the soil or apoplast into cells (Kamiya, Islam, Duan, Uraguchi, & Fujiwara, 2013; Seo et al., 2008). OsPT4 is expressed in epidermal and endodermal cells; it uptakes As(V) from soil to root and may subsequently translocate the toxin into the xylem (Cao et al., 2017). OsPT8 was also shown to play an important role in As(V) uptake (Wang, Zhang, Mao, Xu, & Zhao, 2016; Wu, Ren, McGrath, Wu, & Zhao, 2011). Impacts of mycorrhizal fungi on the expression of OsPT1, 4, and 8 remain to be fully explored, but the fungus *G. intraradices* only specifically upregulated OsPT11in the rice *Oryza sativa*, which was independent of the nutritional status of the plants and phosphate availability in the rhizosphere (Paszkowski, Kroken, Roux, & Briggs, 2002).

In addition to As, mycorrhizal fungi impact the uptake and translocation of other toxic metal(loid)s by plants. Several transporters, including OsNRAMP5 and OsCd1, are involved in cadmium (Cd) uptake and translocation by rice plants (Sasaki, Yamaji, Yokosho, & Ma, 2012; Takahashi et al., 2012; Tan et al., 2019). The impacts of mycorrhizal fungi on their expression also vary according to the combination between fungal and plant species. The fungus *R. intraradices* reduced the expression of the transporters OsNRAMP5 and OsHMA3 in rice roots, lowering Cd content in the roots and shoots. In contrast, the expression levels of these two transporters in the roots were upregulated by the fungus *F. mosseae*; however, the Cd content in roots and shoots was also reduced (Chen et al., 2019). Mycorrhizal fungi were shown to impact the partitioning of methylmercury in rice with less accumulation in grains than in stems and leaves (Li et al., 2023), but the underlying mechanisms remain to be revealed.

Passive migration of metal(loid)s, via apoplast pathway, into xylem is usually not successful due to the hydrophobic apoplastic barrier (Casparian strip and suberin lamella) at the endodermis. However, supported by the force of diffusion and transpiration, toxic metal(loid)s could bypass the hydrophobic barrier at the emerging points of the lateral roots and the root zones where the apoplastic barrier is not mature. The emerging points of the lateral roots are located at the pericycle positioned at the proximal side of the endodermis (Fig. 1), creating a potential leakage of the apoplastic barrier (Santos Teixeira & ten Tusscher, 2019). Mycorrhizal fungal colonization promotes the emergence of lateral roots, increasing the extent of leakage in the apoplast barrier, which could be translated into increased uptake of metal(loid)s by passive migration forced by transpiration and diffusion. The apoplastic barriers Casparian strips and suberin lamella at both the exodermis and endodermis are not well developed in the premature zone of roots, through which toxic metal(loid)s are also able to pass (Ranathunge, Lin, Steudle, & Schreiber, 2011). Mycorrhizal fungal colonization induces more production of roots, resulting in an increase in the total length of premature zone of roots and, thus, more passive migration of metal(loid)s into the xylem (Li, Gao, et al., 2022).

5. Bioremediation by endophytic fungi of toxic metal (loid)s in polluted soil

Compared to mycorrhizal fungi, the involvement of endophytic fungi in bioremediation of soil polluted with toxic metal(loid)s has been less investigated. Among endophytic fungi, *Trichoderma* have been mostly studied. Similar to mycorrhizal fungi, this group of fungi uses constitutively expressed cellular components to adsorb or sequestrate toxic metal(loid)s. They also facilitate phytoextraction of the toxins by modifying plant physiological status (Sun, Karuppiah, & Chen, 2020) (Fig. 1).

Although the ability of the endophytic *Metarhizium* to bioremediate soil and water contaminated by toxic metal(loid)s was only recently explored, a unique specific mechanism to remove mercury pollution was found in this group of fungi (Wu et al., 2022) (Fig. 1). *Metarhizium* are best known for their ability to infect and kill many different arthropods, but most are also saprophytes, rhizosphere colonizers, and beneficial endophytes, with the ability to switch between these different lifestyles (Guo et al., 2017; St Leger & Wang, 2020). Almost all *Metarhizium* species contain a methylmercury demethylase (MMD) and a mercury ion reductase (MIR). The contributions of these two genes to mercury tolerance and the plant–fungus interaction have been genetically and biochemically characterized in the species *M. robertsii* (Wu et al., 2022). *M. robertsii* demethylates methylmercury via MMD to produce divalent mercury, which is then reduced by MIR to volatile elemental mercury. Armed by these two genes, *M. robertsii* can also remove methylmercury and Hg^{2+} from fresh and sea water even in the absence of added nutrients. In methylmercury or Hg^{2+}-contaminated soil, *M. robertsii* associated with maize roots can remove mercury pollutants in the rhizosphere soil, decreasing mercury accumulation in plants and greatly increasing plant growth (Wu et al., 2022). Therefore, MMD and MIR promote the mutually beneficial relationship between plants and *M. robertsii* under mercury stress. This also has important implications for potential bioremediation of mercury-polluted soil. *M. robertsii* can be applied to seeds before planting (Liao et al., 2014). This contrasts favorably with the bacterial sources of MMD (MerB) and MIR (MerA), as these bacteria cannot reproduce in soil, and have yet to be used for bioremediation of mercury-polluted soil (Mahbub et al., 2017). An extra dimension in the quality of the interactions between *Metarhizium* and plants is that this group of fungi can grow and spread through the rhizosphere as hyphal growth (Wu et al., 2022). A number of *Metarhizium* species can efficiently demethylate methylmercury and reduce Hg^{2+}, as well as develop symbiotic relationships with a variety of plants, such as grasses, trees, vegetables, and crops (Ahmad, de Miménez-Gasco, Luthe, Shakeel, & Barbercheck, 2020; Moonjely, Barelli, & Bidochka, 2016). Furthermore, a previous study showed that coevolution occurred between *Metarhizium* and plants, in that *M. robertsii* preferentially associates with grass roots, *M. brunneum* with shrubs, and *M. guizhouense* with trees (Wyrebek, Huber, Sasan, & Bidochka, 2011). Therefore, plant and *Metarhizium* species could be combined in optimal pairs to clean up methylmercury and Hg^{2+} in different types of polluted soil. *Metarhizium* have long been used as biocontrol agents against insect pests, and their safety to humans and the environment has been clearly established through several decades of high-quality research

(Alder–Rangel, 2021, Zhao et al., 2016). In addition, industrial production of *Metarhizium* for insect pest control is highly automated and cost-effective (St Leger, 2021).

In addition to *Mmd* and *Mir*-mediated tolerance to mercury, *M. robertsii* colonization also indirectly reduces toxic metal(loid) accumulation in plants and promotes plant growth. Recently, it was reported that this fungus significantly reduced Cd accumulation in rice roots, stems, leaves, husks, and grains. This is due to *M. robertsii*-induced suppression of Cd intake transporter gene *osNramp5* in the rice roots, and by chemical stabilization of Cd to the residual fraction in the rhizosphere soil (Jiang, Dai, et al., 2022). Production of endogenous hormone levels of indole-3-acetic, gibberellin A3, and brassinolide is also induced by *M. robertsii*, resulting in promotion of photosynthesis and plant growth (Jiang, Fang, et al., 2022). Furthermore, *M. robertsii* protects the plants from Cd-induced oxidative stress by increasing enzyme activities of catalase-peroxidase and the production of GSH, ascorbic acid, and proline in the rice plants (Jiang, Fang, et al., 2022). Similarly, *M. robertsii* colonization suppresses Cd accumulation in *Arabidopsis thaliana* shoots and roots. Different from its impacts on rice, *M. robertsii* elevates Cd efflux capacity via the upregulation of three *PCR* genes in *A. thaliana*. Similar to rice, *M. robertsii* also protects *A. thaliana* by increasing production of catalase-peroxidase, GSH, and multiple HPP proteins for Cd sequestration (Jiang, Fang, et al., 2022).

6. Evolutionary adaptation to toxic metal(loid)s in plant symbiotic fungi

Similar to other stresses, high concentrations of toxic metal(loid)s in soil, whether from natural origins or from anthropogenic activities, may pose considerable challenges to the exposed biota, including plant symbiotic fungi. Since little is currently known on the genetic, molecular, and biochemical basis of tolerance to toxic metal(loid)s in mycorrhizal fungi, few studies have addressed the evolution of their tolerance to the toxins or how such stress tolerance affects the diversification of this group of fungi. In a recent study (Bazzicalupo et al., 2020), substantial genetic and phenotypic differences were identified across *Suillus luteus* individuals in metal-contaminated and noncontaminated soils; however, no evidence was found to support a population differentiation related to metal stress in this species. In contrast, two previous studies on *S. brevipes* (Branco et al., 2015) and *Pisolithus albus* (Jourand et al., 2010) found associations between population structure and stress tolerance but no evidence was provided to show that such population differentiation is a first step for speciation and contributes to adaptation to specific stressful environmental conditions. Mycorrhizal fungi present extremely variable resistance to toxic metal(loid)s, and some strains can survive even in very harsh conditions with metal(loid) toxicities (Riaz et al., 2021). In this scenario, some authors argued that little evolutionary adaptation to stresses have occurred in mycorrhizal fungal communities as there might be sufficient mycorrhizal fungi with a high constitutive tolerance, which are dominant in metal-contaminated environments (Blaudez, Botton, & Chalot, 2000).

Some commonly known endophytic fungi, such as *Metarhizium* and *Trichoderma*, are genetically tractable, and with support from various omics analyses, molecular mechanisms underlying stress tolerance in these fungi can be systematically investigated, which significantly facilitates the investigation of their evolutionary adaptation to stresses. A genome-wide analysis showed that *M. robertsii* has 18 genes that are derived via horizontal gene transfer (HGT). Following genetic and biochemical analyses showed that the acquisition of 12 genes from bacteria or arthropods was necessary for *M. robertsii* to become an insect pathogen (Zhang et al., 2019). This study showed that the methylmercury demethylase encoding gene *Mmd* was also acquired via HGT from bacteria (Wu et al., 2022; Zhang et al., 2019), which contains the *Mer* operon for mercury stress tolerance. MMD homologs are rare and patchily distributed among plant associates and soil fungi, and phylogenetic tracks suggest that fungal *Mmd* genes could have been acquired via HGT through two different evolutionary trajectories (Wu et al., 2022). Homologs of *Mmd* have not been found in mycorrhizal fungi. In contrast to MMD, homologs of the Hg^{2+} reductase MIR were found in many fungi, suggesting Hg^{2+} resistance conferred by MIR is widespread in the fungal kingdom (Wu et al., 2022). Except for tolerance to methylmercury stress, *Mmd* is not involved in tolerance to other toxic metals, including Cd^{2+}, Cr^{2+}, Pb^{2+}, Cu^{2+}, Ag^+, and Fe^{3+}; however, it is not related to pathogenicity, colony growth, and conidiation. In addition, the expression of *Mmd* is specifically upregulated by methylmercury. Therefore, the HGT gene *Mmd* could be specifically involved in tolerance to methylmercury by *Metarhizium* (Wu et al., 2022). However, *Metarhizium* are cosmopolitan and are present in strongly metal-polluted areas as well as environments not contaminated by toxic metal(loid)s (Słaba, Bernat, Różalska, Nykiel, & Długoński, 2013). Hence, the evolutionary mechanisms underlying retention of *Mmd* in worldwide *Metarhizium* populations merit further investigation.

7. Conclusion and future perspectives

Almost all terrestrial plants are associated with fungi, among which the beneficial mycorrhizal and endophytic fungi and their plant hosts facilitate each other's tolerance to various stresses, including toxic metal (loid)s. Mycorrhizal fungi not only sequestrate metal(loid)s by themselves with constitutive tolerance to metal(loid)s but also promote phytoremediation by impacting the physiological status of their hosts. The potential of mycorrhizal fungi to remediate metal(loid) pollution has been extensively explored, but most of these investigations have been conducted on a small scale. In addition to constitutive tolerance to various metal(loid)s, the beneficial endophytic *Metarhizium* acquired specific evolutionary adaptation to methylmercury, resulting in the removal of this metal from soil and water, reduction in their bioaccumulation in plants, and promotion of plant growth. In contrast to the obligate biotrophs mycorrhizal fungi that cannot be massively and automatically produced and are not genetically tractable, the ability of some beneficial endophytic fungi to remediate metal(loid) pollution can be enhanced with genetic engineering (Wu et al., 2022). Furthermore, cost-effective and industrial production technology for these endophytic fungi, such as *Metarhizium* and *Trichoderma*, is available. Therefore, endophytic fungi seem to be well-placed for large-scale field trials for the remediation of metal(loid)-polluted soil and water.

Acknowledgments

This work was funded by the National Natural Science Foundation of China (32172470 and 82261128002). Alene Alder-Rangel reviewed the English in the manuscript.

Conflict of interest

The authors have no conflict of interest to declare.

References

Ahmad, I., de Miménez-Gasco, M. M., Luthe, D. S., Shakeel, S. N., & Barbercheck, M. E. (2020). Endophytic *Metarhizium robertsii* promotes maize growth, suppresses insect growth, and alters plant defense gene expression. *Biological Control, 144*, 104167.
Alder-Rangel, A. (2021). *The adventures of Donald W. Roberts*. Inbioter: International Insect Pathologist.
Al-Sulaiti, M. M., Soubra, L., & Al-Ghouti, M. A. (2022). The causes and effects of mercury and methylmercury contamination in the marine environment: A review. *Current Pollution Reports, 8*, 249–272.
Ashford, A. E., Peterson, R. L., Dwarte, D., & Chilvers, G. A. (1986). Polyphosphate granules in eucalypt mycorrhizas: Determination by energy dispersive x-ray microanalysis. *Canadian Journal of Botany. Journal Canadien de Botanique, 64*, 677–687.

Averill, C., Turner, B., & Finzi, A. (2014). Mycorrhiza-mediated competition between plants and decomposers drives soil carbon storage. *Nature, 505*.

Bazzicalupo, A. L., Ruytinx, J., Ke, Y.-H., Coninx, L., Colpaert, J. V., Nguyen, N. H., et al. (2020). Fungal heavy metal adaptation through single nucleotide polymorphisms and copy-number variation. *Molecular Ecology, 29*, 4157–4169.

Behie, S. W., & Bidochka, M. J. (2014). Nutrient transfer in plant-fungal symbioses. *Trends in Plant Science, 19*, 734–740.

Blaudez, D., Botton, B., & Chalot, M. (2000). Cadmium uptake and subcellular compartmentation in the ectomycorrhizal fungus *Paxillus involutus*. *Microbiology (Reading), 146*(Pt 5), 1109–1117.

Branco, S., Gladieux, P., Ellison, C. E., Kuo, A., LaButti, K., Lipzen, A., et al. (2015). Genetic isolation between two recently diverged populations of a symbiotic fungus. *Molecular Ecology, 24*, 2747–2758.

Cao, Y., Sun, D., Ai, H., Mei, H., Liu, X., Sun, S., et al. (2017). Knocking out OsPT4 gene decreases arsenate uptake by rice plants and inorganic arsenic accumulation in rice grains. *Environmental Science & Technology, 51*, 12131–12138.

Chen, B., & Dong, S. (2022). Mercury contamination in fish and its effects on the health of pregnant women and their fetuses, and guidance for fish consumption—a narrative review. *International Journal of Environmental Research and Public Health, 19*, 15929.

Chen, X., Li, H., Chan, W. F., Wu, C., Wu, F., Wu, S., & Wong, M. H. (2012). Arsenite transporters expression in rice (*Oryza sativa* L.) associated with arbuscular mycorrhizal fungi (AMF) colonization under different levels of arsenite stress. *Chemosphere, 89*, 1248–1254.

Chen, X. W., Wu, L., Luo, N., Mo, C. H., Wong, M. H., & Li, H. (2019). Arbuscular mycorrhizal fungi and the associated bacterial community influence the uptake of cadmium in rice. *Geoderma, 337*, 749–757.

Dhaliwal, S. S., Singh, J., Taneja, P. K., & Mandal, A. (2020). Remediation techniques for removal of heavy metals from the soil contaminated through different sources: a review. *Environmental Science and Pollution Research International, 27*, 1319–1333.

Drijber, R. A., & McPherson, M. R. (2021). 12—Mycorrhizal symbioses. In T. J. Gentry, J. J. Fuhrmann, & D. A. Zuberer (Eds.). *Principles and applications of soil microbiology* (pp. 303–325)(third ed.). Elsevier.

Fomina, M. A., Alexander, I. J., Colpaert, J. V., & Gadd, G. M. (2005). Solubilization of toxic metal minerals and metal tolerance of mycorrhizal fungi. *Soil Biology and Biochemistry, 37*, 851–866.

Geldner, N. (2013). The endodermis. *Annual Review of Plant Biology, 64*, 531–558.

Ghasemi Siani, N., Fallah, S., Pokhrel, L. R., & Rostamnejadi, A. (2017). Natural amelioration of Zinc oxide nanoparticle toxicity in fenugreek (*Trigonella foenum-gracum*) by arbuscular mycorrhizal (*Glomus intraradices*) secretion of glomalin. *Plant Physiology and Biochemistry, 112*, 227–238.

Göhre, V., & Paszkowski, U. (2006). Contribution of the arbuscular mycorrhizal symbiosis to heavy metal phytoremediation. *Planta, 223*, 1115–1122.

Gong, X., & Tian, D. Q. (2019). Study on the effect mechanism of Arbuscular Mycorrhiza on the absorption of heavy metal elements in soil by plants. *IOP Conference Series: Earth andEnvironmental Science, 267*, 052064.

González-Chávez, Ma. del C. A., Ortega-Larrocea, M. del P., Carrillo-González, R., López-Meyer, M., Xoconostle-Cázares, B., Gomez, S. K., et al. (2011). Arsenate induces the expression of fungal genes involved in As transport in arbuscular mycorrhiza. *Fungal Biology, 115*, 1197–1209.

Gonzalez-Guerrero, M., Benabdellah, K., Valderas, A., Azcon-Aguilar, C., & Ferrol, N. (2009). GintABC1 encodes a putative ABC transporter of the MRP subfamily induced by Cu, Cd, and oxidative stress in *Glomus intraradices*. *Mycorrhiza, 20*, 137–146.

González-Guerrero, M., Melville, L. H., Ferrol, N., Lott, J. N. A., Azcón-Aguilar, C., & Peterson, R. L. (2008). Ultrastructural localization of heavy metals in the extraradical mycelium and spores of the arbuscular mycorrhizal fungus *Glomus intraradices*. *Canadian Journal of Microbiology, 54*, 103–110.

González-Guerrero, M., Oger, E., Benabdellah, K., Azcón-Aguilar, C., Lanfranco, L., & Ferrol, N. (2010). Characterization of a CuZn superoxide dismutase gene in the arbuscular mycorrhizal fungus *Glomus intraradices*. *Current Genetics, 56*, 265–274.

Guo, N., Qian, Y., Zhang, Q., Chen, X., Zeng, G., Zhang, X., et al. (2017). Alternative transcription start site selection in Mr-OPY2 controls lifestyle transitions in the fungus *Metarhizium robertsii*. *Nature Communications, 8*, 1565.

Harman, G. E., Howell, C. R., Viterbo, A., Chet, I., & Lorito, M. (2004). *Trichoderma* species—Opportunistic, avirulent plant symbionts. *Nature Reviews. Microbiology, 2*, 43–56.

He, Z., Shentu, J., Yang, X., Baligar, V., Zhang, T., & Stoffella, P. (2015). Heavy metal contamination of soils: Sources, indicators and assessment. *Journal of Environmental Indicators, 9*, 17–18.

Huang, X., Wang, L., Zhu, S., Ho, S.-H., Wu, J., Kalita, P. K., & Ma, F. (2018). Unraveling the effects of arbuscular mycorrhizal fungus on uptake, translocation, and distribution of cadmium in *Phragmites australis* (Cav.) Trin. ex Steud. *Ecotoxicology and Environmental Safety, 149*, 43–50.

Jiang, X., Dai, J., Zhang, X., Wu, H., Tong, J., Shi, J., & Fang, W. (2022). Enhanced Cd efflux capacity and physiological stress resistance: The beneficial modulations of *Metarhizium robertsii* on plants under cadmium stress. *Journal of Hazardous Materials, 437*, 129429.

Jiang, X., Fang, W., Tong, J., Liu, S., Wu, H., & Shi, J. (2022). *Metarhizium robertsii* as a promising microbial agent for rice in situ cadmium reduction and plant growth promotion. *Chemosphere, 305*, 135427.

Jones, D. L. (1998). Organic acids in the rhizosphere—A critical review. *Plant and Soil, 205*, 25–44.

Jourand, P., Ducousso, M., Loulergue-Majorel, C., Hannibal, L., Santoni, S., Prin, Y., & Lebrun, M. (2010). Ultramafic soils from New Caledonia structure *Pisolithus albus* in ecotype. *FEMS Microbiology Ecology, 72*, 238–249.

Kamiya, T., Islam, R., Duan, G., Uraguchi, S., & Fujiwara, T. (2013). Phosphate deficiency signaling pathway is a target of arsenate and phosphate transporter *OsPT1* is involved in As accumulation in shoots of rice. *Soil Science and Plant Nutrition, 59*, 580–590.

Kapoor, A., & Viraraghavan, T. (1995). Fungal biosorption—An alternative treatment option for heavy metal bearing wastewaters: A review. *Bioresource Technology, 53*, 195–206.

Lei, G. J., Yamaji, N., & Ma, J. F. (2021). Two metallothionein genes highly expressed in rice nodes are involved in distribution of Zn to the grain. *The New Phytologist, 229*, 1007–1020.

Li, H., Gao, M., Mo, C., Wong, M., Chen, X. W., & Wang, J. (2022). Potential use of arbuscular mycorrhizal fungi for simultaneous mitigation of arsenic and cadmium accumulation in rice. *Journal of Experimental Botany, 73*, 50–67.

Li, J., Sun, Y., Jiang, X., Chen, B., & Zhang, X. (2018). Arbuscular mycorrhizal fungi alleviate arsenic toxicity to *Medicago* sativa by influencing arsenic speciation and partitioning. *Ecotoxicology and Environmental Safety, 157*, 235–243.

Li, R.-Y., Ago, Y., Liu, W.-J., Mitani, N., Feldmann, J., McGrath, S. P., et al. (2009). The rice aquaporin Lsi1 mediates uptake of methylated arsenic species. *Plant Physiology, 150*, 2071–2080.

Li, W.-W., & Yu, H.-Q. (2014). Insight into the roles of microbial extracellular polymer substances in metal biosorption. *Bioresource Technology, 160*, 15–23.

Li, X., Zhou, M., Shi, F., Meng, B., Liu, J., Mi, Y., et al. (2023). Influence of arbuscular mycorrhizal fungi on mercury accumulation in rice (*Oryza sativa* L.): From enriched isotope tracing perspective. *Ecotoxicology and Environmental Safety, 255*, 114776.

Li, Y., Xu, J., Hu, J., Zhang, T., Wu, X., & Yang, Y. (2022). Arbuscular mycorrhizal fungi and glomalin play a crucial role in soil aggregate stability in Pb-contaminated soil. *International Journal of Environmental Research and Public Health, 19*, 5029.

Liao, X., O'Brien, T. R., Fang, W., & St Leger, R. J. (2014). The plant beneficial effects of *Metarhizium* species correlate with their association with roots. *Applied Microbiology and Biotechnology, 98*, 7089–7096.

Liu, H., Trieu, A. T., Blaylock, L. A., & Harrison, M. J. (1998). Cloning and characterization of two phosphate transporters from *Medicago truncatula* roots: Regulation in response to phosphate and to colonization by arbuscular mycorrhizal (AM) fungi. *Molecular Plant-Microbe Interactions: MPMI, 11*, 14–22.

Liu, L., Li, W., Song, W., & Guo, M. (2018). Remediation techniques for heavy metal-contaminated soils: Principles and applicability. *Science of the Total Environment, 633*, 206–219.

Ma, J. F., Yamaji, N., Mitani, N., Xu, X.-Y., Su, Y.-H., McGrath, S. P., & Zhao, F.-J. (2008). Transporters of arsenite in rice and their role in arsenic accumulation in rice grain. *Proceedings of the National Academy of Sciences, 105*, 9931–9935.

Ma, Y., Rajkumar, M., Oliveira, R. S., Zhang, C., & Freitas, H. (2019). Potential of plant beneficial bacteria and arbuscular mycorrhizal fungi in phytoremediation of metal-contaminated saline soils. *Journal of Hazardous Materials, 379*, 120813.

Mahbub, K. R., Bahar, M. M., Labbate, M., Krishnan, K., Andrews, S., Naidu, R., & Megharaj, M. (2017). Bioremediation of mercury: Not properly exploited in contaminated soils!. *Applied Microbiology and Biotechnology, 101*, 963–976.

Malloch, D. W., Pirozynski, K. A., & Raven, P. H. (1980). Ecological and evolutionary significance of mycorrhizal symbioses in vascular plants (a review). *Proceedings of the National Academy of Sciences, 77*, 2113–2118.

Martin, F., Rubini, P., Côté, R., & Kottke, I. (1994). Aluminium polyphosphate complexes in the mycorrhizal basidiomycete *Laccaria bicolor*: A27Al-nuclear magnetic resonance study. *Planta, 194*, 241–246.

Mishra, S., Tripathi, R. D., Srivastava, S., Dwivedi, S., Trivedi, P. K., Dhankher, O. P., & Khare, A. (2009). Thiol metabolism play significant role during cadmium detoxification by Ceratophyllum demersum L. *Bioresource Technology, 100*, 2155–2161.

Moonjely, S., Barelli, L., & Bidochka, M. J. (2016). Insect pathogenic fungi as endophytes. *Advances in Genetics, 94*, 107–135.

More, T. T., Yadav, J. S. S., Yan, S., Tyagi, R. D., & Surampalli, R. Y. (2014). Extracellular polymeric substances of bacteria and their potential environmental applications. *Journal of Environmental Management, 144*, 1–25.

Pande, V., Pandey, S. C., Sati, D., Bhatt, P., & Samant, M. (2022). Microbial interventions in bioremediation of heavy metal contaminants in agroecosystem. *Frontiers in Microbiology, 13*, 824084.

Parniske, M. (2008). Arbuscular mycorrhiza: The mother of plant root endosymbioses. *Nature Reviews. Microbiology, 6*, 763–775.

Paszkowski, U., Kroken, S., Roux, C., & Briggs, S. P. (2002). Rice phosphate transporters include an evolutionarily divergent gene specifically activated in arbuscular mycorrhizal symbiosis. *Proceedings of the National Academy of Sciences, 99*, 13324–13329.

Ranathunge, K., Lin, J., Steudle, E., & Schreiber, L. (2011). Stagnant deoxygenated growth enhances root suberization and lignifications, but differentially affects water and NaCl permeabilities in rice (*Oryza sativa* L.) roots. *Plant, Cell & Environment, 34*, 1223–1240.

Rangel, D. E. N., Finlay, R. D., Hallsworth, J. E., Dadachova, E., & Gadd, G. M. (2018). Fungal strategies for dealing with environmental and agricultural stress. *Fungal Biology, 122*, 602–612. https://doi.org/10.1016/j.funbio.2018.02.002.

Rascio, N., & Navari-Izzo, F. (2011). Heavy metal hyperaccumulating plants: How and why do they do it? And what makes them so interesting? *Plant Science, 180*, 169–181.

Riaz, M., Kamran, M., Fang, Y., Wang, Q., Cao, H., Yang, G., et al. (2021). Arbuscular mycorrhizal fungi-induced mitigation of heavy metal phytotoxicity in metal contaminated soils: A critical review. *Journal of Hazardous Materials, 402*, 123919.

Rillig, M. C. (2004). Arbuscular mycorrhizae, glomalin, and soil aggregation. *Canadian Journal of Soil Science, 84*, 355–363.

Santos Teixeira, J. A., & ten Tusscher, K. H. (2019). The systems biology of lateral root formation: Connecting the dots. *Molecular Plant, 12*, 784–803.

Sasaki, A., Yamaji, N., Yokosho, K., & Ma, J. F. (2012). Nramp5 is a major transporter responsible for manganese and cadmium uptake in rice. *The Plant Cell, 24*, 2155–2167.

Sayin, F., Khalvati, A., Erdinçler, A., & Soils, M. T. (2019). Effects of sewage sludge application and arbuscular mycorrhizal fungi (*G. mosseae* and *G. intraradices*) interactions on the heavy metal phytoremediation in chrome mine tailings.

Seo, H.-M., Jung, Y., Song, S., Kim, Y., Kwon, T., Kim, D.-H., et al. (2008). Increased expression of *OsPT1*, a high-affinity phosphate transporter, enhances phosphate acquisition in rice. *Biotechnology Letters, 30*, 1833–1838.

Siddiqui, Z. A., & Pichtel, J. (2008). Mycorrhizae: An overview. In Z. A. Siddiqui, M. S. Akhtar, & K. Futai (Eds.). *Mycorrhizae: Sustainable agriculture and forestry* (pp. 1–35). Dordrecht: Springer Netherlands.

Słaba, M., Bernat, P., Różalska, S., Nykiel, J., & Długoński, J. (2013). Comparative study of metal induced phospholipid modifications in the heavy metal tolerant filamentous fungus *Paecilomyces marquandii* and implications for the fungal membrane integrity. *Acta Biochimica Polonica, 60*, 695–700.

Smith, S. E., & Read, D. J. (Eds.). (1997). *Mycorrhizal symbiosis*(second ed.). London: Academic Press.

St Leger, R. (2021). From the lab to the last mile: Deploying transgenic approaches against mosquitoes. *Frontiers in Tropical Diseases, 2*, 804066.

St Leger, R. J., & Wang, J. B. (2020). *Metarhizium*: Jack of all trades, master of many. *Open Biology, 10*, 200307.

Sun, J., Karuppiah, V., & Chen, J. (2020). Chapter 14—The mechanism of heavy metal absorption and biodegradation of organophosphorus pesticides by *Trichoderma*. In V. K. Gupta, S. Zeilinger, H. B. Singh, & I. Druzhinina (Eds.). *New and future developments in microbial biotechnology and bioengineering* (pp. 303–318). Elsevier.

Takahashi, R., Ishimaru, Y., Shimo, H., Ogo, Y., Senoura, T., Nishizawa, N. K., & Nakanishi, H. (2012). The OsHMA2 transporter is involved in root-to-shoot translocation of Zn and Cd in rice. *Plant, Cell Environment, 35*, 1948–1957.

Tan, L., Zhu, Y., Fan, T., Peng, C., Wang, J., Sun, L., & Chen, C. (2019). OsZIP7 functions in xylem loading in roots and inter-vascular transfer in nodes to deliver Zn/Cd to grain in rice. *Biochemical and Biophysical Research Communications, 512*, 112–118.

Tedersoo, L., Bahram, M., Toots, M., Diédhiou, A. G., Henkel, T. W., Kjøller, R., et al. (2012). Towards global patterns in the diversity and community structure of ectomycorrhizal fungi. *Molecular Ecology, 21*, 4160–4170.

Teste, F. P., Jones, M. D., & Dickie, I. A. (2020). Dual-mycorrhizal plants: Their ecology and relevance. *The New Phytologist, 225*, 1835–1851.

Ueno, D., Yamaji, N., Kono, I., Huang, C. F., Ando, T., Yano, M., & Ma, J. F. (2010). Gene limiting cadmium accumulation in rice. *Proceedings of the National Academy of Sciences of the United States of America, 107*, 16500–16505.

Wang, P., Zhang, W., Mao, C., Xu, G., & Zhao, F.-J. (2016). The role of OsPT8 in arsenate uptake and varietal difference in arsenate tolerance in rice. *Journal of Experimental Botany, 67*, 6051–6059.

Wang, Q., Lu, H., Chen, J., Hong, H., Liu, J., Li, J., & Yan, C. (2018). Spatial distribution of glomalin-related soil protein and its relationship with sediment carbon sequestration across a mangrove forest. *Science of the Total Environment, 613–614*, 548–556.

Wang, Q., Mei, D., Chen, J., Lin, Y., Liu, J., Lu, H., & Yan, C. (2019). Sequestration of heavy metal by glomalin-related soil protein: Implication for water quality improvement in mangrove wetlands. *Water Research, 148,* 142–152.

Wu, C., Tang, D., Dai, J., Tang, X., Bao, Y., Ning, J., et al. (2022). Bioremediation of mercury-polluted soil and water by the plant symbiotic fungus *Metarhizium robertsii*. *Proceedings of the National Academy of Sciences of the United States of America, 119,* e2214513119.

Wu, Z., Ren, H., McGrath, S. P., Wu, P., & Zhao, F.-J. (2011). Investigating the contribution of the phosphate transport pathway to arsenic accumulation in rice. *Plant Physiology, 157,* 498–508.

Wyrebek, M., Huber, C., Sasan, R. K., & Bidochka, M. J. (2011). Three sympatrically occurring species of *Metarhizium* show plant rhizosphere specificity. *Microbiology (Reading, England), 157,* 2904–2911.

Zhang, H., Xu, Y., Kanyerere, T., Wang, Y., & Sun, M. (2022). Washing reagents for remediating heavy-metal-contaminated soil: A review. *Frontiers in Earth Science, 10.*

Zhang, Q., Chen, X., Xu, C., Zhao, H., Zhang, X., Zeng, G., et al. (2019). Horizontal gene transfer allowed the emergence of broad host range entomopathogens. *Proceedings of the National Academy of Sciences of the United States of America, 116,* 7982–7989.

Zhao, F. J., Ma, J. F., Meharg, A. A., & McGrath, S. P. (2009). Arsenic uptake and metabolism in plants. *The New Phytologist, 181,* 777–794.

Zhao, F.-J., Tang, Z., Song, J.-J., Huang, X.-Y., & Wang, P. (2022). Toxic metals and metalloids: Uptake, transport, detoxification, phytoremediation, and crop improvement for safer food. *Molecular Plant, 15,* 27–44.

Zhao, H., Lovett, B., & Fang, W. (2016). Genetically engineering entomopathogenic fungi. *Advances in Genetics, 94,* 137–163.

CHAPTER SEVEN

Molecular aspects of copper homeostasis in fungi

Dayane Moraes, Mirelle Garcia Silva-Bailão, and Alexandre Melo Bailão*

Universidade Federal de Goiás (UFG), Goiânia, GO, Brazil
*Corresponding author. e-mail address: ambailao@ufg.br

Contents

Abstract

Copper homeostasis in fungi is a tightly regulated process crucial for cellular functions. Fungi acquire copper from their environment, with transporters facilitating its uptake into the cell. Once inside, copper is utilized in various metabolic pathways, including respiration and antioxidant defense. However, excessive copper can be toxic by promoting cell damage mainly due to oxidative stress and metal displacements. Fungi employ intricate regulatory mechanisms to maintain optimal copper levels. These involve transcription factors that control the expression of genes involved in copper transport, storage, and detoxification. Additionally, chaperone proteins assist in copper trafficking within the cell, ensuring its delivery to specific

Advances in Applied Microbiology, Volume 129
ISSN 0065-2164, https://doi.org/10.1016/bs.aambs.2024.08.001

targets. Furthermore, efflux pumps help remove excess copper from the cell. Altogether, these mechanisms enable fungi to balance copper levels, ensuring proper cellular function while preventing toxicity. Understanding copper homeostasis in fungi is not only essential for fungal biology but also holds implications for various applications, including biotechnology and antifungal drug development.

1. Copper biologic functions

Fungi require transition metals such as copper (Cu), iron (Fe), manganese (Mn), and zinc (Zn) to affect cellular biological events, like other organisms. These metals serve as structural components in proteins or catalytic chemical elements for enzymatic reactions. Cu is an essential metal ion in fungi because it acts as an enzymatic cofactor in biochemical processes, including aerobic respiration, detoxification, signal transduction, and iron transport (reviewed by Smith, Logeman, & Thiele, 2017). These activities are possible because copper transitions between reduced forms (Cu (I) or Cu^+) and oxidized forms (Cu (II) or Cu^{2+}). Organisms exploit this characteristic in redox reactions. Copper also induces conformational changes in proteins (reviewed by Nevitt, Ohrvik, & Thiele, 2012), such as transcription factors, thereby interfering with their activity and consequently modulating gene expression. Copper can bind to proteins due to the affinity of Cu^+ to thiol and thioester groups in cysteine and methionine. In contrast, Cu^{2+} binds to oxygen or the imidazole groups of aspartic acid, glutamic acid, and histidine (Rubino & Franz, 2012).

Despite Cu's essential role in various biological processes, quantitative dysregulation of this compound causes severe cellular damage. Cu is an inorganic element; thus, it is not created or destroyed, so cells require sophisticated regulatory mechanisms to maintain copper homeostasis (reviewed by Festa & Thiele, 2011). Studies consistently suggest that free Cu in the cytoplasm is maintained at controlled and strictly low concentrations (Rae, Schmidt, Pufahl, Culotta, & O'Halloran, 1999), demonstrating the cells' remarkable ability to regulate metal levels precisely. These strategies are crucial for human pathogens because, throughout evolution, host cells have developed defense mechanisms that deprive pathogens of this micronutrient, a feature known as nutritional immunity. On the other hand, immune cells also increase intraphagosomal Cu to exploit its toxicity to control infection (Moraes et al., 2023; Ray & Rappleye, 2022).

Free copper in the cell is toxic because it leads to oxidative stress due to its catalytic activity in Fenton-type reactions. The accumulation of Cu^+ in the cytoplasm promotes the generation of reactive oxygen species (ROS), such as hydroxyl radicals and hydroxyl anions ($Cu^+ + H_2O_2 \rightarrow Cu^{2+} + \bullet OH + OH^-$), by reacting with hydrogen peroxide. ROS can damage membranes by compromising lipid structure due to peroxidation, protein oxidation, and oxidation of nitrogenous bases in nucleic acids, leading to breaks in DNA and RNA chains (Imlay & Linn, 1988). Furthermore, in proteins/enzymes, copper can displace other metals from their natural ligands (Irving & Williams, 1948), compromising the structure of iron-sulfur clusters (Macomber & Imlay, 2009) and even replacing zinc atoms in zinc finger DNA-binding structures. Protein miss-metalation can result in impaired protein activity, as it results in improper structure.

2. Controlling copper availability

Fungi harbor specialized machinery to control the intracellular copper amount. These microorganisms maintain adequate cellular copper levels by modulating the expression of genes that encode components of copper uptake and detoxification machinery in response to changes in environmental levels of this metal. Many of these processes are transcriptionally controlled by copper-sensitive transcription factors, whether under copper scarcity or excess ion conditions. In fungi, copper uptake begins with transporters on the plasma membrane, facilitating the unidirectional import of copper into the cytoplasm. Once inside the cell, copper is bound to copper chaperones or other buffering components that facilitate metal delivery to copper-dependent enzymes or subcellular storage compartments, avoiding toxicity. Therefore, when copper accumulates beyond homeostatic capacities, fungi employ copper detoxification mechanisms, such as storage in the Golgi complex, vacuoles, and vesicles or metal efflux through specialized transporters. Under conditions of copper deficiency, cells mobilize copper deposits or activate copper import machinery, for example, by inducing the production of transporters localized on the plasma membrane and/or on the vacuolar surface (Beaudoin & Labbé, 2007; Festa & Thiele, 2011, Smith et al., 2017).

3. Copper homeostasis in scarcity

The plasma membrane constitutes a barrier to copper distribution in cells due to its impermeability to ions. Therefore, the copper ion uptake system facilitates the import of this metal across the lipid bilayer, delivering it to cytoplasmic proteins that distribute it to Cu-proteins. This system comprises copper transporters, copper-reducing molecules, and copper chaperones. Copper enters the cell through passive transport and independently of ATP (Lee, Peña, Nose, & Thiele, 2002), likely facilitated by a deficient intracellular copper concentration (Rae et al., 1999). Copper ion uptake also appears to be stimulated by acidic pH in the extracellular environment and the extremely high intracellular concentration of K^+, suggesting a $Cu^+/2K^+$ antiport mechanism (De Rome & Gadd, 1987; Lee et al., 2002).

Saccharomyces cerevisiae may also absorb copper by other low-affinity systems. Divalent metal transporters, found on the cell surface, encoded by the low-affinity iron transport protein sequence (FET4) (Dix, Bridgham, Broderius, & Eide, 1997), are capable of transporting iron, cobalt, cadmium, nickel, and copper (Hassett, Dix, Eide, & Kosman, 2000). Additionally, divalent metal ion transporters (SMF 1/2) play an essential role in copper and manganese homeostasis (Liu, Supek, Nelson, & Culotta, 1997), as SMF1 may import Cu into cells (Cohen, Nelson, & Nelson, 2000).

4. High-affinity copper transporters

At low concentrations, copper is internalized from the environment into the cell through high-affinity Cu^+ importers called copper transport proteins (CTR). CTR family members are structurally conserved in fungal, plant, and mammalian species (Puig, Lee, Lau, & Thiele, 2002). CTR importers form a transmembrane complex that typically has three domains associated with the plasma membrane (Dancis, Yuan et al., 1994), requiring assembly/multimerization to create a symmetric homotrimeric or a heterotrimeric protein with a Cu-permeable pore between subunit interfaces (Dancis, Haile, Yuan, & Klausner, 1994; Dancis, Yuan et al., 1994; Peña, Koch, & Thiele, 1998; Puig et al., 2002; Zhou & Thiele, 2001). The membrane-spanning domains contain putative copper-binding motifs, with negatively charged amino acids predominantly hydrophilic, which facilitate metal ion binding. The amino-terminal domain typically contains methionine-rich motifs (Met-X-Met or Met-X_2-Met; "Mets" motifs) embedded in large amino acid repeat elements [SM(N/S)MD(A/E)MSSASKT(V/I/L)S (S/T)MS] forming a hydrophilic region residing in the periplasmic space between the yeast membrane and the cell wall (Dancis, Yuan et al., 1994).

The number of methionine-rich motifs varies depending on the transporter. CTRs without the metal-binding motif in the extracellular domain of the amino-terminal region have an abundance of cysteine residues in this protein and the transmembrane regions (Knight, Labbé, Kwon, Kosman, & Thiele, 1996; Peña, Puig, & Thiele, 2000). In "Mets" motifs, sulfur atoms from cysteines and methionines support metal binding, consequently attracting copper to the region and increasing copper concentration near the pore (Koch, Peña, & Thiele, 1997; Nose, Rees, & Thiele, 2006). The third transmembrane domain contains a motif (Gly-X$_3$-Gly) implicated in intra-molecular interactions of the monomers, which is necessary for CTR molecule assembly/multimerization and proper localization in the plasma membrane (Aller, Eng, De Feo, & Unger, 2004). Supporting this, mutations in the third transmembrane domain of CTR transporters disrupt multi-merization and result in incorrect localization of copper transporters (Aller et al., 2004). The carboxy-terminal portion of CTR proteins is a hydrophilic sequence that includes two cysteine and histidine motifs (Cys-X-Cys and His-X-His) and a sequence rich in charged amino acids, which are potential metal ligands (Dancis, Yuan et al., 1994; Puig et al., 2002). Additionally, in the carboxy-terminal/cytoplasmic portion of CTRs, cysteine residues serve as intracellular supports for Cu$^+$ transfer to chaperones (Nose et al., 2006). Cysteine residues in the carboxy-terminal cytosolic domain are also crucial for CTR oligomerization, suggesting the stabilization of multimerization by disulfide bonds (Eisses & Kaplan, 2005).

S. cerevisiae encodes two functionally redundant plasma membrane proteins of the CTR family, namely CTR1 (Dancis, Yuan et al., 1994) and CTR3 (Knight et al., 1996), responsible for high-affinity copper transport into the cell. Cells expressing either CTR1 or CTR3 alone (Peña et al., 2000) can absorb sufficient amounts of copper to maintain copper-dependent metabolic activities in S. cerevisiae; however, double mutants have defective copper uptake (Knight et al., 1996). Disruption of S. cere-visiae cells for CTR1 leads to the manifestation of copper-scarcity-related phenotypes, including the inability to grow on non-fermentable carbon sources, defective iron uptake, and increased sensitivity to oxidative stress. These phenotypes are likely attributable to defects in copper incorporation into mitochondrial cytochrome c oxidase, FET3 ferroxidase, and Cu-Zn superoxide dismutase, respectively. (Dancis, Yuan et al., 1994). Deleting the CTR3 gene in yeast has also been observed to result in respiratory deficiency and increased sensitivity to oxidative stress (Peña, Puig, & Thiele, 2000).

Furthermore, co-immunoprecipitation studies suggest that CTR1 molecules self-associate to form a homotrimeric complex. The same was observed for CTR3 (Peña, Puig, & Thiele, 2000). However, despite being functionally redundant and having similar association patterns, S. cerevisiae CTR1 and CTR3 transporters are not significantly homologous at the protein sequence level (Labbé & Thiele, 1999). Regarding differences, CTR1 is a 406 amino acids protein with eight copies of the methionine motif in the amino-terminal region. Mutagenesis experiments demonstrate that the "met" close to the first transmembrane domain is essential for metal binding in the extracellular domain. In contrast, the remaining "met" domains in the amino-terminal region are necessary to form the trans-membrane copper channel (Puig et al., 2002). The second transmembrane domain in S. cerevisiae CTR1 presents a conserved methionine motif (Met-X_3-Met) involved in the formation of the Cu^+ passage pore, suggesting a structural similarity with channel proteins (Nose et al., 2006). On the other hand, CTR3 contains 241 amino acids and only one methionine motif (Met-X-Met) but features a cysteine-rich region (Knight et al., 1996; Peña, Puig, & Thiele, 2000). CTR3 undergoes different post-transcriptional regulation compared to CTR1. CTR3 concentrations remain stable on the cell surface under low or high copper availability (Peña, Puig, & Thiele, 2000). Conversely, the level of CTR1 is stable only under deprivation and is degraded in environments with high copper availability (Ooi, Rabinovich, Dancis, Bonifacino, & Klausner, 1996). Metal excess induces ubiquitination of CTR1, which signals the transporter to be internalized by endocytosis. Subsequently, this homotrimer is delivered to the lysosome lumen, where it is degraded by proteases (Liu, Sitaram, & Burd, 2007), a process dependent on the carboxy-terminal region of the monomers (Wu, Sinani, Kim, & Lee, 2009).

Schizosaccharomyces pombe also expresses two CTR proteins on the cell surface named CTR4 (Labbé, Peña, Fernandes, & Thiele, 1999) and CTR5 (Zhou & Thiele, 2001). S. pombe CTR4 is a 289 amino acids protein with five copies of the Met-X_2-Met-X-Met sequence in the extracellular region of the amino-terminal portion, similar to CTR1 in S. cerevisiae. However, the transmembrane domains and residues in the carboxy-terminal portion (residues 111–248) show more significant homology with S. cerevisiae CTR3. This suggests that S. pombe CTR4 may result from an evolutionary event in which the CTR1 and CTR3 from S. cerevisiae were fused into a single coding region (Labbé et al., 1999). Deletion of the CTR4 gene in S. pombe renders identical phenotypes to those observed in S. cerevisiae double mutants for CTR1 and

CTR3, with reduced cell growth in low-copper medium and alteration in activities of copper-dependent enzymes (Labbé et al., 1999). The expression of CTR4 and CTR5 is activated under copper deprivation conditions, while the repression of these genes occurs in conditions of metal abundance (Beaudoin & Labbé, 2001). CTR5 is a 173 amino acid protein that presents two copies of partially overlapping Met-X_2-Met-X-Met motifs and a Cys-X-Met-X-Met sequence in the amino-terminal region. Interestingly, CTR5 is essential for the proper localization of CTR4 on the cell membrane and vice-versa (Zhou & Thiele, 2001). Unlike *S. cerevisiae*, where homomultimerization takes place, functional copper transporters require the assembly of CTR4 and CTR5 to form a heterotrimeric complex (Labbé et al., 1999), built of at least two molecules of CTR4 and one molecule of CTR5 (Ioannoni, Beaudoin, Mercier, & Labbé, 2010). In the amino-terminal portion, *S. pombe* CTR4 transporter shows an almost palindromic sequence that is rich in cysteine and tryptophan residues ($CysX_4YTrpNTrpYX_4Cys$), and CTR5 has a similar amino acid sequence. This region plays a role in the activity/functioning of copper transporters in *S. pombe*. Those CTRs utilize a pair of cysteines from the palindromic region (one cysteine from each subunit), in which one captures two copper ions to form a disulfide bond by thiolate coordination. In contrast, the other subunit stabilizes copper in a tryptophan-dependent way, which prevents Cu^+ oxidation. Therefore, it is assumed that one acts as a metal reductase and the other as a copper receptor (Okada & Miura, 2016).

In *Aspergillus fumigatus*, four proteins (AfCtrA1, AfCtrA2, AfCtrB, and AfCtrC) have been identified with high homology to *S. cerevisiae* CTR1, but experimentally, only AfCtrA2 and AfCtrC functionally complement this homolog in the yeast. Conversely, AfCtrA1 and AfCtrB fail to complement *S. cerevisiae* CTR1, inferring that these proteins seem to operate in other cellular compartments but not in the outer plasma membrane. The double deletion of AfCtrA2 and AfCtrC drastically reduces intracellular copper and results in growth defects, with reduced SOD and catalase activities, but does not alter virulence in a murine model of invasive aspergillosis. Especially, AfCtrA2 and AfCtrC exhibit two transmembrane domains each, but contrasting CtrA2 has only two Met-$X_{(1-5)}$-Met motifs, while CtrC presents four of those copper-binding motifs (Park, Lian, Chang, Kang, & Yun, 2014).

In *Cryptococcus neoformans*, copper uptake is mediated by two functionally redundant Cu transporters, CTR1 and CTR4 (Waterman et al., 2007), which are homologous to the *S. cerevisiae* CTR1. CTR1 or CTR4 successfully reversed the copper-deficient phenotype in a double CTR1 and CTR3 mutant in *S. cerevisiae* (Ding et al., 2011). However, in *C. neoformans*,

CTR1 plays a primary role in high-affinity copper transport under standard growth conditions, while CTR4 appears to function predominantly under more severe copper deficiency conditions (Ding et al., 2011). Deletion of CTR1 in *C. neoformans* leads to a drastic increase in CTR4 expression, while increased levels of CTR1 were not observed in the absence of CTR4 (Ding et al., 2011). Additionally, disruption of CTR4 leads to a copper uptake defect, but this impaired metal acquisition does not abolish Cu-protein activities, including Cu–Zn superoxide dismutase and cytochrome *c* oxidase. Mutants for CTR4 exhibit growth delay in cultures supplemented with menadione and modest effects on non-fermentable carbon sources. These results indicate the partial role of CTR4 in copper transport in *C. neoformans*, while CTR1 has a significant role in copper uptake (Zhang et al., 2016). Nonetheless, mRNA levels of CTR1 and CTR4 are responsive to copper availability, which is repressed under high copper concentration conditions and induced in response to copper deficiency (Ding et al., 2011; Zhang et al., 2016). Structurally, CTR1 and CTR4 contain consensus copper-binding elements/regions, typical of CTR transporters; however, they have only two transmembrane domains. Another difference between CTR1 and CTR4 in *C. neoformans* is the presence of a glycosylation site near the amino-terminal extracellular binding motifs in CTR4, as in mammalian CTR1; however, this site is not found in fungal CTR1. This modification is essential for the copper uptake activity of CTR4 since mutant cells with impaired glycosylation have a low copper growth-defective phenotype, similar to CTR1/CTR4 double mutants (Sun et al., 2014).

The copper uptake in *Candida albicans* involves the high-affinity copper transporter gene *Ca*CTR1 (Marvin, Williams, & Cashmore, 2003). *Ca*CTR1 is a 251 amino acid protein with two or three transmembrane segments without an exposed amino-terminal sequence to the periplasmic space. Methionine residues are abundantly found in this molecule, with one motif Met-X_2-Met and eight methionine pairs separated by an undefined amino acid residue (Met-X-Met), localized in the first 71 amino acid residues of *Ca*CTR1. This region also has four conserved domains: M(A/X)M(S/A)(S/A/X)(S/A/T)(S/A/T)(S/A/T/X)(S/X) (Marvin et al., 2003). *Ca*CTR1 is important for survival in low copper, growth in non-fermentable carbon sources, fighting oxidative stress, and proper iron acquisition. Intriguingly, CaCTR1 mutants exhibit aberrant morphology (Marvin, Mason, & Cashmore, 2004).

The search for CTR orthologs in dimorphic fungi genomes identified three Ctr sequences encoded by *Blastomyces dermatitidis*, *Coccidioides* species, *Histoplasma capsulatum*, *Paracoccidioides brasiliensis* (while *Paracoccidioides lutzii* coding for two Ctr transporters), and *Talaromyces marneffei*. Only *Sporothrix* species produce four CTRs. Bioinformatics and phylogenetic analyses suggest the existence of CTR1 (homologous to CTR1 *S. cerevisiae*, CTR1 *C. neoformans,* and CTRA2 *A. fumigatus*), CTR2 (homologous to CTR2 *S. cerevisiae* and CTR6 *S. pombe,* described above) and CTR3 (homologous to CTR3 *S. cerevisiae*, CTR4 *C. neoformans,* and CTR4 *S. pombe*) in these species. All dimorphic fungal CTR orthologs harbor the copper transporter domain, typically containing three transmembrane domains (Moraes et al., 2023). Functional studies revealed that CTR3 is essential for copper acquisition and virulence in *H. capsulatum*. This transporter is necessary for fungal survival in macrophages treated with interferon-gamma. Furthermore, the CTR3 transporter responds to changes in copper availability and is not influenced by the supply of other ions, such as zinc (Assunção et al., 2020; Moraes et al., 2023). Additionally, the loss of CTR3 function does not affect *Histoplasma* growth under restricted iron conditions (Shen, Beucler, Ray, & Rappleye, 2018). The *P. lutzii* ortholog (PAAG_05251) is repressed by copper overload (Portis et al., 2020), while the *P. brasiliensis* (Pb18) CTR3-ortholog is induced by copper deprivation (Petito et al., 2020).

5. Ferrireductases

Copper exists in two valence states (Cu^+ and Cu^{2+}), but the uptake of this metal is limited to the cuprous form. In *S. cerevisiae*, the reduction of the most commonly found form (Cu^{2+}) occurs with the assistance of the ferrireductases FRE1 and FRE2. The ferrireductase FRE1 contains 686 amino acids with a molecular weight of 78.8 kDa. The sequence presents two regions rich in hydrophobic amino acids in the amino-terminal portion, compatible with transmembrane domains (Dancis, Roman, Anderson, Hinnebusch, & Klausner, 1992). This reducing molecule was initially described as influential in the reduction and transport of iron in *S. cerevisiae*, and it exhibits a highly similar sequence to a plasma membrane cytochrome *b* (X-CGD protein) that has redox activity and transfers electrons from cytoplasmic NADPH to molecular oxygen outside the cell (Dancis et al., 1992). Further, it was discovered that this enzyme had an influence on Cu-reduction and acquisition in *S. cerevisiae* (Hassett & Kosman, 1995). Ferrireductase expression is regulated by the transcription factor MAC1, which binds to the consensus sequence (A/T)TTTG CTCA (required for the transcriptional activation of Cu-acquisition genes) in the promoter region (Yamaguchi-Iwai et al., 1997). FRE2 has copper reduction activity similar to FRE1, but copper does not influence the expression of the gene. This finding was explained by the presence of a pyrimidine-rich sequence close to the TATA box, which could prevent the formation of the transcriptional complex in the FRE2 promoter (Georgatsou, Mavrogiannis, Fragiadakis, & Alexandraki, 1997).

6. Mobilization of copper stock

The existence of an alternative copper transporter in *S. cerevisiae* yeasts was suggested based on the discreet growth of CTR1/CTR3 double mutants in metal scarcity and non-fermentable carbon sources (Knight et al., 1996). The CTR2 was identified/characterized (Kampfenkel, Kushnir, Babiychuk, Inzé, & Van Montagu, 1995; Knight et al., 1996) as a transporter with a central role in mobilizing copper stocks in *S. cerevisiae* during deprivation.

CTR2 encodes a low-affinity copper transporter in the vacuolar membrane, specifically designed to mobilize copper from the vacuolar lumen to cytosolic copper chaperones, enhancing yeast growth under metal deprivation conditions (Rees, Lee, & Thiele, 2004). The mechanism used by the CTR2 protein is mechanically similar to the copper transporter by CTR1 in *S. cerevisiae* yeasts (Kampfenkel et al., 1995). The low-affinity transporter CTR2 is a membrane protein with three transmembrane domains. These transporters function as a homotrimer. Their structural characteristics are similar to CTR1 and CTR2. Still, they share less than 23% of the amino acid sequence, and sequence conservation is found in hydrophobic transmembrane regions, including the 20 methionines in the first transmembrane domain. CTR2 features an amino-terminal region rich in methionine and a carboxy-terminal region rich in cysteine/histidine, the latter is exposed to the cytoplasmic environment similar to CTR1. Additionally, the methionine motif (Met-X_3-Met) in the second transmembrane domain is essential for copper transport to the cytoplasm. Deletion of CTR2 causes vacuolar copper hyperaccumulation, even under copper scarcity, confirming that this protein supplies copper from the vacuole to the cytoplasm during metal scarcity (Rees et al., 2004).

In *S. pombe*, CTR6 mediates the low-affinity transport of intracellular copper from the vacuolar lumen to the cytoplasm, and its expression is regulated in a copper-dependent manner (Bellemare et al., 2002). CTR6 is an integral membrane protein containing three transmembrane regions, including a Met-X_3-Met motif in the second transmembrane sequence and a Gly-X_3-Gly motif in the third transmembrane sequence, the latter being the candidate responsible for the assembly of the CTR6 homotrimer complex in the vacuolar membrane. CTR6 in *S. pombe* is orthologous to *S. cerevisiae* CTR2. The amino-terminal region (33 amino acids) of CTR6 is located in the vacuole lumen and contains a Met-X-His-Cys-X-Met-X-Met sequence (residues 9–14). This domain seems involved in the copper capture inside the organelle due to its similarity to methionine motifs in the CTR transporter family; therefore, it is described as a copper-binding motif (Bellemare et al., 2002).

A homolog of the *S. cerevisiae* CTR2 gene has been identified in *C. neoformans*. The cryptococcal CTR2 gene encodes a 228 amino acid protein containing three Met-X_2-Met motifs in the amino-terminal domain. Like other low-affinity copper transporters, *C. neoformans* CTR2 contains only one predicted transmembrane domain (residues 15–35). The sequence similarities are located mainly in the first 50 residues of the amino-terminal sequence, which shares 46% sequence identity with CTR in *A. fumigatus* and 42% with predicted CTRs in *H. capsulatum*. Functionally, CTR2-silenced mutants in *C. neoformans* show growth sensitivity under copper-limiting conditions, with observed defects related to polysaccharide capsule formation, melanization, and increased phago-cytosis by macrophages (Chun & Madhani, 2010).

7. Copper detoxification machinery

Eukaryotic cells, such as fungi, exhibit mechanisms to counteract intracellular copper overload, aiming to decrease the concentration of free metal in the cytoplasm. These cells can: (1) synthesize metallothioneins, which act as copper buffering molecules, (2) store the metal in vesicles/vacuoles, (3) promote metal efflux, and (4) inactivate copper uptake.

8. Metallothioneins

Metallothioneins (MTs) are a family of proteins found in a wide range of eukaryotes. They exhibit a remarkable affinity for metal binding, mainly because of their cysteine-rich composition. MTs play an essential role in the intracellular homeostasis of heavy metals by binding to various metal ions through clusters/bonds to thiolate residues, thereby acting as metal ion storage molecules and preventing metal toxicity (reviewed by Hamer, 1986).

In *S. cerevisiae*, intracellular copper resistance is primarily mediated by the CUP1 region in the genome (Brenes-pomales, Lindegren, & Lindegren, 1955, Butt, Sternberg, Herd, & Crooke, 1984, Ecker et al., 1986). This region consists of tandem coding sequences (5 to 15 copies) of copper metallothionein, also called CUP1 (Fogel, Welch, Cathala, & Karin, 1983). The CUP1 protein contains 61 amino acids with 12 cysteine residues and eight serine residues (Butt et al., 1984), able to bind to eight copper ions (Winge, Nielson, Gray, & Hamer, 1985). The upstream sequence of CUP1 is critical for gene expression regulation. The fragment between -105 and -180 in the 5' flanking region of CUP1 is necessary to initiate copper-induced transcription. This region contains two copper-regulatory elements (UAS: upstream activator sequence TCTTTTCCG CTGAACCGTTCCAGCAAAAAAGA and TCTTTTGCTGGCATTT CTTCTAGAAGCAAAAAGA), necessary for transcription factor binding. The two regulatory elements act synergistically to promote CUP1 expression, and mutants lacking one of these regions have half the expression levels as the wild type (Thiele & Hamer, 1986). CUP1 shows a 20-fold induction when *S. cerevisiae* cells are exposed to high copper (2 mM), which increases the resistance to metal excess (Karin et al., 1984). The induction of CUP1 in high copper relies on the transcription factor ACE1 (discussed below) (Hamer, Thiele, & Lemontt, 1985; Gorman, Clark, Lee, Debouck, & Rosenberg, 1986; Welch, Fogel, Buchman, & Karin, 1989). ACE1 recognizes the motifs of the UAS sequence within the promoter region of the CUP1 coding gene and activates gene transcription (Buchman, Skroch, Dixon, Tullius, & Karin, 1990; Huibregtse, Engelke, & Thiele, 1989; Szczypka & Thiele, 1989). A study suggests the presence of small amounts of ACE1 in the UAS region even under copper-free conditions, but when Cu rises, more ACE1 binds to the CUP1 promoter, which results in more RNA polymerase II recruitment, consequently increasing mRNA levels (Wimalarathna, Pan, & Shen, 2012). The presence of a large amount of ACE1 induces the repositioning of nucleosomes and remodeling of chromatin structure in the open reading frame of CUP1 and flanking sequences, potentially via recruitment of a chromatin remodeling enzyme, which facilitates the formation of a transcription complex at the CUP1 promoter and the passage of RNA polymerase II (Shen, Leblanc, Alfieri, & Clark, 2001). At low copper, CUP1 apoproteins display negative autoregulation that binds to the promoter region, repressing gene expression (reviewed by Butt & Ecker, 1987).

Defense against copper toxicity in *S. cerevisiae* can also be accomplished by the protein CRS5 (copper resistance suppressor YO4031W) (Culotta, Howard, & Liu, 1994), an efficient metallothionein in buffering high intracellular concentrations of copper ions. The CRS5 locus harbors a single-copy gene encoding a 69 amino acid polypeptide with 19 cysteine residues arranged in Cys-Cys, Cys-X-Cys, and Cys-X$_2$-Cys motifs. However, CRS5 shares low sequence similarity with CUP1 (Culotta et al., 1994). Besides that, CUP1 is much more effective than CRS5 in protecting *S. cerevisiae* yeast cells against copper toxicity, possibly due to a single UAS in the promoter region or the lower Cu-affinity of CRS5 (Jensen, Howard, Strain, Winge, & Culotta, 1996).

C. neoformans encodes two metallothioneins (CMT1: CNAG_05449 and CMT2: CNAG_00306) that are copper-responsive and precisely function in Cu-detoxification (Ding et al., 2011). CMT1 and CMT2 are induced under copper excess, ensuring the growth of *C. neoformans* in high micronutrient levels (1 mM). This result evidences cryptococcal metallothioneins as critical components in the copper-detoxifying machinery (Ding et al., 2011). Phylogenetic analyses suggest that CMT1 and CMT2 are distantly related to *S. cerevisiae* CRS5 and CUP1, as well as human metallothioneins (Ding et al., 2013). Structurally, Cmt1 and Cmt2 present typical copper-binding motifs (Cys-X-Cys) essential for forming copper ion clusters by fungal metallothioneins. However, the distribution of these elements appears peculiar, as these motifs are grouped in several segments separated by three spacer regions in Cmt1 and four in Cmt2. Comparisons of the spacer regions showed that the first share a high level of similarity, while the second and third regions are identical (Ding et al., 2013).

C. albicans presents the metallothionein CRD2 (copper resistance determinant). CRD2 has 76 amino acids with 12 cysteine residues, 10 of which are part of Cys-X-Cys motifs. This metallothionein has been identified as an element for copper tolerance because mutations in this gene result in reduced growth rates at elevated copper levels. However, copper does not induce CRD2 expression. Mutants for CRD2 showed sensitivity only up to the first 18 h of incubation, suggesting an initial buffering activity of this metallothionein, preventing metal toxicity before the production of other Cu-detoxifying elements such as CRD1/CRP (Riggle & Kumamoto, 2000).

A. fumigatus contains AfCDRA, a protein homologous to CRD2 in *C. albicans*. AfCRDA is upregulated not only by copper but also by zinc.

However, gene knockout does not affect fungal sensitivity to those metals (Cai et al., 2018). *Aspergillus nidulans* also has a CRDA metallothionein induced by high copper, silver, or cadmium, but it does not play a role in copper tolerance. Moreover, in *A. nidulans,* CRDA expression is not regulated by the high copper regulator ACE1 (Antsotegi-Uskola, Markina-Iñarrairaegui, & Ugalde, 2017).

In *S. pombe*, the PCCS protein is associated with increased copper resistance. Fission yeasts use the carboxy-terminal portion for copper buffering. The domain (residues 223–297) has high similarity with metallothioneins, contains seventeen cysteine residues arranged in eight pairs of Cys-Cys configuration, is rich in lysine and serine amino acids, and has no aromatic amino acids. Thus, PCCS in *S. pombe* seems to act as a potential candidate for copper ion capture, protecting the cell against toxicity, since PCCS mutants show compromised growth high copper (starting at 420 uM of $CuSO_4$; Laliberté et al., 2004).

9. Storage and distribution

Copper distribution is a complex process because of its affinity for thiol and thioester groups present in amino acids cysteines and methionines, as well as imidazole groups observed in aspartic and glutamic acid and histidines. Therefore, this metal can be associated with a variety of proteins that do not require copper for proper activity. As a result, copper moves both in extracellular fluids and within cells with the assistance of various transport proteins. These proteins contribute to processes related to both the quantitative control of intracellular copper and the viability of fungal cells, as they ensure the delivery of copper to cuproenzymes and mediate the storage of this micronutrient (reviewed by Inesi, 2017, reviewed by Hatori & Lutsenko, 2016). Fungal cells have chaperones, small metal-binding proteins, to execute copper sequestration and delivery to desired compartments/proteins. This maintains intracellular free copper at low concentrations, preventing miss-metalation mediated damages and providing adequate metal supply to cuproenzymes. For example, in *S. cerevisiae,* (1) ATX1 mediates intracellular copper transport from CTRs to P-type ATPase transporters on the membrane of the Golgi complex and secretory vesicles; (2) chaperones SCO1, COX11, and COX17 deliver copper to cytochrome c oxidase in the mitochondria; and (3) CCS1 (Lys7p) provides copper to Cu-Zn superoxide dismutase.

10. ATX1 and CCC2

The antioxidant metallochaperone 1 (ATX1) in *S. cerevisiae* (Lin & Culotta, 1995) was initially identified as a suppressor of oxidative damage in cells of delete mutants for superoxide dismutase 1 (SOD1). The over-expression of ATX1 compensated SOD1 deletion, preventing the harmful effects of superoxide anions and hydrogen peroxide. The ATX1 was shown to consume superoxide anions (Portnoy et al., 1999). However, ATX1 has been primarily recognized as an essential element mediating copper transport to the Golgi complex and secretory vesicles based on sequence similarity to copper carrier proteins in humans and bacteria (Lin & Culotta, 1995). Regarding the main activity exerted by ATX1, X-ray absorption analyses showed that this protein binds to copper ions and these chaperone cytosolic domains of P-type ATPase copper transporters on the membrane of the Golgi complex and secretory vesicles (the CCC2 described later; Pufahl et al., 1997). ATX1 is indirectly associated with the proper development of *S. cerevisiae* yeast cells under iron deprivation, as the copper available in the Golgi complex serves for incorporation into cuproenzymes, such as iron multicopper oxidases (the FET3 importers), which is an essential component of the high-affinity iron uptake complex (Lin, Pufahl, Dancis, O'Halloran, & Culotta, 1997). ATX1 assists in copper transport and distribution in *S. cerevisiae* cells. ATX1 encodes a cytosolic polypeptide with 73 amino acids with a conserved motif composed of methionine and cysteines ($MetXCysX_2Cys$) in the amino-terminal region (Lin & Culotta, 1995) that binds to copper ions. The Cu-binding to ATX1 apoprotein leads to the forming of a positively charged surface, resulting from the exposure/proximity of lysine residues. This process influences this chaperone's tertiary structure and activity (Arnesano et al., 2001; Banci et al., 2006).

The Ca^{2+}-sensitive cross-complementer (CCC2; Fu, Beeler, & Dunn, 1995) was initially described in *S. cerevisiae* as a calcium-transporter of the P-type ATPase family and was also characterized as an intracellular Cu translocator. CCC2 copper transport activity was suggested because gene-knockout mutants required extra copper for growth (Fu et al., 1995). The CCC2 transporter is localized in membranes of secretory vesicles of the Golgi complex (trans-Golgi network) (Yuan, Dancis, & Klausner, 1997) and presents four cytoplasmic domains separated by eight transmembrane domains. Additionally, CCC2 contains four ATP-binding domains, one phosphorylation domain, and a phosphatase activity domain, similar to those found in metal transporters. The cytoplasmic portion has two metal-binding domains (MDBs) identical to the one found in ATX1 (Met-TreCysX2Cys; Fu et al., 1995).

Protein-protein interaction assays (two-hybrid) indicate that ATX1 interacts with the amino-terminal region of CCC2 in a copper-dependent fashion (van Dongen, Klomp, & Merkx, 2004). ATX1 delivers copper (Cu^+) to CCC2 through direct interactions between the $CysX_2Cys$ in the chaperone and the MDBs of the CCC2 (Pufahl et al., 1997). The exposed site of ATX1, when bound to copper and together with the acidic residues on the surface of the CCC2 transporters, facilitates electrostatic interactions between the proteins (Arnesano et al., 2001). Copper binding to the MDBs in CCC2 triggers ATP hydrolysis, which forms an acyl-phosphate inter-mediate in a conserved DKTGT motif, which is essential to generate the force required to pump copper into the organelle (Lowe et al., 2004). Consistently, mutation of conserved lysines on the surface of ATX1 sig-nificantly reduces the copper-dependent interaction between ATX1 and CCC2; in particular, mutation of Lys65 in ATX1 abolishes the function of this protein (Portnoy et al., 1999). Similarly, mutation of 583CysPCys565 (a component of the phosphorylation domain) in the CCC2 transporter results in impaired yeast growth because it compromises the transport/storage of copper supply for cuproenzymes (Lowe et al., 2004).

The proteins ATX1 and CCC2 have been described in *C. neoformans*, primarily focusing on their indirect activity in melanin production. Cryptococcal strains with mutations in CCC2 and ATX1 show reduced virulence due to compromised melanization and phenotypes related to copper deficiency, such as the inability to grow in low Cu or Fe (Walton, Idnurm, & Heitman, 2005), as observed in *S. cerevisiae*.

A copper transporter CTPA (copper transporter *Aspergillus*) has been described in *A. fumigatus*. Still, like in *Cryptococcus*, it has an indirect function in conidial melanization, being necessary for conidial pigmentation (Upadhyay, Torres, & Lin, 2013). There are homologs of the chaperones Atx1, Ccs1, and Cox17 in the genome of *A. fumigatus* (AFUB_008300, AFUB_025550, and AFUB_041410). Their function has not yet been confirmed, requiring further studies to determine the role of these proteins in copper trafficking in *A. fumigatus* (Song, Li, & Jiang, 2019).

Based on a comparative analysis of amino acid sequences, the existence of a copper transporter for the Golgi complex in *C. albicans* has been suggested, named CaCCC2, due to its similarity (32%) to *S. cerevisiae* CCC2. The CaCCC2 sequence contains three motifs with ATPase activity/function (CPC, DKTGT, and GDGIND), eight transmembrane motifs, and metal-binding motifs GMTCXXC in the amino-terminal region of the protein. In *C. albicans*, the CaCCC2 transporter has little influence on virulence in mouse models, as a moderate reduction in the ability to kill the host was detected when animals were infected with mutant strains (Weissman, Shemer, & Kornitzer, 2002).

Data mining in the *S. pombe* genome database suggests that this species' open reading frame (ORF) SPBC1709.10c encodes a putative protein orthologous to *S. cerevisiae* ATX1. This sequence in *S. pombe* shares a high identity (56%) with ATX1 of *S. cerevisiae* and has the conserved MetXCysX$_2$Cys motif, suggesting a role in copper distribution (reviewed by Peter, Laliberté, Beaudoin, & Labbé, 2008). The activity of this protein in delivering copper to a possible transporter in *S. pombe* orthologous to *S. cerevisiae* CCC2 has yet to be elucidated. Still, it is crucial to supply copper to copper amine oxidase 1 (CAO1) (Peter et al., 2008; Laliberté & Labbé, 2006).

11. COX11, COX17, SCO1 and cytochrome *c* oxidase

The cytochrome *c* oxidase (COX) acts as an essential enzyme for the electron transport chain because it catalyzes the electron transfer from cytochrome *c* to molecular oxygen, promoting the generation of a chemiosmotic potential (proton gradient) necessary for ATP synthase activity. The activity of the COX enzyme in yeast depends on three copper ions inserted into two of the 12 subunits of the complex found in eukaryotes. The COX2 subunit requires two copper ions at the binuclear site called CuA, while the COX1 subunit requires a single copper ion at the mononuclear site CuB (Tsukihara et al., 1995). COX1 and COX2 are encoded by the mitochondrial genome (as well as COX3), while nuclear sequences synthesize the other subunits. However, all subunits of the COX complex are found in mitochondria in eukaryotes (in the plasma membrane in prokaryotes), specifically anchored in the inner mitochondrial membrane. For this reason, the copper atoms that need to be coupled to these COX subunits must be imported from the cytoplasm. Three proteins (COX11, COX17, and SCO1) are implicated in copper delivery/insertion into the mitochondria and, consequently, to the COX subunits.

The metalation of COX involves the metallochaperone COX17, a small 69-residue protein soluble in the cytoplasm (hydrophilic) (Glerum, Shtanko, & Tzagoloff, 1996). Hence, it is easily observed in the intermembrane space of the mitochondria (Beers, Glerum, & Tzagoloff, 1997). The activity of *S. cerevisiae* COX17 in supplying copper to cytochrome *c* oxidase occurs essentially due to the presence of a copper-binding sequence (CCXC) in this metallochaperone, which facilitates the transient trapping of copper until it is delivered to the target chaperones (Beers et al., 1997). The dependency on this sequence was confirmed since a simple substitution in any of the cysteine residues (Cys23, 24, 26) by the amino acid serine results in a non-functional cytochrome c oxidase complex. The point mutation prevented yeasts from growing on non-fermentable carbon sources (Heaton, Nittis, Srinivasan, & Winge, 2000). Nuclear magnetic resonance spectroscopy and X-ray absorption spectroscopy showed that two of the cysteine residues in the metal-binding motif (Cys23 and Cys26) bind to a single copper ion. Copper binds trigonally to the thiolate groups of cysteines, which stabilize the tertiary structure of COX17 and facilitate the interaction of this protein with target metallochaperones (Abajian, Yatsunyk, Ramirez, & Rosenzweig, 2004; Srinivasan, Posewitz, George, & Winge, 1998). Positively charged lysine and arginine residues (Lys19, Lys21, Lys30, and Arg33) in the COX17 sequence appear to contribute to this interaction (Abajian et al., 2004). This was demonstrated by functional analysis, which found that substituting the amino acid Arg33 abolishes protein function (Punter & Glerum, 2003). Once copper is bound to COX17, the protein mediates the donation of the ions to the COX 1 and 2 subunits, and this process depends on two accessory factors called COX11 and SCO1, which load the metal ions on the CuB site in COX1 and the CuA site in COX2, respectively.

The SCO1 protein (synthesis of cytochrome c oxidase 1) of *S. cerevisiae* is intimately involved in the activity of the COX complex. SCO1 promotes the assembly of the CuA binuclear site of the COX2 subunit in mitochondrial cytochrome c oxidase, acting precisely in post-translational steps of COX2 synthesis. SCO1 deletion does not induce quantitative changes in mRNA levels of COX2 but decreases COX activity due to degradation of the COX1 and COX2 subunits (Krummeck & Rödel, 1990). Sequence analysis of SCO1 indicates the presence of a metal-binding region (CXXXCP) essential for protein function because single or double mutations replacing the cysteines result in defective aerobic respiration in *S. cerevisiae* due to inefficient COX complex activity (Rentzsch et al., 1999). X-ray spectroscopy combined with functional studies has suggested that the two cysteines in the CXXXCP motif and the conserved His135 at the C-terminal end bind to a single copper atom (Balatri, Banci, Bertini, Cantini, & Ciofi-Baffoni, 2003, Nittis, George, &

Winge, 2001). Nevertheless, crystallographic studies have also identified an unexpected copper-binding site built up by the residues Cys181, Cys216, and His239. The latter can adopt positions close to both pairs of Cu-binding cysteines (Abajian & Rosenzweig, 2006). The association between SCO1 and COX17 in *S. cerevisiae* appears to depend on the Cys57 residue in COX17 (Horng et al., 2005). Evaluations of structural dynamics suggest that conserved residues 217KKYRVYF223 are critical for copper transfer from COX17 to the SCO1 apoprotein since mutants in this region have deficient COX activity (Rigby, Cobine, Khalimonchuk, & Winge, 2008). Little is known about the formation of complexes between SCO1 and COX2, but immunoprecipitation experiments and affinity chromatography have indicated the interaction of these two proteins (Lode, Kuschel, Paret, & Rödel, 2000).

In fungi, the metalation of the CuB site of COX1 requires the activity of COX11, which receives copper from COX17 (Hiser, Di Valentin, Hamer, & Hosler, 2000). COX11 is an essential chaperone for accumulating the COX1 mutant strains, which have decreased activity of the COX complex and reduced amounts of COX1 (Carr, George, & Winge, 2002). COX11 from *S. cerevisiae* is a 28 kDa polypeptide anchored in the inner mitochondrial membrane by a single transmembrane domain (residues 85–107; Carr, Maxfield, Horng, & Winge, 2005). The carboxy-terminal region of this protein is projected into the intermembrane space and contains isolated cysteines (Cys111, Cys208, and Cys210) or conserved CXC motif able to bind to a single copper ion. Substitution mutations in Cys residues decrease copper affinity and confer respiratory incompetence in *S. cerevisiae*. This protein is built up in dimers, which is essential because COX1 also occurs as a dimeric protein and a single COX11 dimer can provide sufficient copper for both CuB sites in COX1 (Carr et al., 2002).

12. CCS/Lys7 and Cu/Zn superoxide dismutase

Superoxide dismutase (SOD) enzymes act in combating superoxide radicals by catalyzing the dismutation reaction of this compound into molecular oxygen and hydrogen peroxide. Redundantly, Cu/Zn SOD, Mn SOD, or Fe/Mn SOD operate in this process (reviewed by Beyer, Imlay, & Fridovich, 1991). The metals in these enzymes act as cofactors and are essential for the redox reaction in ROS breakdown. The concentration of free copper in the intracellular niche remains undetectable under normal physiological conditions (Rae et al., 1999), a situation that makes copper acquisition by SOD via passive diffusion unfeasible. To deliver copper to Cu/Zn superoxide dismutase (SOD1), *S. cerevisiae* yeast utilizes LYS7 (Culotta et al., 1997), which was initially identified as involved in the lysine biosynthesis pathway (Horecka, Kinsey, & Sprague, 1995).

The CCS gene in *S. cerevisiae* encodes a 26–30 kDa chaperone with three distinct domains, two of which are copper bindings, while the central domain is crucial for interaction with SOD1. The amino-terminal domain of CCS exhibits notable homology with the copper chaperone ATX1 and contains the copper-binding motif MetXCysX$_2$Cys, essential for CCS function under copper-limiting conditions (Schmidt et al., 1999). Mutation in the central domain abrogates CCS-SOD1 binding (Schmidt et al., 1999; Schmidt, Kunst, & Culotta, 2000). The carboxy-terminal domain carries a cysteine motif (CysXCys) that transfers the metal to Cu/Zn SOD (Schmidt et al., 1999).

SOD1 and CCS exist in the cytoplasm as homodimers (CCS only when copper is present) and form a heterodimer/heterotetramer (dimer of dimers) for copper transfer (Lamb, Torres, O'Halloran, & Rosenzweig, 2001); however, a significant fraction of these proteins are also found in the intermembrane space of the mitochondria. Although these proteins are located in both compartments, the absence of CCS avoids SOD accumulation in the mitochondria. The localization of CCS in the inter-membrane space is assured by the interaction of the amino-terminal domain of CCS with the protein system/complex Mia40, which is anchored in the mitochondrial membrane. This system transfers energy to form a disulfide bond between the residues of the CysX$_2$Cys motif in the first domain of CCS, preventing the return of CCS to the cytoplasm through the translocase system, which only transports unfolded proteins.

In the fission yeast *S. pombe*, the protein PCCS (SPAC22E12.04) appears to be a candidate molecule for copper trafficking within the cell, as experimental characterization of this element demonstrates an equivalent function to CCS in *S. cerevisiae*. PCCS in has four domains, unlike CCS in *S. cerevisiae*. However, only three domains in *S. pombe* are necessary and sufficient for copper delivery to SOD1. *S. pombe*, PCCS lacks the Met-X-Cys-X2-Cys motif, which is necessary for copper binding under copper deprivation in *S. cerevisiae*. Fission yeast presents only one cysteine residue at position 11 that aligns with the last Cys residue present in the Met-X-Cys-X$_2$-Cys motif in *S. cerevisiae*. However, *S. pombe* protein sequence specificities in the carboxy-terminal region result in activity peculiarities of this chaperone. This region is rich in cysteine residues that carry copper to SOD and facilitates the action of this protein as metallothionein. Furthermore, PCCS lacks three histidines and the aspartic acid found in the second domain, which is essential for zinc loading in SOD1 in *S. cerevisiae*. The residues necessary for PCCS-protein interaction with SOD1 are highly conserved when comparing the two proteins, and the effectors necessary for copper delivery to SOD1 are localized in the third domain (Laliberté et al., 2004).

13. Copper efflux pump

C. albicans is copper resistant and can grow in a medium containing up to 25 mM of $CuSO_4$, while *S. cerevisiae* has satisfactory growth up to 2 mM of copper. The increased Cu resistance observed in *C. albicans* cells led to the discovery of the copper resistance-associated P-type ATPase *Ca*CRP1 (Weissman, Berdicevsky, Cavari, & Kornitzer, 2000). *Ca*CRP1 sequence has 1197 amino acids with metal-binding regions CXXC and DGMXCXXC in the N-terminus and three motifs in the ATPase functional domain (CPC, DKTCT, GDGINDC) at the C-terminus (Weissman et al., 2000). Deleting the CRP1 gene decreases the *C. albicans* grown in high copper by tenfold (Weissman et al., 2000). The growth defect likely occurs due to Cu-extrusion deficiency as mutants accumulate 40 times more copper inside the cell than wild cells (Weissman et al., 2000). CRP1 in *C. albicans* is linked to virulence. The copper efflux pump is necessary for fungal survival in macrophage cultures (Mackie et al., 2016). The levels of CRP1 transcripts are upregulated by high copper (Mackie et al., 2016; Weissman et al., 2000). This copper-detoxifying gene seems more important at early infection stages since CRP1 increases during early infection and then declines later (Mackie et al., 2016).

In *A. nidulans*, copper detoxification occurs primarily through copper efflux mediated by the efflux pump CRPA. The CRPA mRNA levels are upregulated in response to prolonged copper exposure. This phenomenon depends on the transcription factor ACEA. This copper transporter likely pumps Cu out of the cells, avoiding metal poisoning. CRPA has eight transmembrane domains, an auxiliary motif for copper translocation (CPC), and five metal-binding motifs, including three GlyMetXCysX$_2$Cys (heavy metal-associated domains) and two CXXC motifs in the N-terminus region. Additionally, CRPA contains an aspartyl kinase domain (DKTG) in the cytoplasmic loop, an aspartate residue that is transiently phosphorylated during the ATP catalytic cycle, a phosphatase domain (TGES), and a consensus domain for ATP binding (GDGVNDSP); thus, the protein contains conserved motifs found in heavy metal P-type ATPase transporters necessary for proper copper transport (Antsotegi-Uskola et al., 2017).

The response of *Histoplasma* to copper was first demonstrated in a study that utilized the CRP1 promoter as a tool for driving the transcription of heterologous genes. Gebart and colleagues (2006) showed that treatment of *H. capsulatum* yeasts with copper (1 to 100 μM copper sulfate) was sufficient to allow the expression of heterologous reporter genes (lacZ and GFP) under the control of the CRP1 promoter, with this process being dose-dependent. Moraes and colleagues (2023) characterized CRP1 in *H. capsulatum*. *In silico* analyses showed that this transporter contains a metal-binding site (CXXC and GMXCXXC), conserved P-type ATPase pump catalytic domains, and transmembrane domains responsible for building up pores for copper ions. A CPC sequence was found in the sixth transmembrane domain. CRP1 composes the virulence arsenal of *H. capsulatum*. CRP1-silenced strains were more susceptible to macrophage killing and had decreased survival in the mouse histoplasmosis model (Moraes et al., 2024).

14. Regulation of copper homeostasis

The regulation of copper homeostasis in fungi is maintained by the fine-tuning adjustments at the transcriptional level in genes encoding for transporters, metallothioneins, chaperones, and transcription factors. The regulators are copper sensors that perceive metal levels by Cu-binding/releasing, promoting structural changes. Hence, they coordinate the availability of proteins to ensure proper Cu levels in the cell. Many fungi contain two copper-responsive transcription factors, which reciprocally activate genes in response to deprivation or excess, while others present a single global regulatory molecule of copper homeostasis.

15. ACE1/AMT1

Copper increasing in *S. cerevisiae* is detected by the transcription factor ACE1 (Thiele, 1988), also known as CUP2 (Welch et al., 1989). ACE1 was identified during screening of a collection of copper-sensitive mutants. In this study, the ACE1-deleted strain exhibited increased Cu sensitivity and delayed CUP1 mRNA accumulation during metal exposure ($\geq$25 μM $CuSO_4$; Thiele, 1988).

ACE1 in *S. cerevisiae* is a nuclear polypeptide with 225 amino acid residues (Fürst, Hu, Hackett, & Hamer, 1988; Szczypka & Thiele, 1989). The N-terminus portion contains a zinc-binding subdomain and a copper-regulatory binding subdomain (CuRD), while the carboxy-terminal half features the transactivation region (Farrell, Thorvaldsen, & Winge, 1996). The first sub-domain (residues 1–40) is termed the zinc-binding module, as direct muta-genesis experiments at the site prevented zinc atom binding (Farrell et al., 1996). The zinc-ion ligands include the thiolate groups on the amino acids Cys11, Cys14, and Cys23 and the imidazole of His25 (Farrell et al., 1996). Experiments with Cys11 substitution demonstrate a reduction in ACE1's ability to bind to the CUP1 DNA, suggesting the role of this region in DNA-binding activity (Buchman et al., 1990). Additionally, the short sequence within the zinc module (R/K)GRP (residues 36–39) appears essential for DNA binding and for establishing contact with the A/T-rich region in the minor groove of the target gene promoter sequence. A Gly37Glu substitution in the (R/K)GRP sequence reduces ACE1 binding affinity to the CUP1 promoter DNA by 7-fold (Buchman et al., 1990). Adjacent to the zinc-binding subdomain is the copper-regulatory binding region (CuRD sub-domain, residues 41–100), a region containing a cysteine residue that binds to four Cu^{1+} atoms (Fürst et al., 1988). Copper ion binding promotes a con-formational change in the ACE1 protein, which enables it to bind to the promoter of the target genes. In *S. cerevisiae*, ACE1 has a cluster of tet-racopper thiolate that stabilizes the tertiary structure, which interacts with upstream activating sequences (UAS) in the promoters (Fürst et al., 1988). A single amino acid change at cysteine 11 compromises the interaction between ACE1 and UAS, although it does not lead to a complete loss of DNA-binding activity. However, the interaction between the Cys-11 mutant ACE1 and targets does not result in transcriptional activation, probably due to the inability to interact with other elements that compose the transcriptional machinery (Buchman et al., 1990). Comparative analyses of *S. cerevisiae* ACE1 sequences with other fungal copper-responsive transcription factors show a consensus sequence ($CysX_2CysX_{12-14}CysXCysX_{10-27}CysXCysX_5CysXCys$) within the CuRD subdomain. This consensus sequence is predicted to consist of two lobes containing four cysteines, each spaced by a varying region (Graden et al., 1996). Together, the sequence of the zinc-binding module and the CuRD subdomain, present in the amino-terminal portion, constitutes a rich region in basic residues, as it contains twelve cysteine motifs (CysXCys or $CysX_2Cys$) (Fürst et al., 1988), eleven of which have been shown to play a critical role for DNA binding (Hu, Fürst, & Hamer, 1990). The carboxy-terminal half of ACE1 is highly acidic and harbors the transactivation domain (Szczypka & Thiele, 1989). When ACE1 is bound to copper and consequently to specific promoter regions in target genes, the transactivation domain influences the transcription of genes under ACE1 control, likely by recruiting elements that compose the transcription activation machinery. For example, the essentiality of this region has been demonstrated for the proper activity of ACE1 in inducing CUP1 transcription in *S. cerevisiae* (Peña et al., 1998).

ACE1 contacts the gene to be regulated through interactions with the minor grooves in the target DNA between two major/greater groove interaction sites. Minor groove interactions are considered critical for forming a stable CuACE1-DNA complex in the nucleus (Szczypka & Thiele, 1989). Activated ACE1 (copper-bound) recognizes specific metal-responsive elements (MREs) in the promoters of target genes (HTHNNGCTGD). Mutations in the MREs reduced ACE1 activity, which results in loss of expression of the copper-responsive genes (Huibregtse et al., 1989; Evans, Engelke, & Thiele, 1990).

The ACE1 protein of *S. cerevisiae* is constitutively expressed in the absence or presence of copper (Szczypka & Thiele, 1989). This fact ensures adequate levels of the transcription factor to respond to toxic Cu levels rapidly. Studies have shown that *S. cerevisiae* cells grown under low copper levels contain a substantial amount of CUP2 (ACE1) (Culotta, Hsu, Hu, Fürst, & Hamer, 1989; Szczypka & Thiele, 1989). However, under these conditions, this transcription factor is inactive apoprotein and unable to bind to the UAS of the CUP1 DNA sequence (Buchman, Skroch, Welch, Fogel, & Karin, 1989). In response to copper increasing (>1 mM), ACE1 activates the expression of CUP1 (Thiele, 1988; Welch et al., 1989) and CRS5 (Culotta et al., 1994), as well as SOD1 (Gralla, Thiele, Silar, & Valentine, 1991). Cells with ACE1 deletion fail to induce MT genes and exhibit copper hypersensitivity phenotypes (Thiele, 1988; Welch et al., 1989).

AMT1 is an orthologous protein of ACE1 that responds to copper rising in *C. glabrata* (now renamed to *Nakaseomyces glabratus*; (Zhou & Thiele, 1991). The AMT1 polypeptide has 265 amino acids with an increased copper-binding CuRD region (residues 41–110) in the amino-terminal portion, as in *S. cerevisiae* ACE1. Nevertheless, AMT1 shows high homology with the first 100 amino acids of the N-terminus of *S. cerevisiae* ACE1. This region includes eleven cysteine residues similar to ACE1, with the first seven being strictly conserved (Zhou & Thiele, 1991). Despite structural similarities with ACE1, *C. glabrata* AMT1 is subject to positive transcriptional autoregulation mediated by the binding of Cu-activated AMT1 to a single copper-responsive element in the AMT1 promoter. *C. glabrata* cells treated with copper increase AMT1 mRNA, followed by accumulation of metallothionein (Zhou & Thiele, 1993).

In *A. fumigatus*, copper excess is detected by the copper-responsive transcription factor ACEA (an ACE1 orthologous). The ACEA presents, in the amino-terminal region, the zinc-binding module and the copper-regulatory domain, rich in cysteine residues, arranged in CysXCys and $CysX_2Cys$ motifs. Interestingly, ACEA in *A. fumigatus* has two conserved consensus sequences in the N-terminus, one within the zinc-binding module ($CysX_2Cys$-XRGHRX$_3$-CysXHis) and another KGRP close to the copper-binding domain portion, and both are essential for copper resistance (Cai et al., 2018). When *A. fumigatus* cells encounter high copper concentrations, ACEA induces the expression of the copper efflux pump CRPA and the metallothionein CRDA, ROS detoxification proteins such as SOD1, CAT1, and CAT2, as well as the transcription factors AFTA and YAP1 that promote oxidative stress control (Wiemann et al., 2017). However, understanding the ACEA binding motif in target genes in *A. fumigatus* is still limited.

16. MAC1

The transcription factor encoded by the metal-activated binding protein (MAC1) mediates the homeostatic control in *S. cerevisiae* cells under copper scarcity by regulating the expression of the high-affinity copper transport system (Yamaguchi-Iwai et al., 1997). Structurally, MAC1 exhibits notable homology with the amino-terminal region of ACE1, with 53% identity (Jungmann et al., 1993). Analysis of the MAC1 417 amino acids (46 kDa polypeptide) revealed the presence of a zinc-binding module at the N-terminus (residues 1–40; Jensen, Posewitz, Srinivasan, & Winge, 1998). At this portion, MAC1 has the Zn-binding domain that ties two Zn atoms, although the second zinc ion-binding motif remains uncharacterized (Jensen et al., 1998). The zinc module also displays a conserved (R/K)GRP motif (residues 36–39; Jamison McDaniels, Jensen, Srinivasan, Winge, & Tullius, 1999; Yamaguchi-Iwai et al., 1997), known to form a concave surface inserted into the A/T-rich minor groove in the promoter region of target genes, where the two arginines (R) likely interact with the A/T base pair in the DNA, while the proline residue appears to contact the peptide chain outside the minor DNA groove (Jamison McDaniels et al., 1999). The minimal DNA-binding domain of MAC1 consists of the first 159 residues of the amino-terminal portion of the polypeptide sequence. Therefore, part of the sequence in the amino-terminal half (residues 41–159) forms the MAC1 DNA-binding structure (Jensen et al., 1998). In MAC1, only three of the eleven cysteines necessary for ACE1 function are conserved; however, the protein has additional cysteine-rich domains in the C-terminus region that might participate in copper metal coordination (Jungmann et al., 1993). Two cysteine-rich motifs are found in the C-terminus. Those motifs contain five cysteines and one histidine ($CysXCysX_4CysXCysX_2CysX_2His$), designated REP-I and REP-II (residues 322–337; Zhu, Labbé, Peña, & Thiele, 1998) and are responsible for MAC1 activity (Graden & Winge, 1997). REP-I and REP-II bind to eight Cu1 + atoms, resulting in MAC1 activity inhibition (Jensen & Winge, 1998). Point mutations at multiple positions in REP-I and REP-II revealed that these elements play independent, distinct roles in MAC1 function (Keller, Gross, Kelleher, & Winge, 2000; Zhu et al., 1998). Point mutations in the two REP sequences cause different levels of strain impairment. REP-II mutation completely prevents CTR transporter expression and causes growth defects due to inefficient respiration, while mutants in REP-I can maintain at least partial CTR production (Zhu et al., 1998). Both domains participate in intramolecular interaction (interaction of MAC1 domains) that represses MAC1 activity. Specifically, at high copper, cysteine-rich motifs (carboxy-terminal activation domain) bind to the DNA-binding domain of the N-terminus, promoting structural changes that directly inhibit MAC1 activity (Jensen & Winge, 1998). Additionally, immuno-fluorescence data indicate that MAC1 is a nuclear protein (Jungmann et al., 1993) that corroborates with the two nuclear localization signal (NSL) sequences (Serpe, Joshi, & Kosman, 1999). Sequences consistent with NSL appear in the amino-terminal half $KKXRX_{15}KKXK$ (residues 155–177) of MAC1 and in the C-terminus (Serpe et al., 1999).

The MAC1 protein binds to copper-responsive elements (CuREs) $TTTGC(T/G)C(A/G)$ in the promoters of the MAC1-regulated gene, such as the CTR1 and CTR3 transporters and the ferroreductases FRE1 and FRE7 (Labbé, Zhu, & Thiele, 1997, Martins, Jensen, Simon, Keller, & Winge, 1998, Yamaguchi-Iwai et al., 1997). Electrophoretic mobility shift assays (EMSA) demonstrated direct and specific binding of MAC1 (residues 1–159) to a CTR1 DNA duplex containing the copper-responsive element

(Jensen et al., 1998). CuRE elements in promoters are found in at least two copies in each gene and are critical for efficient transcriptional activation (Jensen et al., 1998). These elements seem to work synergistically as an increase in their number directly enhances CTR1 expression. The sequence between CuRE sites is variable, and changes in this region have a limited effect on expression (Jensen et al., 1998). Alternatively, deletion (Dancis et al., 1992; Jensen et al., 1998; Labbé et al., 1997; Martins et al., 1998; Yamaguchi-Iwai et al., 1997) or mutation (Labbé et al., 1997; Martins et al., 1998) of any of these elements impairs MAC1 binding, demonstrating that tandem repetition is necessary for the full activation of target gene promoters. Supporting this, nucleotide substitution experiments showed the require-ment of two copies of the GCTC element in CuRE for proper MAC1 function, suggesting MAC1 binds to DNA as a dimer (Yamaguchi-Iwai et al., 1997). Two-hybrid analyses indicated a Cu-independent intramole-cular MAC1p-MAC1p interaction (Serpe et al., 1999).

MAC1 in *S. cerevisiae* is constitutively expressed, although transcrip-tional activation depends on copper and a transactivation domain (Jensen & Winge, 1998). At elevated intracellular copper levels, MAC1 becomes inactive through an intramolecular interaction between the DNA-binding domain in the amino-terminal region and REP-I and REP-II. The interaction promotes the dissociation of MAC1 from the promoter ele-ments, which blocks the transcriptional activity (Graden & Winge, 1997; Keller, Bird, & Winge, 2005). MAC1 in *S. cerevisiae* is rapidly degraded in a sufficient copper environment to eliminate the copper uptake machinery (Zhu et al., 1998). The degradation is accompanied by a decrease in the expression of target genes, such as the CTR1 gene product (Yonkovich, McKenndry, Shi, & Zhu, 2002).

In the filamentous fungus *A. fumigatus*, the transcription factor AfMAC1 controls copper acquisition by inducing the expression of three copper transporters (CTRA1, CTRA2, and CTRC) and two metallor-eductases (Kusuya et al., 2017). AfMAC1 shows a similar amino acid sequence to *S. cerevisiae* MAC1, with DNA and copper-binding sequences in the amino-terminal region. Intriguingly, AfMAC1 appears to exert control over a broader range of activities in *Aspergillus* cells and not only acts in copper homeostasis. *A. fumigatus* cells with AfMAC1 deletion have a severe growth defect in low copper, but metal supplementation rescues the phenotype. Deletion mutants exhibit reduced laccase and SOD activity and are more sensitive to H_2O_2 (1 mM). The severe growth defects in low copper observed in AfMAC1 mutants indicate that the fungus does not have sufficient metal stocks to ensure adequate copper supply in poor Cu-environments (Park, Kim, & Yun, 2017). The influence of MAC1, rather than CUF (discussed below), on copper homeostasis in *A. fumigatus* emerged due to the presence of the 5'-GCTCG-3' sequence (similar to those used for MAC1 binding to promoters of CTR transporter genes in *S. cerevisiae*) in the promoters of the copper transporters CTRA2 and CTRC (Park et al., 2014). EMSA analyses showed the binding of AfMAC1 to the 5'-TGTGCTCA-3' sequence in the promoter of CTRC in *A. fumigatus* (Park et al., 2017).

Histoplasma genome encodes an ortholog of MAC1. This transcription factor is necessary for the positive regulation of CTR3 under low copper conditions, and MAC1-deficient strains are attenuated in their ability to infect host macrophages. MAC1 regulates CTR3 by binding in CuRE motifs as a promoter of the high-affinity copper transporter. MAC1 also regulates CRD2 and CCC1, identified as copper-binding factors and vacuolar metal ion transporters. *Histoplasma* MAC1 regulates genes related to the homeostasis of other metals, such as iron and zinc, and ROS degradation, such as catalases and superoxide dismutase (Ray & Rappleye, 2022). Two copper regulators have also been pointed out in the dimorphic fungi *Blastomyces*, *Paracoccidioides*, and *Coccidioides* (Moraes et al., 2023; Ray & Rappleye, 2022).

17. CUF1

Using two distinct transcription factors involved in copper homeostasis regulation in some fungi is not conserved throughout the fungal kingdom; for example, *S. pombe* and *C. neoformans* use a single transcription factor responsive to Cu factor 1 (CUF1). On the other hand, only in *C. neoformans* does CUF1 share strong similarities with MAC1 and ACE1 (Labbé et al., 1999; Waterman et al., 2007). This observation in phylogenetically distant fungi indicates the existence of a common ancestor, which possibly underwent recombination events leading to the emergence of a chimeric protein (Nevitt et al., 2012).

The CUF1 gene of *S. pombe* (Labbé et al., 1999) consists of sequences similar to those essential for the activity of the transcription factors ACE1/ AMT1 and MAC1. CUF1 exhibits 51% and 45% identity with the amino-terminal region of ACE1 in *S. cerevisiae* and AMT1 in *C. glabrata*, respectively. Specifically, CUF1 harbors extensive homology with the amino acid residues 41 to 61 region of ACE1 in *S. cerevisiae*. Regarding similarity with MAC1, residues 1 to 61 in CUF1 exhibit low identity with this transcription factor. However, CUF1 presents a cysteine-rich domain containing five residues and one histidine residue in the C-terminus region (^{328}CysGlnCysGlyAspAsnCysGluCysLeuGlyCysLeuThrGis^{342}), which is similar to MAC1 (Labbé et al., 1999). This sequence appears to play a critical role in the quantitative monitoring of copper ions in *S. pombe* cells, as when this domain is disrupted, CUF1 fails to bind copper, resulting in high levels of CTR4 mRNAs (Beaudoin, Mercier, Langlois, & Labbé, 2003).

CUF1 exhibits a non-canonical nuclear localization signal sequence (NLS; 11 to 53 residues) in the N-terminus. Nevertheless, CUF1 is mainly localized in the cytoplasm of cells growing under copper-rich conditions, while it accumulates in the nucleus under copper scarcity. This suggests that Cu-loading CUF1 induces conformational changes that mask the NLS, blocking its translocation to the nucleus (Beaudoin & Labbé, 2006). CUF1 moves from the nucleus to the cytoplasm at sufficient copper concentrations, likely due to the leucine-rich nuclear export signal (NES; 349LAALNHISAL358) located at the C-terminal. The nuclear export is mediated by CRM1 (Beaudoin & Labbé, 2006).

CUF1 deletion in *S. pombe* results in the inability to utilize non-fermentable carbon sources impairs superoxide dismutase activity and causes defects in iron accumulation at copper deprivation (Labbé et al., 1999). These phenotypes suggest the involvement of CUF1 in the control of genes necessary for high-affinity copper transport. CUF1 deletion avoids the induction of high-affinity copper transporter genes (Beaudoin et al., 2003). CUF1 of *S. pombe* is activated during copper limitation when it induces the expression of the plasma membrane copper transporters CTR4 (Labbé et al., 1999) and CTR5 (Zhou & Thiele, 2001) and the vacuolar copper transporter designated CTR6 (Bellemare et al., 2002). In response to copper deficiency, CUF1 binds to the DNA sequence 5'-D(T/A)DDHGCTGD-3' (CuSE) in the promoter region of target genes (Beaudoin & Labbé, 2001).

C. neoformans also encodes a transcription factor CnCUF1 (Waterman et al., 2007) with structural similarity to *S. pombe* CUF1 and other transcription factors. The CnCUF1 gene encodes for a 591 amino acid peptide. The amino-terminal region (63 amino acids) of CnCUF1 shows high identity with *S. cerevisiae* ACE1 and *C. glabrata* AMT1, and this region contains seven cysteine residues and a conserved KGRP motif (Waterman et al., 2007). Additionally, cryptococcal CUF1 contains two copper-binding cysteine motifs, and the first one ($CXCX_3CXCX_2CX_2H$) is identical to REP-1 found in *S. cerevisiae* MAC1 (Jiang et al., 2011). Despite the similarity, *C. neoformans* CUF1 contains an additional cysteine-rich domain ($CCX_3CXCX_4CXCX_3CCXCCX_2$ CXC) and an extensive proline-serine-rich motif (Kosman, 2018).

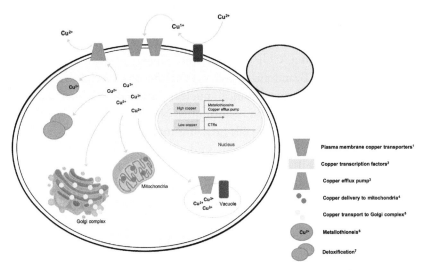

Fig. 1 The figure depicts the schematic representation of the functionally characterized proteins that influence copper homeostasis in *Aspergillus* sp., *Candida* sp., *C. neoformans*, *S. cerevisiae*, and *S. pombe*. [1]In the plasma membrane: *A. fumigatus AfCtrA2* and *AfCtrC* (Park et al., 2014), *C. albicans CaCTR1* (Marvin et al., 2003), *C. neoformans* CTR1 and CTR4 (Waterman et al., 2007), *S. cerevisiae* CTR1 and CTR3 (Knight et al., 1996), *S. pombe* CTR4 (Labbé et al., 1999) and CTR5 (Zhou & Thiele, 2001). In vacuoles: *C. neoformans* CTR2 (Chun & Madhani, 2010), *S. cerevisiae* CTR2 (Kampfenkel et al., 1995, Knight et al., 1996), *S. pombe* CTR6 (Bellemare et al., 2002). [2]*A. fumigatus* ACEA (Cai et al., 2018), *A. fumigatus* AfMAC1 (Park et al., 2017), *C. glabrata* AMT1 (Zhou, Szczypka, Sosinowski, & Thiele, 1992), *C. albicans* CaMAC1, *C. neoformans* CnCUF1 (Waterman et al., 2007), ACE1 (*activation of CUP1 expression*) (Thiele, 1988), *S. cerevisiae* MAC1 (*metal binding activator*) (Yamaguchi-Iwai et al., 1997), *S. pombe* CUF1 (Labbé et al., 1999). [3]*A. nidulans* CRPA (Antsotegi-Uskola et al., 2017), *A. fumigatus* CRPA (Wiemann et al., 2017), *C. albicans CaCRP1* (*copper resistance-associated P-type ATPase*) (Weissman et al., 2000). [4] *S. cerevisiae* COX11 (Hiser et al., 2000), *S. cerevisiae* COX17 (Glerum et al., 1996), SCO1 (*synthesis of cytochrome c oxidase* 1) (Schulze & Rödel, 1988). [5]*A. fumigatus* CTPA (*copper transporter Aspergillus*) (Upadhyay et al., 2013), *C. albicans CaCCC2* (Weissman et al., 2002), *C. neoformans* ATX1 and CCC2 (Walton et al., 2005), ATX1 *S. cerevisiae* (Lin & Culotta, 1995), *S. cerevisiae* CCC2 (*Ca²⁺-sensitive cross-complementer*) (Fu et al., 1995). [6]*A. fumigatus* CRDA (Cai et al., 2018), *C. albicans* CRD2 (*copper resistance determinant*) (Riggle & Kumamoto, 2000), *C. neoformans* CMT1 and CMT2 (Ding et al., 2011), *S. cerevisiae* CUP1 (Butt et al., 1984, Ecker et al., 1986, Fogel et al., 1983), *S. cerevisiae* CRS5 (Culotta et al., 1994). [7]*S. cerevisiae* CCS1 (*copper chaperone for* SOD) (Culotta et al., 1997), *S. pombe* PCCS (Laliberté et al., 2004).

Deletion of the cryptococcal copper-dependent transcription factor leads to growth defects under conditions with scarce copper (Waterman et al., 2007), similar to mutants for *S. cerevisiae* MAC1 and *S. pombe* CUF1. On the other hand, cryptococcal CUF1 protein is also required for copper detoxification under copper-rich conditions. Atomic absorption spectrometry measurements demonstrated intracellular Cu accumulation in strains deleted for CnCUF1, associated with insufficient metallothionein levels in a copper-enriched medium. This suggests the importance of this transcription factor in promoting Cu-detoxification (Jiang et al., 2011). Together, these experiments highlight the dual role of *C. neoformans* CUF1 as a transcription factor in controlling both copper deprivation and excess in this species.

18. Conclusion

Copper homeostasis represents a pivotal biological process in both pathogenic and model fungi. These organisms have evolved sophisticated systems to regulate copper levels, which involve a complex interplay of mechanisms. This ensures that this essential micronutrient is available for vital processes while preventing potential toxicity. The maintenance of copper balance in fungi depends upon membrane transporters and a variety of proteins regulated by specific transcription factors. This process has been the subject of extensive study in the yeast model *S. cerevisiae* and pathogenic fungi, including the genera *Aspergillus*, *Candida*, and *Cryptococcus*. The process of copper uptake into cells is facilitated by transporters that exhibit high affinity and specificity for copper ions. Once inside the cytoplasm, fungi must meticulously regulate copper levels to avoid the potential toxicity of elevated intracellular concentrations. The mechanisms employed to address the accumulation of intracellular copper include the synthesis of metallothioneins, which act as copper chelators; the storage of copper in vesicles; the promotion of copper efflux; and the inhibition of copper uptake. The transcriptional regulation of genes involved in copper uptake, buffering, efflux, and detoxification is orchestrated by fungal transcription factors that control the expression of these genes. These regulators detect fluctuations in copper levels and activate detoxification or acquisition mechanisms in response to excess or deficiency (Fig. 1).

Understanding these mechanisms may provide insights into fungal biology and new targets for developing strategies for combating fungal infections, which could ultimately benefit human health. By targeting the unique components of fungal copper homeostasis, such as specific transporters or regulatory proteins, it may be possible to disrupt the pathogen's ability to thrive within the host. This approach could lead to the design of new drugs that are more selective for fungal pathogens, reducing the risk of side effects and improving treatment outcomes.

Acknowledgments

We are grateful to Alene Alder-Rangel from Alder's English Services for the review of the English. This work was supported by the Conselho Nacional de Desenvolvimento Científico e Tecnológico - CNPq (grant numbers 407706/2021-6; 312888/2022-8; 405299/2023-0; 314289/2023-2); the Fundação de Amparo a Pesquisa do Estado de Goiás – FAPEG (grant numbers 202310267000240; 202310267000213; 202310267000747); the Instituto Nacional de Ciência e Tecnologia da Interação Patógeno-Hospedeiro - INCT-IPH/FAPEG (grant number 201810267000022). DM received a fellowship from FAPEG. MGSB and AMB are CNPq productivity fellows.

References

Abajian, C., & Rosenzweig, A. C. (2006). Crystal structure of yeast Sco1. *Journal of Biological Inorganic Chemistry: JBIC a Publication of the Society of Biological Inorganic Chemistry, 11*(4), 459–466.

Abajian, C., Yatsunyk, L. A., Ramirez, B. E., & Rosenzweig, A. C. (2004). Yeast cox17 solution structure and Copper(I) binding. *The Journal of Biological Chemistry, 279*(51), 53584–53592.

Aller, S. G., Eng, E. T., De Feo, C. J., & Unger, V. M. (2004). Eukaryotic CTR copper uptake transporters require two faces of the third transmembrane domain for helix packing, oligomerization, and function. *The Journal of Biological Chemistry, 279*(51), 53435–53441.

Antsotegi-Uskola, M., Markina-Iñarrairaegui, A., & Ugalde, U. (2017). Copper resistance in *Aspergillus nidulans* Relies on the P$_I$-type ATPase CrpA, regulated by the transcription factor AceA. *Frontiers in Microbiology, 8*, 912.

Arnesano, F., Banci, L., Bertini, I., Cantini, F., Ciofi-Baffoni, S., Huffman, D. L., & O'Halloran, T. V. (2001). Characterization of the binding interface between the copper chaperone Atx1 and the first cytosolic domain of Ccc2 ATPase. *The Journal of Biological Chemistry, 276*(44), 41365–41376.

Assunção, L. D. P., Moraes, D., Soares, L. W., Silva-Bailão, M. G., de Siqueira, J. G., Baeza, L. C., ... Bailão, A. M. (2020). Insights into *Histoplasma capsulatum* behavior on zinc deprivation. *Frontiers in Cellular and Infection Microbiology, 10*, 573097.

Balatri, E., Banci, L., Bertini, I., Cantini, F., & Ciofi-Baffoni, S. (2003). *Solution structure of Sco1: A thioredoxin-like protein involved in cytochrome c oxidase assembly. Structure (London, England: 1993), 11*, 1431–1443.

Banci, L., Bertini, I., Cantini, F., Felli, I. C., Gonnelli, L., Hadjiliadis, N., ... Voulgaris, P. (2006). The Atx1-Ccc2 complex is a metal-mediated protein-protein interaction. *Nature Chemical Biology, 2*(7), 367–368.

Beaudoin, J., & Labbé, S. (2001). The fission yeast copper-sensing transcription factor Cuf1 regulates the copper transporter gene expression through an Ace1/Amt1-like recognition sequence. *The Journal of Biological Chemistry, 276*(18), 15472–15480.

Beaudoin, J., & Labbé, S. (2006). Copper induces cytoplasmic retention of fission yeast transcription factor cuf1. *Eukaryot Cell, 5*(2), 277–292.

Beaudoin, J., & Labbé, S. (2007). Crm1-mediated nuclear export of the *Schizosaccharomyces pombe* transcription factor Cuf1 during a shift from low to high copper concentrations. *Eukaryotic Cell, 6*(5), 764–775.

Beaudoin, J., Mercier, A., Langlois, R., & Labbé, S. (2003). The *Schizosaccharomyces pombe* Cuf1 is composed of functional modules from two distinct classes of copper metalloregulatory transcription factors. *The Journal of Biological Chemistry, 278*(16), 14565–14577.

Beers, J., Glerum, D. M., & Tzagoloff, A. (1997). Purification, characterization, and localization of yeast Cox17p, a mitochondrial copper shuttle. *The Journal of Biological Chemistry, 272*(52), 33191–33196.

Bellemare, D. R., Shaner, L., Morano, K. A., Beaudoin, J., Langlois, R., & Labbe, S. (2002). Ctr6, a vacuolar membrane copper transporter in *Schizosaccharomyces pombe*. *The Journal of Biological Chemistry, 277*(48), 46676–46686.

Beyer, W., Imlay, J., & Fridovich, I. (1991). Superoxide dismutases. *Progress in Nucleic Acid Research and Molecular Biology, 40*, 221–253.

Brenes-pomales, A., Lindegren, G., & Lindegren, C. C. (1955). Gene control of copper sensitivity in Saccharomyces. *Nature, 176*(4487), 841–842.

Buchman, C., Skroch, P., Dixon, W., Tullius, T. D., & Karin, M. (1990). A single amino acid change in CUP2 alters its mode of DNA binding. *Molecular and Cellular Biology, 10*(9), 4778–4787.

Buchman, C., Skroch, P., Welch, J., Fogel, S., & Karin, M. (1989). The CUP2 gene product, regulator of yeast metallothionein expression, is a copper-activated DNA-binding protein. *Molecular and Cellular Biology, 9*(9), 4091–4095.

Butt, T. R., & Ecker, D. J. (1987). Yeast metallothionein and applications in biotechnology. *Microbiological Reviews, 51*(3), 351–364.

Butt, T. R., Sternberg, E., Herd, J., & Crooke, S. T. (1984). Cloning and expression of a yeast copper metallothionein gene. *Gene, 27*(1), 23–33.

Cai, Z., Du, W., Zhang, Z., Guan, L., Zeng, Q., Chai, Y., ... Lu, L. (2018). The *Aspergillus fumigatus* transcription factor AceA is involved not only in Cu but also in Zn detoxification through regulating transporters CrpA and ZrcA. *Cellular Microbiology, 20*(10), e12864.

Carr, H. S., George, G. N., & Winge, D. R. (2002). Yeast Cox11, a protein essential for cytochrome c oxidase assembly, is a Cu(I)-binding protein. *J Biol Chem, 277*(34), 31237–31242.

Carr, H. S., Maxfield, A. B., Horng, Y. C., & Winge, D. R. (2005). Functional analysis of the domains in Cox11. *J Biol Chem, 280*(24), 22664–22669.

Chun, C. D., & Madhani, H. D. (2010). Ctr2 links copper homeostasis to polysaccharide capsule formation and phagocytosis inhibition in the human fungal pathogen *Cryptococcus neoformans*. *PLoS One, 5*(9), e12503.

Cohen, A., Nelson, H., & Nelson, N. (2000). The family of SMF metal ion transporters in yeast cells. *The Journal of Biological Chemistry, 275*(43), 33388–33394.

Culotta, V. C., Howard, W. R., & Liu, X. F. (1994). CRS5 encodes a metallothionein-like protein in *Saccharomyces cerevisiae*. *The Journal of Biological Chemistry, 269*(41), 25295–25302.

Culotta, V. C., Hsu, T., Hu, S., Fürst, P., & Hamer, D. (1989). Copper and the ACE1 regulatory protein reversibly induce yeast metallothionein gene transcription in a mouse extract. *Proceedings of the National Academy of Sciences of the United States of America, 86*(21), 8377–8381.

Culotta, V. C., Klomp, L. W., Strain, J., Casareno, R. L., Krems, B., & Gitlin, J. D. (1997). The copper chaperone for superoxide dismutase. *The Journal of Biological Chemistry, 272*(38), 23469–23472.

Dancis, A., Haile, D., Yuan, D. S., & Klausner, R. D. (1994). The *Saccharomyces cerevisiae* copper transport protein (Ctr1p). Biochemical characterization, regulation by copper, and physiologic role in copper uptake. *The Journal of Biological Chemistry, 269*(41), 25660–25667.

Dancis, A., Yuan, D. S., Haile, D., Askwith, C., Eide, D., Moehle, C., ... Klausner, R. D. (1994). Molecular characterization of a copper transport protein in *S. cerevisiae*: An unexpected role for copper in iron transport. *Cell, 76*(2), 393–402.

Dancis, A., Roman, D. G., Anderson, G. J., Hinnebusch, A. G., & Klausner, R. D. (1992). Ferric reductase of *Saccharomyces cerevisiae*: Molecular characterization, role in iron uptake, and transcriptional control by iron. *Proceedings of the National Academy of Sciences of the United States of America, 89*(9), 3869–3873.

De Rome, L., & Gadd, G. M. (1987). Measurement of copper uptake in *Saccharomyces cerevisiae* using a Cu2+-selective electrode. *FEMS Microbiology Letters, 43*(3), 283–287.

Ding, C., Festa, R. A., Chen, Y. L., Espart, A., Palacios, Ò., Espín, J., ... Thiele, D. J. (2013). *Cryptococus neoformans* copper detoxification machinery is critical for fungal virulence. *Cell Host & Microbe, 13*(3), 265–276.

Ding, C., Yin, J., Tovar, E. M., Fitzpatrick, D. A., Higgins, D. G., & Thiele, D. J. (2011). The copper regulon of the human fungal pathogen *Cryptococcus neoformans* H99. *Molecular Microbiology, 81*(6), 1560–1576.

Dix, D., Bridgham, J., Broderius, M., & Eide, D. (1997). Characterization of the FET4 protein of yeast. Evidence for a direct role in the transport of iron. *The Journal of Biological Chemistry, 272*(18), 11770–11777.

Ecker, D. J., Butt, T. R., Sternberg, E. J., Neeper, M. P., Debouck, C., Gorman, J. A., & Crooke, S. T. (1986). Yeast metallothionein function in metal ion detoxification. *The Journal of Biological Chemistry, 261*(36), 16895–16900.

Eisses, J. F., & Kaplan, J. H. (2005). The mechanism of copper uptake mediated by human CTR1: A mutational analysis. *The Journal of Biological Chemistry, 280*(44), 37159–37168.

Evans, C. F., Engelke, D. R., & Thiele, D. J. (1990). ACE1 transcription factor produced in *Escherichia coli* binds multiple regions within yeast metallothionein upstream activation sequences. *Mol Cell Biol, 10*(1), 426–429.

Farrell, R. A., Thorvaldsen, J. L., & Winge, D. R. (1996). Identification of the Zn(II) site in the copper-responsive yeast transcription factor, AMT1: A conserved Zn module. *Biochemistry, 35*(5), 1571–1580.

Festa, R. A., & Thiele, D. J. (2011). Copper: An essential metal in biology. *Current Biology: CB, 21*(21), R877–R883.

Fogel, S., Welch, J. W., Cathala, G., & Karin, M. (1983). Gene amplification in yeast: CUP1 copy number regulates copper resistance. *Current Genetics, 7*(5), 347–355.

Fu, D., Beeler, T. J., & Dunn, T. M. (1995). Sequence, mapping and disruption of CCC2, a gene that cross-complements the Ca(2+)-sensitive phenotype of csg1 mutants and encodes a P-type ATPase belonging to the Cu(2+)-ATPase subfamily. *Yeast (Chichester, England), 11*(3), 283–292.

Fürst, P., Hu, S., Hackett, R., & Hamer, D. (1988). Copper activates metallothionein gene transcription by altering the conformation of a specific DNA binding protein. *Cell, 55*(4), 705–717.

Georgatsou, E., Mavrogiannis, L. A., Fragiadakis, G. S., & Alexandraki, D. (1997). The yeast Fre1p/Fre2p cupric reductases facilitate copper uptake and are regulated by the copper-modulated Mac1p activator. *The Journal of Biological Chemistry, 272*(21), 13786–13792.

Glerum, D. M., Shtanko, A., & Tzagoloff, A. (1996). Characterization of COX17, a yeast gene involved in copper metabolism and assembly of cytochrome oxidase. *The Journal of Biological Chemistry, 271*(24), 14504–14509.

Gorman, J. A., Clark, P. E., Lee, M. C., Debouck, C., & Rosenberg, M. (1986). Regulation of the yeast metallothionein gene. *Gene, 48*(1), 13–22.

Graden, J. A., & Winge, D. R. (1997). Copper-mediated repression of the activation domain in the yeast Mac1p transcription factor. *Proceedings of the National Academy of Sciences of the United States of America, 94*(11), 5550–5555.

Graden, J. A., Posewitz, M. C., Simon, J. R., George, G. N., Pickering, I. J., & Winge, D. R. (1996). Presence of a copper(I)-thiolate regulatory domain in the copper-activated transcription factor Amt1. *Biochemistry, 35*(46), 14583–14589.

Gralla, E. B., Thiele, D. J., Silar, P., & Valentine, J. S. (1991). ACE1, a copper-dependent transcription factor, activates expression of the yeast copper, zinc superoxide dismutase gene. *Proc Natl Acad Sci U S A, 88*(19), 8558–8562.

Hamer, D. H. (1986). Metallothionein. *Annual Review of Biochemistry, 55*, 913–951.

Hassett, R., & Kosman, D. J. (1995). Evidence for Cu(II) reduction as a component of copper uptake by *Saccharomyces cerevisiae*. *The Journal of Biological Chemistry, 270*(1), 128–134.

Hamer, D. H., Thiele, D. J., & Lemontt, J. E. (1985). Function and autoregulation of yeast copperthionein. *Science, 228*(4700), 685–690.

Hassett, R., Dix, D. R., Eide, D. J., & Kosman, D. J. (2000). The Fe(II) permease Fet4p functions as a low affinity copper transporter and supports normal copper trafficking in *Saccharomyces cerevisiae*. *The Biochemical Journal, 351*(Pt 2), 477–484.

Hatori, Y., & Lutsenko, S. (2016). The role of copper chaperone Atox1 in coupling redox homeostasis to intracellular copper distribution. *Antioxidants (Basel, Switzerland), 5*(3), 25.

Heaton, D., Nittis, T., Srinivasan, C., & Winge, D. R. (2000). Mutational analysis of the mitochondrial copper metallochaperone Cox17. *The Journal of Biological Chemistry, 275*(48), 37582–37587.

Hiser, L., Di Valentin, M., Hamer, A. G., & Hosler, J. P. (2000). Cox11p is required for stable formation of the Cu(B) and magnesium centers of cytochrome c oxidase. *The Journal of Biological Chemistry, 275*(1), 619–623.

Horecka, J., Kinsey, P. T., & Sprague, G. F., Jr (1995). Cloning and characterization of the *Saccharomyces cerevisiae* LYS7 gene: Evidence for function outside of lysine biosynthesis. *Gene, 162*(1), 87–92.

Horng, Y. C., Leary, S. C., Cobine, P. A., Young, F. B., George, G. N., Shoubridge, E. A., & Winge, D. R. (2005). Human Sco1 and Sco2 function as copper-binding proteins. *The Journal of Biological Chemistry, 280*(40), 34113–34122.

Hu, S., Fürst, P., & Hamer, D. (1990). The DNA and Cu binding functions of ACE1 are interdigitated within a single domain. *The New Biologist, 2*(6), 544–555.

Huibregtse, J. M., Engelke, D. R., & Thiele, D. J. (1989). Copper-induced binding of cellular factors to yeast metallothionein upstream activation sequences. *Proceedings of the National Academy of Sciences of the United States of America, 86*(1), 65–69.

Imlay, J. A., & Linn, S. (1988). DNA damage and oxygen radical toxicity. *Science (New York, N. Y.), 240*(4857), 1302–1309.

Inesi, G. (2017). Molecular features of copper binding proteins involved in copper homeostasis. *IUBMB Life, 69*(4), 211–217.

Ioannoni, R., Beaudoin, J., Mercier, A., & Labbé, S. (2010). Copper-dependent trafficking of the Ctr4-Ctr5 copper transporting complex. *PLoS One, 5*(8), e11964.

Irving, H., & Williams, R. (1948). Order of stability of metal complexes. *Nature, 162*, 746–747.

Jamison McDaniels, C. P., Jensen, L. T., Srinivasan, C., Winge, D. R., & Tullius, T. D. (1999). The yeast transcription factor Mac1 binds to DNA in a modular fashion. *The Journal of Biological Chemistry, 274*(38), 26962–26967.

Jensen, L. T., & Winge, D. R. (1998). Identification of a copper-induced intramolecular interaction in the transcription factor Mac1 from *Saccharomyces cerevisiae*. *The EMBO Journal, 17*(18), 5400–5408.

Jensen, L. T., Howard, W. R., Strain, J. J., Winge, D. R., & Culotta, V. C. (1996). Enhanced effectiveness of copper ion buffering by CUP1 metallothionein compared with CRS5 metallothionein in *Saccharomyces cerevisiae*. *The Journal of Biological Chemistry, 271*(31), 18514–18519.

Jensen, L. T., Posewitz, M. C., Srinivasan, C., & Winge, D. R. (1998). Mapping of the DNA binding domain of the copper-responsive transcription factor Mac1 from *Saccharomyces cerevisiae*. *The Journal of Biological Chemistry, 273*(37), 23805–23811.

Jiang, N., Liu, X., Yang, J., Li, Z., Pan, J., & Zhu, X. (2011). Regulation of copper homeostasis by Cuf1 associates with its subcellular localization in the pathogenic yeast *Cryptococcus neoformans* H99. *FEMS Yeast Research, 11*(5), 440–448.

Jungmann, J., Reins, H. A., Lee, J., Romeo, A., Hassett, R., Kosman, D., & Jentsch, S. (1993). MAC1, a nuclear regulatory protein related to Cu-dependent transcription factors is involved in Cu/Fe utilization and stress resistance in yeast. *The EMBO Journal, 12*(13), 5051–5056.

Kampfenkel, K., Kushnir, S., Babiychuk, E., Inzé, D., & Van Montagu, M. (1995). Molecular characterization of a putative *Arabidopsis thaliana* copper transporter and its yeast homologue. *The Journal of Biological Chemistry, 270*(47), 28479–28486.

Karin, M., Najarian, R., Haslinger, A., Valenzuela, P., Welch, J., & Fogel, S. (1984). Primary structure and transcription of an amplified genetic locus: The CUP1 locus of yeast. *Proceedings of the National Academy of Sciences of the United States of America, 81*(2), 337–341.

Keller, G., Bird, A., & Winge, D. R. (2005). Independent metalloregulation of Ace1 and Mac1 in *Saccharomyces cerevisiae*. *Eukaryotic Cell, 4*(11), 1863–1871.

Keller, G., Gross, C., Kelleher, M., & Winge, D. R. (2000). Functional independence of the two cysteine-rich activation domains in the yeast Mac1 transcription factor. *The Journal of Biological Chemistry, 275*(38), 29193–29199.

Knight, S. A., Labbé, S., Kwon, L. F., Kosman, D. J., & Thiele, D. J. (1996). A widespread transposable element masks expression of a yeast copper transport gene. *Genes & Development, 10*(15), 1917–1929.

Koch, K. A., Peña, M. M., & Thiele, D. J. (1997). Copper-binding motifs in catalysis, transport, detoxification and signaling. *Chem Biol, 4*(8), 549–560.

Kosman, D. J. (2018). For *Cryptococcus neoformans*, responding to the copper status in a colonization niche is not just about copper. *Molecular Microbiology, 108*(5), 463–466.

Krummeck, G., & Rödel, G. (1990). Yeast SCO1 protein is required for a post-translational step in the accumulation of mitochondrial cytochrome c oxidase subunits I and II. *Current Genetics, 18*(1), 13–15.

Kusuya, Y., Hagiwara, D., Sakai, K., Yaguchi, T., Gonoi, T., & Takahashi, H. (2017). Transcription factor Afmac1 controls copper import machinery in *Aspergillus fumigatus*. *Current Genetics, 63*(4), 777–789.

Labbé, S., & Thiele, D. J. (1999). Pipes and wiring: The regulation of copper uptake and distribution in yeast. *Trends in Microbiology, 7*(12), 500–505.

Labbé, S., Peña, M. M., Fernandes, A. R., & Thiele, D. J. (1999). A copper-sensing transcription factor regulates iron uptake genes in *Schizosaccharomyces pombe*. *The Journal of Biological Chemistry, 274*(51), 36252–36260.

Labbé, S., Zhu, Z., & Thiele, D. J. (1997). Copper-specific transcriptional repression of yeast genes encoding critical components in the copper transport pathway. *The Journal of Biological Chemistry, 272*(25), 15951–15958.

Laliberté, J, & Labbé, S. (2006). Mechanisms of copper loading on the Schizosaccharomyces pombe copper amine oxidase 1 expressed in Saccharomyces cerevisiae. *Microbiology (Reading), 152*(9), 2819–2830.

Laliberté, J., Whitson, L. J., Beaudoin, J., Holloway, S. P., Hart, P. J., & Labbé, S. (2004). The *Schizosaccharomyces pombe* Pccs protein functions in both copper trafficking and metal detoxification pathways. *The Journal of Biological Chemistry, 279*(27), 28744–28755.

Lamb, A. L., Torres, A. S., O'Halloran, T. V., & Rosenzweig, A. C. (2001). Heterodimeric structure of superoxide dismutase in complex with its metallochaperone. *Nature Structural Biology, 8*(9), 751–755.

Lee, J., Peña, M. M., Nose, Y., & Thiele, D. J. (2002). Biochemical characterization of the human copper transporter Ctr1. *The Journal of Biological Chemistry, 277*(6), 4380–4387.

Lin, S. J., & Culotta, V. C. (1995). The ATX1 gene of *Saccharomyces cerevisiae* encodes a small metal homeostasis factor that protects cells against reactive oxygen toxicity. *Proceedings of the National Academy of Sciences of the United States of America, 92*(9), 3784–3788.

Lin, S. J., Pufahl, R. A., Dancis, A., O'Halloran, T. V., & Culotta, V. C. (1997). A role for the *Saccharomyces cerevisiae* ATX1 gene in copper trafficking and iron transport. *The Journal of Biological Chemistry, 272*(14), 9215–9220.

Liu, J., Sitaram, A., & Burd, C. G. (2007). Regulation of copper-dependent endocytosis and vacuolar degradation of the yeast copper transporter, Ctr1p, by the Rsp5 ubiquitin ligase. *Traffic (Copenhagen, Denmark), 8*(10), 1375–1384.

Liu, X. F., Supek, F., Nelson, N., & Culotta, V. C. (1997). Negative control of heavy metal uptake by the *Saccharomyces cerevisiae* BSD2 gene. *The Journal of Biological Chemistry, 272*(18), 11763–11769.

Lode, A., Kuschel, M., Paret, C., & Rödel, G. (2000). Mitochondrial copper metabolism in yeast: Interaction between Sco1p and Cox2p. *FEBS Letters, 485*(1), 19–24.

Lowe, J., Vieyra, A., Catty, P., Guillain, F., Mintz, E., & Cuillel, M. (2004). A mutational study in the transmembrane domain of Ccc2p, the yeast Cu(I)-ATPase, shows different roles for each Cys-Pro-Cys cysteine. *The Journal of Biological Chemistry, 279*(25), 25986–25994.

Mackie, J., Szabo, E. K., Urgast, D. S., Ballou, E. R., Childers, D. S., MacCallum, D. M., ... Brown, A. J. (2016). Host-imposed copper poisoning impacts fungal micronutrient acquisition during systemic *Candida albicans* infections. *PLoS One, 11*(6), e0158683.

Macomber, L., & Imlay, J. A. (2009). The iron-sulfur clusters of dehydratases are primary intracellular targets of copper toxicity. *Proceedings of the National Academy of Sciences of the United States of America, 106*(20), 8344–8349.

Martins, L. J., Jensen, L. T., Simon, J. R., Keller, G. L., & Winge, D. R. (1998). Metalloregulation of FRE1 and FRE2 homologs in *Saccharomyces cerevisiae*. *The Journal of Biological Chemistry, 273*(37), 23716–23721.

Marvin, M. E., Mason, R. P., & Cashmore, A. M. (2004). The CaCTR1 gene is required for high-affinity iron uptake and is transcriptionally controlled by a copper-sensing transactivator encoded by CaMAC1. *Microbiology (Reading, England), 150*(Pt 7), 2197–2208.

Marvin, M. E., Williams, P. H., & Cashmore, A. M. (2003). The *Candida albicans* CTR1 gene encodes a functional copper transporter. *Microbiology (Reading, England), 149*(Pt 6), 1461–1474.

Moraes, D., Rodrigues, J. G. C., Silva, M. G., Soares, L. W., Soares, C. M. D., Bailao, A. M., & Silva-Bailao, M. G. (2023). Copper acquisition and detoxification machineries are conserved in dimorphic fungi. *Fungal Biology Reviews, 44*.

Moraes, D., Tristão, G. B., Rappleye, C. A., Ray, S. C., Ribeiro-Dias, F., Gomes, R. S., ... Bailão, A. M. (2024). The influence of a copper efflux pump in *Histoplasma capsulatum* virulence. *The FEBS Journal, 291*(4), 744–760.

Nevitt, T., Ohrvik, H., & Thiele, D. J. (2012). Charting the travels of copper in eukaryotes from yeast to mammals. *Biochimica et Biophysica Acta, 1823*(9), 1580–1593.

Nittis, T., George, G. N., & Winge, D. R. (2001). Yeast Sco1, a protein essential for cytochrome c oxidase function is a Cu(I)-binding protein. *The Journal of Biological Chemistry, 276*(45), 42520–42526.

Nose, Y., Rees, E. M., & Thiele, D. J. (2006). Structure of the Ctr1 copper trans'PORE'ter reveals novel architecture. *Trends in Biochemical Sciences, 31*(11), 604–607.

Okada, M., & Miura, T. (2016). Copper(I) stabilization by cysteine/tryptophan motif in the extracellular domain of Ctr4. *Journal of Inorganic Biochemistry, 159*, 45–49.

Ooi, C. E., Rabinovich, E., Dancis, A., Bonifacino, J. S., & Klausner, R. D. (1996). Copper-dependent degradation of the *Saccharomyces cerevisiae* plasma membrane copper transporter Ctr1p in the apparent absence of endocytosis. *The EMBO Journal, 15*(14), 3515–3523.

Park, Y. S., Kim, T. H., & Yun, C. W. (2017). Functional characterization of the copper transcription factor AfMac1 from *Aspergillus fumigatus. The Biochemical Journal, 474*(14), 2365–2378.

Park, Y. S., Lian, H., Chang, M., Kang, C. M., & Yun, C. W. (2014). Identification of high-affinity copper transporters in *Aspergillus fumigatus. Fungal Genetics and Biology, 73*, 29–38.

Peña, M. M., Koch, K. A., & Thiele, D. J. (1998). Dynamic regulation of copper uptake and detoxification genes in *Saccharomyces cerevisiae. Molecular and Cellular Biology, 18*(5), 2514–2523.

Peña, M. M., Puig, S., & Thiele, D. J. (2000). Characterization of the Saccharomyces cerevisiae high affinity copper transporter Ctr3. *J. Biol. Chem. 275*, 33244–33251. https://doi.org/10.1074/jbc.m005392200.

Peter, C., Laliberté, J., Beaudoin, J., & Labbé, S. (2008). Copper distributed by Atx1 is available to copper amine oxidase 1 in *Schizosaccharomyces pombe. Eukaryotic Cell, 7*(10), 1781–1794.

Petito, G., de Curcio, J. S., Pereira, M., Bailão, A. M., Paccez, J. D., Tristão, G. B., ... de Almeida Soares, C. M. (2020). Metabolic adaptation of *Paracoccidioides brasiliensis* in response to in vitro copper deprivation. *Frontiers in Microbiology, 11*, 1834.

Portis, I. G., de Sousa Lima, P., Paes, R. A., Oliveira, L. N., Pereira, C. A., Parente-Rocha, J. A., ... de Almeida Soares, C. M. (2020). Copper overload in Paracoccidioides lutzii results in the accumulation of ergosterol and melanin. *Microbiol Res, 239*, 126524.

Portnoy, M. E., Rosenzweig, A. C., Rae, T., Huffman, D. L., O'Halloran, T. V., & Culotta, V. C. (1999). Structure-function analyses of the ATX1 metallochaperone. *The Journal of Biological Chemistry, 274*(21), 15041–15045.

Pufahl, R. A., Singer, C. P., Peariso, K. L., Lin, S. J., Schmidt, P. J., Fahrni, C. J., ... O'Halloran, T. V. (1997). Metal ion chaperone function of the soluble Cu(I) receptor Atx1. *Science (New York, N. Y.), 278*(5339), 853–856.

Puig, S., Lee, J., Lau, M., & Thiele, D. J. (2002). Biochemical and genetic analyses of yeast and human high affinity copper transporters suggest a conserved mechanism for copper uptake. *The Journal of Biological Chemistry, 277*(29), 26021–26030.

Punter, F. A., & Glerum, D. M. (2003). Mutagenesis reveals a specific role for Cox17p in copper transport to cytochrome oxidase. *The Journal of Biological Chemistry, 278*(33), 30875–30880.

Rae, T. D., Schmidt, P. J., Pufahl, R. A., Culotta, V. C., & O'Halloran, T. V. (1999). Undetectable intracellular free copper: The requirement of a copper chaperone for superoxide dismutase. *Science (New York, N. Y.), 284*(5415), 805–808.

Ray, S. C., & Rappleye, C. A. (2022). Mac1-dependent copper sensing promotes histoplasma adaptation to the phagosome during adaptive immunity. *mBio, 13*(2), e0377321.

Rees, E. M., Lee, J., & Thiele, D. J. (2004). Mobilization of intracellular copper stores by the ctr2 vacuolar copper transporter. *The Journal of Biological Chemistry, 279*(52), 54221–54229.

Rentzsch, A., Krummeck-Weiss, G., Hofer, A., Bartuschka, A., Ostermann, K., & Rödel, G. (1999). Mitochondrial copper metabolism in yeast: mutational analysis of Sco1p involved in the biogenesis of cytochrome c oxidase. *Curr Genet, 35*(2), 103–108.

Rigby, K., Cobine, P. A., Khalimonchuk, O., & Winge, D. R. (2008). Mapping the functional interaction of Sco1 and Cox2 in cytochrome oxidase biogenesis. *The Journal of Biological Chemistry, 283*(22), 15015–15022.

Riggle, P. J., & Kumamoto, C. A. (2000). Role of a *Candida albicans* P1-type ATPase in resistance to copper and silver ion toxicity. *Journal of Bacteriology, 182*(17), 4899–4905.

Rubino, J. T., & Franz, K. J. (2012). Coordination chemistry of copper proteins: How nature handles a toxic cargo for essential function. *Journal of Inorganic Biochemistry, 107*(1), 129–143.

Schmidt, P. J., Kunst, C., & Culotta, V. C. (2000). Copper activation of superoxide dismutase 1 (SOD1) in vivo. Role for protein-protein interactions with the copper chaperone for SOD1. *The Journal of Biological Chemistry, 275*(43), 33771–33776.

Schmidt, P. J., Rae, T. D., Pufahl, R. A., Hamma, T., Strain, J., O'Halloran, T. V., & Culotta, V. C. (1999). Multiple protein domains contribute to the action of the copper chaperone for superoxide dismutase. *The Journal of Biological Chemistry, 274*(34), 23719–23725.

Schulze, M., & Rödel, G. (1988). SCO1, a yeast nuclear gene essential for accumulation of mitochondrial cytochrome c oxidase subunit II. *Molecular & General Genetics: MGG, 211*(3), 492–498.

Serpe, M., Joshi, A., & Kosman, D. J. (1999). Structure-function analysis of the protein-binding domains of Mac1p, a copper-dependent transcriptional activator of copper uptake in *Saccharomyces cerevisiae*. *The Journal of Biological Chemistry, 274*(41), 29211–29219.

Shen, C. H., Leblanc, B. P., Alfieri, J. A., & Clark, D. J. (2001). Remodeling of yeast CUP1 chromatin involves activator-dependent repositioning of nucleosomes over the entire gene and flanking sequences. *Molecular and Cellular Biology, 21*(2), 534–547.

Shen, Q., Beucler, M. J., Ray, S. C., & Rappleye, C. A. (2018). Macrophage activation by IFN-γ triggers restriction of phagosomal copper from intracellular pathogens. *PLoS Pathogens, 14*(11), e1007444.

Smith, A. D., Logeman, B. L., & Thiele, D. J. (2017). Copper acquisition and utilization in fungi. *Annual Review of Microbiology, 71*, 597–623.

Song, J., Li, R., & Jiang, J. (2019). Copper homeostasis in *Aspergillus fumigatus*: Opportunities for therapeutic development. *Frontiers in Microbiology, 10*, 774.

Srinivasan, C., Posewitz, M. C., George, G. N., & Winge, D. R. (1998). Characterization of the copper chaperone Cox17 of *Saccharomyces cerevisiae*. *Biochemistry, 37*(20), 7572–7577.

Sun, T. S., Ju, X., Gao, H. L., Wang, T., Thiele, D. J., Li, J. Y., ... Ding, C. (2014). Reciprocal functions of *Cryptococcus neoformans* copper homeostasis machinery during pulmonary infection and meningoencephalitis. *Nature Communications, 5*, 5550.

Szczypka, M. S., & Thiele, D. J. (1989). A cysteine-rich nuclear protein activates yeast metallothionein gene transcription. *Molecular and Cellular Biology, 9*(2), 421–429.

Thiele, D. J. (1988). ACE1 regulates expression of the *Saccharomyces cerevisiae* metallothionein gene. *Molecular and Cellular Biology, 8*(7), 2745–2752.

Thiele, D. J., & Hamer, D. H. (1986). Tandemly duplicated upstream control sequences mediate copper-induced transcription of the *Saccharomyces cerevisiae* copper-metallothionein gene. *Molecular and Cellular Biology, 6*(4), 1158–1163.

Tsukihara, T., Aoyama, H., Yamashita, E., Tomizaki, T., Yamaguchi, H., Shinzawa-Itoh, K., ... Yoshikawa, S. (1995). Structures of metal sites of oxidized bovine heart cytochrome c oxidase at 2.8 A. *Science (New York, N. Y.), 269*(5227), 1069–1074.

Upadhyay, S., Torres, G., & Lin, X. (2013). Laccases involved in 1,8-dihydroxynaphthalene melanin biosynthesis in Aspergillus fumigatus are regulated by developmental factors and copper homeostasis. *Eukaryot Cell, 12*(12), 1641–1652.

van Dongen, E. M., Klomp, L. W., & Merkx, M. (2004). Copper-dependent protein-protein interactions studied by yeast two-hybrid analysis. *Biochemical and Biophysical Research Communications, 323*(3), 789–795.

Walton, F. J., Idnurm, A., & Heitman, J. (2005). Novel gene functions required for melanization of the human pathogen *Cryptococcus neoformans*. *Molecular Microbiology, 57*(5), 1381–1396.

Waterman, S. R., Hacham, M., Hu, G., Zhu, X., Park, Y. D., Shin, S., ... Williamson, P. R. (2007). Role of a CUF1/CTR4 copper regulatory axis in the virulence of *Cryptococcus neoformans*. *The Journal of Clinical Investigation, 117*(3), 794–802.

Weissman, Z., Berdicevsky, I., Cavari, B. Z., & Kornitzer, D. (2000). The high copper tolerance of *Candida albicans* is mediated by a P-type ATPase. *Proceedings of the National Academy of Sciences of the United States of America, 97*(7), 3520–3525.

Weissman, Z., Shemer, R., & Kornitzer, D. (2002). Deletion of the copper transporter CaCCC2 reveals two distinct pathways for iron acquisition in *Candida albicans*. *Molecular Microbiology, 44*(6), 1551–1560.

Welch, J., Fogel, S., Buchman, C., & Karin, M. (1989). The CUP2 gene product regulates the expression of the CUP1 gene, coding for yeast metallothionein. *The EMBO Journal, 8*(1), 255–260.

Wiemann, P., Perevitsky, A., Lim, F. Y., Shadkchan, Y., Knox, B. P., Landero Figueora, J. A., ... Keller, N. P. (2017). *Aspergillus fumigatus* copper export machinery and reactive oxygen intermediate defense counter host copper-mediated oxidative antimicrobial offense. *Cell Reports, 19*(5), 1008–1021.

Wimalarathna, R. N., Pan, P. Y., & Shen, C. H. (2012). Chromatin repositioning activity and transcription machinery are both recruited by Ace1p in yeast CUP1 activation. *Biochemical and Biophysical Research Communications, 422*(4), 658–663.

Winge, D. R., Nielson, K. B., Gray, W. R., & Hamer, D. H. (1985). Yeast metal-lothionein. Sequence and metal-binding properties. *J Biol Chem, 260*(27), 14464–14470.

Wu, X., Sinani, D., Kim, H., & Lee, J. (2009). Copper transport activity of yeast Ctr1 is down-regulated via its C terminus in response to excess copper. *The Journal of Biological Chemistry, 284*(7), 4112–4122.

Yamaguchi-Iwai, Y., Serpe, M., Haile, D., Yang, W., Kosman, D. J., Klausner, R. D., & Dancis, A. (1997). Homeostatic regulation of copper uptake in yeast via direct binding of MAC1 protein to upstream regulatory sequences of FRE1 and CTR1. *The Journal of Biological Chemistry, 272*(28), 17711–17718.

Yonkovich, J., McKenndry, R., Shi, X., & Zhu, Z. (2002). Copper ion-sensing transcription factor Mac1p post-translationally controls the degradation of its target gene product Ctr1p. *The Journal of Biological Chemistry, 277*(27), 23981–23984.

Yuan, D. S., Dancis, A., & Klausner, R. D. (1997). Restriction of copper export in *Saccharomyces cerevisiae* to a late Golgi or post-Golgi compartment in the secretory pathway. *The Journal of Biological Chemistry, 272*(41), 25787–25793.

Zhang, P., Zhang, D., Zhao, X., Wei, D., Wang, Y., & Zhu, X. (2016). Effects of CTR4 deletion on virulence and stress response in *Cryptococcus neoformans*. *Antonie Van Leeuwenhoek, 109*(8), 1081–1090.

Zhou, H., & Thiele, D. J. (2001). Identification of a novel high affinity copper transport complex in the fission yeast *Schizosaccharomyces pombe*. *The Journal of Biological Chemistry, 276*(23), 20529–20535.

Zhou, P., & Thiele, D. J. (1993). Rapid transcriptional autoregulation of a yeast metal-loregulatory transcription factor is essential for high-level copper detoxification. *Genes & Development, 7*(9), 1824–1835.

Zhou, P., Szczypka, M. S., Sosinowski, T., & Thiele, D. J. (1992). Expression of a yeast metallothionein gene family is activated by a single metalloregulatory transcription factor. *Molecular and Cellular Biology, 12*(9), 3766–3775.

Zhou, P. B., & Thiele, D. J. (1991). Isolation of a metal-activated transcription factor gene from Candida glabrata by complementation in *Saccharomyces cerevisiae. Proc Natl Acad Sci U S A, 88*(14), 6112–6116.

Zhu, Z., Labbé, S., Peña, M. M., & Thiele, D. J. (1998). Copper differentially regulates the activity and degradation of yeast Mac1 transcription factor. *The Journal of Biological Chemistry, 273*(3), 1277–1280.

Printed and bound by CPI Group (UK) Ltd, Croydon, CR0 4YY

08/05/2025

01864966-0001